Biotechnological Advances in Agriculture, Healthcare, Environment and Industry

There has been a surge in critical situations such as malnutrition, agricultural productivity, viral diseases, noncommunicable diseases, antibiotic resistance, and depletion of natural resources threatening human welfare during the last decade. Fortunately, a parallel unprecedented growth and development have also taken place in various biotechnological sectors, giving hope and direction to meet these challenges.

This book has been written with a very clear purpose of highlighting the contributions of modern biotechnology in solving emerging problems in various sectors. It has been divided into four major sections: Agricultural Biotechnology, Healthcare Biotechnology, Environmental Biotechnology, and Industrial Biotechnology. The chapters under each section are dedicated to the latest developments with comprehensive information about background, prospects, and applications. We have tried to incorporate the latest innovative biotechnological solutions such as biofortification, exome sequencing, genome sequencing, RNAi, drug discovery, nutrigenomics, bioremediation, nanobiotechnology, and many more, which have contributed remarkably in recent years to address the modern challenges in various sectors.

The key highlights of this book are as follows:

- Multidisciplinary perspective of biotechnological approaches.
- Extensive investigation of hot topics.
- Updated information with relevant data and statistics.
- Application-based study of biotechnology.
- Profusely illustrated and labeled diagrams to supplement the descriptions.

This fully updated book will serve as a high-value resource material for students, teachers, researchers, and industry personnel in various disciplines of life sciences, agricultural sciences, medicine, environment, and industrial biotechnology. It will serve as a comprehensive resource book to develop an in-depth understanding of the emerging challenges in the areas of human welfare and analyze the latest biotechnological advancements in the direction of solving these concerns.

Biotechnological Advances in Agriculture, Healthcare, Environment and Industry

Edited by
Priyanka Siwach, Suresh Kumar Gahlawat,
Raj Kumar Salar, Joginder Singh Duhan, and
Pardeep Kumar Sadh

CRC Press
Taylor & Francis Group
Boca Raton London New York

CRC Press is an imprint of the
Taylor & Francis Group, an **informa** business

Designed cover image: Shutterstock

First edition published 2025
by CRC Press
2385 NW Executive Center Drive, Suite 320, Boca Raton FL 33431

and by CRC Press
4 Park Square, Milton Park, Abingdon, Oxon, OX14 4RN

CRC Press is an imprint of Taylor & Francis Group, LLC

ISBN: 9781032642543 (hbk)
ISBN: 9781032642550 (pbk)
ISBN: 9781032642567 (ebk)

DOI: 10.1201/9781032642567

Typeset in Times
by codeMantra

Contents

Authors

Priyanka Siwach, Ph.D. (Biotechnology & Molecular Biology), is serving as Professor in the Department of Biotechnology, Chaudhary Devi Lal University, Sirsa, Haryana, India, with more than 23 years of regular teaching and research experience. Her specialization is Genomics and Plant Biotechnology. She has actively contributed to academic administration through her distinguished roles as Dean, Faculty of Life Sciences, Dean Research, Chairperson of the Department of Biotechnology and Nodal Officer, Intellectual Property Rights. During her career, she has completed two R&D projects and has guided 10 Ph.D. and 15 M.Phil. students. She has published more than 80 research/review papers/articles in various national and international journals. She has three edited books and one authored book to her credit, and has authored seven book chapters. She has delivered numerous invited talks in her area of specialization in many conferences/workshops. Recently, in collaboration with the Supercomputing Facility for Bioinformatics and Computational Biology, IIT, New Delhi, she developed a novel methodology for the annotation of the genome sequence, and this research was published in *Nucleic Acids Research*, a journal of high repute (with an impact factor of 19.25). Recently, she has been granted one patent by the Indian Patent Office for a novel methodology of genetic transformation in *Ficus religiosa* L. She has been awarded the Silver Jubilee Gold Medal for her Ph.D. thesis. She is also a reviewer for several international journals of reputed publishing houses like Wiley-VCH, Springer, Elsevier and Taylor & Francis.

Suresh Kumar Gahlawat, Ph.D., is a dedicated teacher with more than 26 years of experience. Presently, he is working as a Professor of Biotechnology & Dean of Academic Affairs at Chaudhary Devi Lal University, Sirsa, Haryana, India, and recipient of prestigious fellowships, namely BOYSCASTS and Biotechnology Overseas Associateship funded by Ministry of Science & Technology, Government of India, availed at FRS Marine Laboratory, Aberdeen, Scotland. He has published more than 90 research papers in national and international reputed journals, in addition to review articles and book chapters. He has also authored and edited nine books. He has completed four major research projects and supervised 14 Ph.D. and M.Phil. research scholars. He has delivered more than 45 invited talks in the field of specialisation in various academic staff colleges and conferences/seminars. He has visited several laboratories, such as theCentre for Environment, Fisheries and Aquaculture Science (CEFAS), Weymouth; University of Stirling; BBSRC; and The Pirbright Institute during his post-doctoral fellowships. He has developed molecular diagnostic methods for detection of bovine viruses and filed four Indian patents. He has developed easy and inexpensive methods for isolation of macrophages, a sensitive non-destructive method for the detection of IPNV carrier from Atlantic salmon, a loop-mediated isothermal amplification method for the detection of *Renibacterium salmoninarum*, *Aeromonas hydrophila* and *Pseudomonas florescens* His major research work was carried out on infectious pancreatic necrosis virus (IPNV) carriers, Type-I interferon signaling, expression of interferon and interferon-induced genes, disease diagnosis, probiotics and nanotechnology on marine/freshwater fish and shellfish species. He has served as an academic administrator, contributed as Dean of Colleges, Dean of Student's Welfare, Dean of Research, Dean, Faculty of Life Sciences, Chairperson, Department of Biotechnology, among others. He has attended several trainings including the Leadership for Academician Programme (LeAP-2019) at Aligarh Muslim University, Aligarh, and Monash University, Melbourne, Australia, under the MHRD initiative.

Raj Kumar Salar, Ph.D., is a Professor in the Department of Biotechnology, Chaudhary Devi Lal University, Sirsa, Haryana, India, with more than 25 years of experience. He served as Dean of the Faculty of Life Sciences, Chairperson of the Department of Biotechnology and Controller of Examination at Chaudhary Devi Lal University. He is the recipient of a postdoctoral fellowship from the Ministry of Education, Slovak Republic, to pursue advanced research at the Department of Biochemical Technology, Slovakia, and visited several other countries, including Japan, Norway, Austria, and Hungary, for academic pursuits. Besides thermophilic fungi, his current research interests include fermentation technology and harnessing the medicinal value of plants and microbes. He has supervised several Ph.D. and M.Phil. students guiding their research work. He has published more than 100 research papers and reviews in national and international reputed journals and edited four books. He has also authored a book titled *Thermophilic Fungi: Basic Concepts and Biotechnological Applications*, published by CRC Press. He has also received several research and development projects from the University Grants Commission (UGC), New Delhi, and the Haryana State Council for Science and Technology (DST), Ministry of Science & Technology, New Delhi. He is the academic editor of the prestigious journal *PLOS ONE*. He is also a reviewer for several journals published by Wiley-VCH, Springer, Elsevier, and Taylor & Francis. He has received the prestigious King Abdulaziz City of Science & Technology Award for the best paper published in *3 Biotech* journal. He is also an elected Fellow of the Mycological Society of India.

Joginder Singh Duhan, Ph.D. (Microbiology), is a Professor and Chairperson in the Department of Biotechnology, Microbiology and Zoology, Chaudhary Devi Lal University, Sirsa, India with more than 25 years of experience in teaching and research. He is the recipient of the Young Scientist Award given by the Association of Microbiologists of India and the K.K. Nanda Award for the best research and several other best paper awards. He has published more than 100 research papers in national and international reputed journals. He has edited four books: three are published by Springer and one by Narosa. He has received R&D projects from UGC and CSIR, New Delhi. He has supervised the research work of 15 M. Phil. and 11 Ph.D. students. His specialization includes Microbial Biotechnology, Fermentation Technology, and Nanotechnology. He is a reviewer of several journals published by Elsevier, Springer, MDPI, Wiley and Taylor and Francis. He is a life member of several prestigious academic bodies/associations such as the Indian Science Congress, Association of Microbiologists of India, SOCDAB, and so on.

Pardeep Kumar Sadh, Ph.D., is currently serving as an Assistant Professor (part-time) in the Department of Biotechnology at Chaudhary Devi Lal University, Sirsa, Haryana, India. He completed his MSc, M.Phil., and Ph.D. in the same department at Chaudhary Devi Lal University. Dr. Sadh has authored more than 35 research and review articles in national and international reputed journals, contributed 4 book chapters and edited one book published by Wiley. His research interests encompass Bioresource Technology, Functional Foods, Food Biotechnology, and Agricultural Waste Management. During his Ph.D., he was awarded the Rajiv Gandhi National Fellowship by the University Grants Commission under the Government of India. He has received best poster awards at various conferences and certificates of excellence for his work as a reviewer for internationally reputed journals. Additionally, Dr. Sadh is a recognized reviewer and referee for several international and national journals.

Contributors

Amanjyoti
Department of Food Science and Technology
Chaudhary Devi Lal University
Sirsa, Haryana, India

Kiran Bala
Om Sterling Global University
Hisar, Haryana, India

Priyanka Balyan
Department of Plant Science
Mahatma Jyotiba Phule Rohilkhand University
Bareilly, India

Upendra Balyan
Department of Molecular Biology & Biotechnology, College of Biotechnology
CCS Haryana Agricultural University
Hisar, Haryana, India

Manjeet Bansal
Department of Civil Engineering
Maharaja Ranjit Singh Punjab Technical University
Bathinda, Haryana, India

Kanisht Batra
Department of Animal Biotechnology
Lala Lajpat Rai University of Veterinary and Animal Sciences
Hisar, Haryana, India

Manju Bernela
ICAR-National Research Centre on Equines
Hisar, Haryana, India

Basanti Brar
Om Sterling Global University
Hisar, Haryana, India

Savita Budania
Department of Animal Biotechnology
Lala Lajpat Rai University of Veterinary and Animal Sciences
Hisar, Haryana, India

Shiwani Chahal
Department of Biotechnology
Chaudhary Devi Lal University
Sirsa, Haryana, India

Deepika Chaudhary
Department of Animal Biotechnology
Lala Lajpat Rai University of Veterinary and Animal Sciences
Hisar, Haryana, India

Asha Choudhary
Department of Biotechnology
Chaudhary Devi Lal University
Sirsa, India

Rita Choudhary
National Centre of Excellence for Advanced Research in Agriculture Nanotechnology TERI-Deakin Nanobiotechnology Centre (TDNBC)
The Energy and Resources Institute (TERI)
New Delhi, India

Deepika
Department of Biotechnology
Chaudhary Devi Lal University
Sirsa, Haryana, India

Rani Devi
Department of Energy and Environmental Sciences
Chaudhary Devi Lal University
Sirsa, Haryana, India

Sunena Dhania
Department of Bio and Nanotechnology
Guru Jambheshwar University of Science and Technology
Hisar, Haryana, India

Sanju Bala Dhull
Department of Food Science and Technology
Chaudhary Devi Lal University
Sirsa, Haryana, India

Natoavina Faliarizao
Department of Food Science and Human Nutrition
Michigan State University
East Lansing, Michigan

Bharat Garg
Department of Molecular Biology & Biotechnology, Bioinformatics
College of Biotechnology, CCS Haryana Agricultural University
Hisar, Haryana, India

Anita Rani Gill
Department of Bio & Nanotechnology
Guru Jambheshwar University of Science and Technology
Hisar, Haryana, India

Gulden Goksen
Department of Food Technology, Vocational School of Technical Sciences at Mersin Tarsus Organized Industrial Zone
Tarsus University
Mersin, Turkey

Vinita Hooda
Maharshi Dayanand University
Rohtak, Haryana, India

Heena Jain
Department of Biotechnology
Chaudhary Devi Lal University
Sirsa, Haryana, India

Seema Joshi
Om Sterling Global University
Hisar, Haryana, India

Avneet Kaur
Department of Chemistry, University Institute of Science
Chandigarh University, Gharuan
Mohali, India

Pawan Kaur
National Centre of Excellence for Advanced Research in Agriculture Nanotechnology TERI-Deakin Nanobiotechnology Centre (TDNBC)
The Energy and Resources Institute (TERI)
New Delhi, India

Mohd. Kashif Kidwai
Department of Energy and Environmental Sciences
Chaudhary Devi Lal University
Sirsa, Haryana, India

Anita Singh Kirrolia
Department of Environmental Science & Engineering
Guru Jambheshwar University of Science & Technology
Hisar, Haryana, India

Aman Kumar
Department of Animal Biotechnology
Lala Lajpat Rai University of Veterinary and Animal Sciences
Hisar, Haryana, India

Anuj Kumar
Laboratory of Immunity
University Medical College
Shantou, China
Department of Microbiology and Immunology, IWK Health Center, Canadian Centre for Vaccinology, Faculty of Medicine
Dalhousie University
Halifax, Canada

Dharmender Kumar
Department of Biotechnology
Deenbandhu Chhotu Ram University of Science and Technology
Murthal, India

Sushila Maan
Department of Animal Biotechnology
Lala Lajpat Rai University of Veterinary and Animal Sciences
Hisar, Haryana, India

Sachin Malik
Department of Biotechnology
Deenbandhu Chhotu Ram University of Science and Technology
Murthal, India

Shikha Malik
Department of Biotechnology
Deenabandhu Chhotu Ram University of Science & Technology
Murthal, India

Anju Manuja
ICAR-National Research Centre on Equines
Hisar, Haryan, India

Sumnil Marwaha
ICAR-National Research Centre on Camel
Bikaner, Rajasthan

Akhilesh Mishra
Yu lab Computational Biology Department
St. Jude Children's Research Hospital
Memphis, Tn

Chanchal Mony
National Centre of Excellence for Advanced Research in Agriculture Nanotechnology TERI-Deakin Nanobiotechnology Centre (TDNBC)
The Energy and Resources Institute (TERI)
New Delhi, India
and
School of Life and Environmental Sciences
Geelong, Australia

Sunil Mor
Animal Disease Research & Diagnostic Laboratory, Department of Animal Disease Research
Veterinary and Biomedical Sciences
Brookings, South Dakota

Vikash Nain
Department of Food Science and Technology
Chaudhary Devi Lal University
Sirsa, Haryana. India

Pierre Navarro
Lycée Jacquard BTS Métiers de l'Eau Rue Jacquard
Lavelanet, France

Kiran Nehra
Department of Biotechnology
Deenabandhu Chhotu Ram University of Science & Technology
Murthal, Haryana, India

Manju Nehra
Department of Food Science and Technology
Chaudhary Devi Lal University
Sirsa, Haryana, India

Anamika Pal
National Centre of Excellence for Advanced Research in Agriculture Nanotechnology TERI-Deakin Nanobiotechnology Centre (TDNBC), The Energy and Resources Institute (TERI)
New Delhi, India
and
School of Life and Environmental Sciences
Geelong, Australia

Sourav Panigrahi
Department of Molecular Biology & Biotechnology
College of Biotechnology, CCS Haryana Agricultural University
Hisar, Haryana, India

Parul
Maharshi Dayanand University
Rohtak, Haryana, India

Minakshi Prasad
ICAR-National Research Centre on Equines
Hisar, Haryana, India

Sukhvinder Singh Purewal
University Centre for Research & Development
Chandigarh University
Gharuan, India

Anita Rani
Chaudhary Devi Lal University
Sirsa, Haryana, India

Jyoti Rani
Department of Botany
Chaudhary Devi Lal University
Sirsa, Haryana, India

Ritu Rani
Department of Biotechnology
Deenbandhu Chhotu Ram University of Science and Technology
Murthal, Haryana, India

Ruma Rani
ICAR-National Research Centre on Equines
Hisar, Haryana, India

Suman Rani
SIES Indian Institute of Environmental Management
University of Mumbai
Navi Mumbai, India

Kaushlesh Ranjan
Department of Veterinary Physiology and Biochemistry
Sardar Vallabhbhai Patel University of Agriculture & Technology
Meerut, India

Nishu Rawat
Department of Biotechnology
Deenabandhu Chhotu Ram University of Science & Technology
Murthal, Haryana, India

Neeru S. Redhu
Department of Bioinformatics & Computational Biology
College of Biotechnology, CCS Haryana Agricultural University
Hisar, Haryana, India

Pawan Kumar Rose
Department of Energy and Environmental Sciences
Chaudhary Devi Lal University
Sirsa, Haryana, India

Pardeep Kumar Sadh
Department of Biotechnology
Ch. Devi Lal University
Sirsa, Haryana, India

Satish Kumar Sain
Central Institute of Cotton Research
Sirsa, Haryana, India

Raj Kumar Salar
Department of Biotechnology
Chaudhary Devi Lal University
Sirsa, Haryana, India

Kawaljeet Sandhu
Department of Food Science and Technology
Maharaja Ranjit Singh Punjab Technical university
Bathinda, Punjab

Alok Kumar Saran
Department of Energy and Environmental Sciences
Chaudhary Devi Lal University
Sirsa, Haryana, India

Khushboo Sethi
ICAR-National Research Centre on Equines
Hisar, Haryana, India

Anubha Sharma
Department of Animal Biotechnology
Lala Lajpat Rai University of Veterinary and Animal Sciences
Hisar, Haryana, India

Shashi
Maharshi Dayanand University
Rohtak, Haryana, India

Inderjeet Singh
Department of Biotechnology
Chaudhary Devi Lal University
Sirsa, Haryana, India

Vikram Singh
Wheat & Barley Section, Department of Genetics & Plant Breeding, College of Agriculture
CCS Haryana Agricultural University
Hisar, Haryana, India

Yogita Singh
Department of Molecular Biology & Biotechnology, College of Biotechnology
CCS Haryana Agricultural University
Hisar, Haryana, India

Anil Kumar Siroha
College of Agriculture
Chaudhary Charan Singh Haryana Agricultural University
Bawal, India

Priyanka Siwach
Department of Biotechnology
Chaudhary Devi Lal University
Sirsa, Haryana, India

Sayed Sartaj Sohrab
Department of Medical Laboratory Sciences, Faculty of Applied Medical Sciences
King Abdulaziz University
Jeddah, Saudi Arabia

Meenakshi Suhag
Institute of Environmental Studies
Kurukshetra University
Kurukshetra, Haryana, India

Jyoti Taunk
Department of Soil Chemistry, Plant Nutrition & Microbiology Institute of Soil Water and Environmental Sciences
Agricultural Research Organization — The Volcani Institute
Rishon LeZion, Israel

Rajesh Thakur
Department of Bio and Nano Technology
Guru Jambheshwar University of Science and Technology
Hisar, Haryana, India

Rahul Thory
Department of Food Science & Technology
Guru Nanak Dev University
Amritsar, India

Jayanti Tokas
Department of Biochemistry, College of Basic Sciences & Humanities
CCS Haryana Agricultural University
Hisar, Haryana, India

Shikha Yashveer
Department of Molecular Biology & Biotechnology, Bioinformatics, College of Biotechnology
CCS Haryana Agricultural University
Hisar, India

Part I

Agricultural Biotechnology

1 Wheat Biofortification
A Solution to Hunger and Malnutrition

Upendra Balyan, Yogita Singh, Priyanka Balyan, Sourav Panigrahi, and Anuj Kumar

1.1 INTRODUCTION

One of the most important components of nutrition security is mineral nutrition. Around three billion individuals globally suffer from zinc deficiency. Over 47% of preschoolers worldwide exhibit a range of Fe-deficient symptoms, including stunted physical and mental development and decreased learning capacity. Micronutrient deficiencies, particularly Zn and Fe, result from the scant quantity of micronutrients available in various plant products such as cereals, pulses, and so on, which form a part of the staple diet for humans. In developing countries, cereals, particularly rice, maize, and wheat, are the essential sources of calories and sustenance. These grains provide a significant portion of total energy (up to 30.4%) and protein (20%–30%) in the typical diet. However, more than a third of the world's populace lacks essential vitamins, minerals, and micronutrients, with iron and zinc deficiencies affecting 60% and 30% of individuals, respectively. Cereal grains also contain trace amounts of other elements like iodine, selenium, and more, which are vital for supporting overall health and nutrition. This deficiency in key micronutrients is a significant global health concern and contributes to the widespread issue of malnutrition. For instance, more than 200 million people in India were reported to be malnourished in 2019 (FAO, 2021). Common deficiencies include vitamin A, zinc, iodine, folate, and iron, which lead to hidden hunger and hinder the growth and development of children (Zou et al., 2019). Hidden hunger can invisibly damage the health and development of populations, driven by factors like poor diet, illness, reduced nutrient absorption, and increased micronutrient requirements during specific life stages such as pregnancy, lactation, and infancy (Jeje et al., 2023).

Micronutrients like vitamins and minerals, although required in small quantities, are vital for cellular functions. They form the basis for vitamins, proteins, and amino acids. In developing countries, imbalances in food and nutrients often lead to health issues such as anemia, stunted growth, cognitive and physical developmental problems, blindness, and higher rates of illness and death, especially in women and infants (Stewart et al., 2010).

Micronutrient malnutrition is particularly widespread in underdeveloped and developing nations where diets primarily consist of staple foods such as rice, wheat, maize, and millet (Khush et al., 2012). Current agricultural systems are attempting to

DOI: 10.1201/9781032642567-2

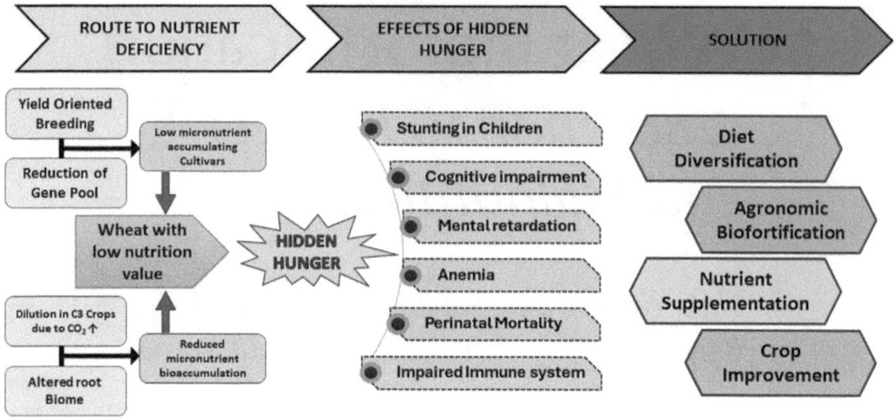

FIGURE 1.1 Causes and effects of hidden hunger and various approaches toward its solution.

replace nutrient-deficient food crops with nutrient-rich cultivars to address this problem. However, cereals remain the dominant food crops worldwide, posing challenges for replacing these staples due to economic and cultural factors. Figure 1.1 shows an overview of the causes and effects of hidden hunger and the possible solutions to combat it. Therefore, increasing the micronutrient content of these staple food crops by a process known as biofortification is seen as a practical solution to combat nutrient deficiencies without altering the established dietary patterns.

Trace metals such as zinc and iron, vital for many enzymatic processes, are abundant in wheat, a widely consumed crop. These metals support functions like auxin and carbohydrate metabolism, protein synthesis, membrane integrity, and various metabolic processes (Rehman et al., 2018). However, their absorption is limited by antinutrients. While numerous strategies have been developed to enhance their bioavailability, issues of cost and sustainability persist. Possible solutions include diversifying diets, increasing mineral intake, fortifying food and crops, promoting regular consumption through horticultural interventions, and implementing public health measures to combat micronutrient deficiency-related diseases.

1.2 WHEAT BIOFORTIFICATION

Wheat stands out as one of the most frequently grown and consumed cereals and is regarded as a staple food in the majority of the world. Cereals are the primary source of food and feed worldwide. Cereal grains are especially significant because of their biochemical makeup, which contains substances that have preventive and therapeutic effects for humans and natural plant protection (Malaguti et al., 2014; Shewry and Hey, 2015).

According to the FAO, around three billion people lack a healthy diet, and over 720 million suffer from hunger. The COVID-19 pandemic has exacerbated these issues, leading to a rise in "hidden hunger" caused by insufficient intake of vital micronutrients like provitamin A, iron, zinc, and selenium. A sustainable, economical, and long-term strategy to address micronutrient insufficiency is biofortification, which aims to improve the nutritional quality of food crops using either traditional breeding, agronomic techniques, or biotechnological tools. Staple crops, due to their dietary significance, are often the focus of biofortification research. Specifically, wheat contributes to about 20% of global protein and calorie consumption and roughly 30% of Fe and Zn intake. However, many wheat-based foods fail to meet daily micronutrient needs, particularly in less affluent areas. Hence, the continuation of wheat biofortification is vital to produce nutrient-rich, sustainable foods and mitigate micronutrient deficiencies (Ibba et al., 2022). Biofortification enriches food crops with essential nutrients, improving diet quality, especially for underprivileged populations in developing countries. Diversifying diets can yield positive results. Economically, biofortification requires a one-time investment, offering a cost-effective solution to micronutrient deficiencies. Once biofortified crops are processed, no additional funding is needed (Pfeiffer and McClafferty, 2007).

The main objective of biofortification is to reduce micronutrient malnutrition in underdeveloped nations. For this purpose, enriched foods containing the essential nutrients should be included in the primary diet instead of a separate supplement. Plants benefit from biofortification in terms of production, vitality, quality, and stress tolerance. It also improves the physical and chemical properties of the soil. Due to the depletion of numerous minerals in monoculture and incorrect fertilizer application, the health of the soil on farmed land deteriorates (Bouis and Saltzman, 2017).

The concept of biofortification is the practice of economically and sustainably breeding crops to increase their nutritional value. Either the traditional method of selective breeding or genetic engineering can be used for this. Because of certain obstacles to the potential uptake of soil nutrients, cereal crops like wheat typically lack certain minerals, necessitating fortification (Fageria and Baligar, 2008). The availability of nutrients in wheat is negatively impacted by the ongoing use of weak fertilizers in fields with low mineral concentrations (Fageria et al., 2002). Wheat is the second most-produced cereal after maize and is traded internationally more than any other crop. Furthermore, annual wheat production has almost tripled since the 1940s, and this trend is expected to continue as the population grows (Iizumi et al., 2017). Wheat that has been biofortified may help lessen malnutrition caused by famine, which typically affects low-income nations (Shewry and Hey, 2015).

For biofortified crops to be successful, the following factors must be considered:

1. The biofortified crop needs to be lucrative for the farmer and produce large yields.
2. Customers in target locations, where micronutrient malnutrition is a problem, must also find the crop appetizing.
3. Age and gender-specific requirements for the daily consumption of the main food must be met.
4. After the crop is harvested, nutrients shouldn't be wasted because they can drastically deteriorate during processing, cooking, or storage.
5. The consumer needs to absorb the nutrients (Meenakshi, 2009; Bouis and Welch, 2010).

1.3 METHODS OF WHEAT BIOFORTIFICATION

Micronutrient deficiency-related malnutrition in populations is being addressed through various approaches, including food supplementation, dietary diversity, industrial fortification, and nutritional fortification. Industrial fortification, while cost-effective, can raise product prices, making them unaffordable for the most deprived (Nestel, 1993). For a significant section of the population, food supplementation is not sustainable, despite its reliability. Biofortification, which enhances nutrient concentrations in crop plants, is considered the most viable strategy. It involves ensuring high-yield, cost-effective crops accepted by farmers and consumers, with retained nutrient content (Dhaliwal et al., 2022).

Winkler (2011) outlines four stages of biofortification: discovery, development, dissemination, and outcomes. These stages involve identifying target populations, setting nutrient levels, screening germplasm, breeding crops, testing performance, measuring nutrient retention, evaluating absorption, developing dissemination strategies, and improving nutritional status.

Biofortification uses three main approaches: agronomic, breeding, and molecular/transgenic. Staple crops like rice, wheat, and maize are targeted using these methods, as shown in Figure 1.2 (Garg et al., 2018). Agronomic biofortification involves nutrient application to soil or plant leaves and Nanotechnology, which enhances nutrient absorption (El-Saadony et al., 2019). Breeding biofortified crops is successful, with genomic section, marker-assisted selection (MAS), quantitative trait loci (QTL) mapping, and gene-editing techniques like CRISPR/Cas. Challenges include trade-offs between nutrients and interactions among multiple nutrients. The transgenic approach to biofortification plays a significant role, particularly in situations where there is limited genetic diversity, low heritability, and concerns about linkage drag.

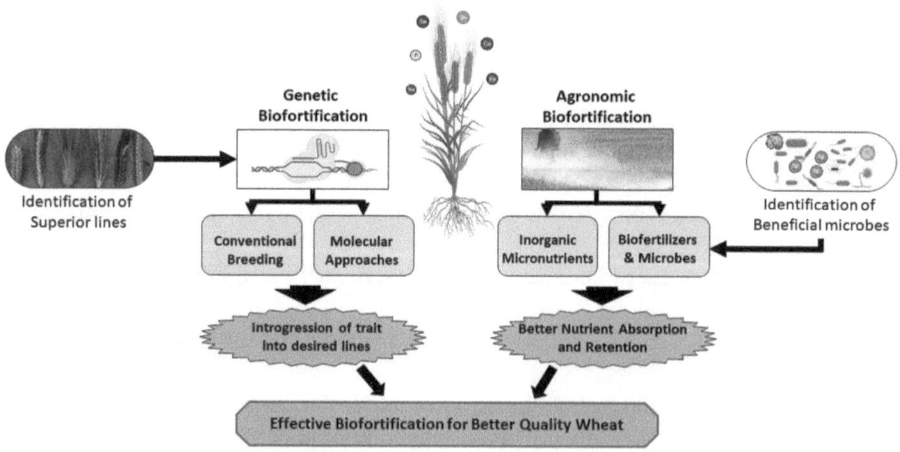

FIGURE 1.2 Various approaches for biofortification of wheat.

1.3.1 Agronomic Biofortification

Applying micronutrient-rich fertilizers to plants in a readily phytoavailable form, adjusting the pH of the soil, rotating crops, or adding beneficial soil microorganisms are some agronomic techniques to increase the mineral contents of cereal crops. (White and Broadley, 2011; Rawat et al., 2013; Rengel et al., 1999) Micronutrient fertilizers can be applied to the soil or applied topically to the leaves of cereal crops to implement agronomic biofortification. Some agronomic measures of biofortification include crop rotation, soil moisture management, tillage, application of green manures, such as *Rhizobium*, *Azotobacter*, *Azospirillum*, and proper application of biofertilizers and fertilizers containing N, P, K, and S and applying soil amendments to troubled soils (e.g., gypsum in alkaline and saline soils and lime in acidic soils).

Farmers traditionally use mineral fertilizers to improve the health and growth of their crops. Similarly, applying fertilizers can enhance crop mineral content and elevate nutrient concentration to a certain extent (Rengel et al., 1999). Zinc fertilizers added to the soil can enhance the uptake and accumulation of zinc up to 60 mg/kg in whole wheat grain (Cakmak, 2008). Agronomic biofortification using iron and foliar nitrogen fertilizers has shown a positive correlation with high concentrations of iron uptake in cases of iron and zinc (Aciksoz et al., 2011). Foliar zinc application effectively increases zinc uptake and improves bioavailability by reducing the quantity of phytic acid, an antinutritional component (Yang et al., 2011). Different parameters including soil moisture content its interaction with other components, soil organic matter content, soil pH, and soil aeration, further influence the degree of nutrient bioavailability to crops (Alloway, 2009). Agronomic biofortification aids in enhancing the chemical and physical characteristics of the soil, thereby increasing crop yields.

It has been found that applying chemical and organic fertilizers conjugately increases the uptake of micronutrients in wheat (White and Broadley, 2005). The addition of organic manure to wheat grains worked well with the necessary elements. Arbuscular mycorrhizal fungi (AMF) and fertilizers used together were also found to have a synergistic effect on the micronutrient absorption of wheat. Microorganisms play an important part in modern agriculture because they provide plants with a variety of benefits, including nutritional support. *Bacillus aryabhattai*, the Zn-solubilizing bacteria, greatly aids in the growth of plants in Central India by eliminating nutrients from the vertisols (Ramesh et al., 2014). Moreover, by combining organic and artificial fertilizers, wheat grain iron enrichment has been accomplished (Ramzani et al., 2016).

The effectiveness of agronomic biofortification is largely determined by fertilizer application. Different techniques impact the micronutrient content in grains in various ways. Micronutrient fertilizers, while sometimes boosting grain yield, can alter plant physiology and potentially reduce crop output. Among the various techniques, foliar application is the most efficient. However, for optimal results, it should be applied with an understanding of the crop's growth stage. Foliar treatment may not be as beneficial at other stages, as nutrients are not typically stored in leaves as they are in seeds and fruits. Despite successful nutrient absorption from the soil, crops may not always be able to access these nutrients due to certain restrictions (Frossard et al., 2000). This strategy, while effective, is not sustainable as it requires consistent fertilizer application and can be costly.

1.3.2 Genetic Biofortification

In addition to nutrient management in soil, conventional plant breeding has proven more successful for current crop releases biofortified with specific nutrients (Garg et al., 2018). This strategy involves crossbreeding plants with desired traits and selecting offspring with preferred agronomic properties, thereby enhancing nutrient content in food grains. Genetic variations in grain micronutrient content can be identified using marker-assisted breeding, gene discovery, or traditional breeding techniques (Grusak, 2002). Effective genetic biofortification enhances nutrient bioavailability, promotes micronutrient translocation to grains, and improves plant nutrient uptake (Mulualem, 2015). Recent advancements in genetic biofortification have helped reduce micronutrient deficiencies and provide a long-term, diet-based solution to complement other therapies (Ludwig and Slamet-Wedin, 2019). Plant breeders select and breed nutrient-rich wheat cultivars to enhance the micronutrient concentration of iron and zinc in the resulting cultivars. As a result of targeted improvements for nutritional traits, 393 nutrient-rich crop varieties have been released or tested in 63 countries worldwide, benefiting 48 million people (Virk et al., 2021). The creation of wheat cultivars high in Fe and Zn has accelerated in India over the past several years. Fourteen wheat cultivars that were biofortified and younger than 5 years were added to seed chains during 2021, with DBW187 (2315q), HD8759 (420q), and DDW 47 (269.5q), producing the most breeder seed. Although the 14 biofortified wheat varieties yielded 3890.60 quintals of breeder seed for later multiplication, they only accounted for 19.54% of the total amount of breeder seed produced during 2020–2021 (Kamble et al., 2022).

The most effective and widely used technique for biofortifying wheat is MAS. The primary concern in wheat is the grain protein content (GPC). Numerous studies worldwide have mapped several quantitative trait loci (QTLs). The QTL Gpc-B1 is largely responsible for the grain protein and mineral contents inherited by hexaploid cultivar species from their parent dicoccoides species. However, the nature of QTLs depends on the absorption, transportation, and storage of nutrients in the grain. Owing to the complexity of these QTLs, conventional breeding has not been entirely successful. Mapping research indicates that iron and zinc are linked at the same loci. QTLs for iron, zinc, and selenium have been identified. The introgression of these QTLs can undoubtedly enhance these features through MAS. However, MAS faces challenges due to the complex polygenic nature of QTLs and their interactions with the environment. Recently, the "genomic selection" or genome-wide selection (GWS) approach was introduced. This method can address these issues as it estimates the genetic breeding value.

The breeding for Zn-enriched wheat in India is headed by CIMMYT and other national organizations. A total of four varieties with a Zn level of 40%–80% were made available in India (Saltzman et al., 2016). According to Velu et al. (2015), the newly created cultivars have a Zn content level in grain that is 20%–40% higher (+8–12 ppm extra level). The first zinc hyperaccumulating variety (PBW1Zn) was recently created by the Punjab Agricultural University and made available for growing in the North Western Plain Zone. The Indian Institute of Wheat and Barley Research (IIWBR), Karnal, released one more variety (WB2), which has high Zn and Fe in grains.

TABLE 1.1
Genes Involved in Uptake, Transport, and Storage of Micronutrients in Wheat

Metals	Transporter Gene Family	References
Fe	Deoxymugineic acid synthase (DMAS), natural resistance-associated macrophage protein (NRAMP), ferric chelate reductases (FRO), iron-regulated/ferroportin (IREG/FPN), ferritin (FER), and iron-deficiency-specific clone 3 (IDS3)	Bashir and Nishizawa (2006), Borg et al. (2012), Borrill et al. (2014), Wang et al. (2019), Mathpal et al. (2018)
Zn/Fe	Zinc–induced facilitator like (ZIFL), yellow stripe or yellow-stripe-like (YS/YSL), zinc-regulated transporter (ZRT)/iron-regulated transporter-like protein (IRT) (ZIP)	Kumar et al. (2019), Sharma et al. (2019), Wang et al. (2019)
Cd/Zn	Heavy metal atpases (HMA)	Tan et al. (2013)
Mn/Zn/Fe	Vacuolar iron transporter (VIT), metal tolerance protein (MTP)	Connorton et al. (2017), Wang et al. (2019), Sharma et al. (2020), Vatansever et al. (2017)
Heavy metals	Vacuolar iron transporter-like (VITL)	Sharma et al. (2020)
	Phytosiderophores	
DMA/MA/NA for transporting Fe/Zn/Mn/Cu/Ni	Transporter of mugineic acid (TOM), Nicotianamine aminotransferase (NAAT), Nicotianamine synthase (NAS)	Bonneau et al. (2016), Wang et al. (2019), Gautam et al. (2023)

Increasing the bioaccumulation of micronutrients, increasing the amount of the enzyme needed to produce the desired nutrition, downregulating the genes involved in nutrient degradation, downregulating the branching pathway-related genes, enhancing the bioavailability, and using a transgenic approach are all examples of molecular breeding for biofortification, the details of which are given in Table 1.1.

Researchers have mapped recombinant inbred populations with a wide segregation range in the genomic region associated with grain zinc (GZn) and grain iron (GFe) to identify QTLs to enhance the nutritional quality of wheat (Crespo-Herrera et al., 2017). Using diversity arrays technology (DArT) markers, MAS increased the GZn efficiency of wheat to 31%–90% after separating two QTLs on chromosomes 1B and 6B among recombinant inbred lines (RILs). By combining the genic sequence from RNA-seq with the genomic data from whole-genome shotgun sequencing of polyploid wheat, it may be possible to identify genome-specific markers linked to genes controlling the accumulation of grain iron and zinc. Furthermore, mining alleles and combining genomic tools with breeding strategies, gene knockdown (RNA interference), oligo-directed mutagenesis, high-throughput SNP genotyping, transcriptome analysis, next-generation sequencing, and gene-editing methods like CRISPR/Cas, TALENs, and MegaN are being used for trait-targeted biofortification in rice, wheat, and maize (McMullen et al., 2009; Abdallah et al., 2015).

Biofortification has been successful in various crops, including rice and wheat, where the targeted crop's gene pool contains exploitable genetic diversity. However, trade-offs between micronutrients remain a significant barrier to producing crops biofortified with several minerals. One way to find a trade-off between iron and zinc accumulation in whole grains is to analyze a mapping population designed for iron and zinc biofortification of wheat. Similar findings have been observed (Haydon et al., 2012; Briat et al., 2015), where increased zinc accumulation inhibits iron accumulation and vice versa.

Additionally, a tripartite interaction between phosphorus, zinc, and iron may impact the accumulation of micronutrients, leading to an overabundance of zinc and iron when phosphorus is lacking (Zheng et al., 2009). This shows the complexity of tissue-specific accumulation, remobilization, and absorption of nutrients in agricultural plants, indicating that the accumulation of one nutrient may be at the expense of another. Further physiological and molecular research is necessary due to the interplay of multiple nutrients that accumulate in edible components in addition to these two.

In situations with limited genetic diversity, poor heritability, and linkage drag the transgenic method becomes crucial for biofortification. For instance, there's no genetic diversity in the iron content of the endosperms of the three most commonly farmed cereal grains: rice, wheat, and maize (Wang et al., 2011; Jhinjer et al., 2018). Yet, expressing the soybean's ferritin gene has enhanced the Fe content of winter wheat (Xiaoyan et al., 2012). Moreover, upregulating the phytochrome gene (*phyA*) has been shown to increase phytase activity, thereby increasing iron's bioavailability in wheat (Brinch-Pedersen et al., 2000). Silencing the ATP-binding cassette transporter subfamily C member 13 (*ABCC13*) transporter gene in wheat reduces the concentration of phytic acid, a prominent antinutritional factor in seeds.

While the transgenic approach is seen as a potential future method for biofortification, its acceptance among consumers remains low due to legislative issues and lack of awareness. Genetic engineering techniques like CRISPR/Cas9, TALEN, and Zinc Finger Nuclease (ZFN) can alter how phenotypes express unwanted genetic sequences. Mutagenic breedias also offered advantages by creating novel mutants with traits such as high iron content, low phytase content, and so on.

However, the use of molecular methods for biofortification has its drawbacks. These include the presence of nutrients like phytic acid and polyphenols, a decrease in yield with an increase in micronutrients, low bioavailability of micronutrients, and complex interactions between QTLs and the environment.

The combination of genomic and genetic engineering techniques has improved biofortification in various cereal crops, including rice, wheat, maize, and oats. It has addressed every genetic complexity, transforming the challenges of biofortification into a smoother process with plenty of room for progress. In urgent cases, it also offers time-saving benefits.

1.4 MICROBES IN BIOFORTIFICATION

Plant growth-promoting rhizobia (PGPRs) and specific endophytes (non-pathogenic microorganisms that colonize the internal tissues of plants and the rhizosphere) are known for their nitrogen-fixing abilities and their capacity to absorb other micronutrients.

They have been shown to promote plant growth, enhance yields, and increase nutrient levels in grains across various crop plants (Lobo et al., 2019). While there is a solid understanding of the physiological and molecular processes via which plants get nutrients through microbes, the extensive use of microbes in nutrient management for field crops is constrained by practical challenges (Jacoby et al., 2017).

On the contrary, it has been shown that microorganisms that promote plant development not only improve soil fertility, plant growth, and yield but also raise the concentrations of several micronutrients in crop grains (Rana et al., 2012; Ramesh et al., 2014).

Symbiotic relationships depend heavily on metabolic exudates, which are mostly secreted by plant roots. The transfer of fatty acids via ATP-binding cassette transporter subfamily G (ABCG) proteins, including oleic, stearic, and linoleic acid, is necessary to promote these connections (Jiang et al., 2017). Different root exudates include significant concentrations of specific plant-based secondary metabolites, including hormones, coumarins, sterols, and glucosinolates. These chemicals are probably transferred to the mycorrhizal zone by ABCG transporters. In turn, the endophytes can improve the uptake and retention of micronutrients such as boron, zinc, iron, flavonoids, vitamins B and C, sapogenins, and saponins (Wu et al., 2013). They also contribute to nitrogen fixation (N_2) and the uptake of essential minerals in crops.

Given the undesirable environmental effects of synthetic N_2 fertilizers (Commoner, 1977), microbial enrichment presents a potentially "greener" way to provide N_2 to crop roots. Research has shown that certain microbes such as *Sedum alfredii*, the zinc hyper-accumulating plant, specifically *Sphinogomonas sp.* SaMR12 and *Enterobacter sp.* ScCS20, can effectively colonize the rhizosphere and cause greater zinc bioavailability, leading to an increase in zinc levels in rice grains (Wang et al., 2014). Moreover, co-inoculation of wheat plants with these bacteria showed better effectiveness than bacterial infections alone. Grain selenium levels rose in wheat plants treated with *Bacillus* sp. R12 or *Pseudomonas* sp. R8 in conjunction with *Glomus clarideum* (Durán et al., 2013).

Injecting wheat roots with mycorrhizal fungi enhances root structure and nutrient absorption, including iron, zinc, and phosphorus. In sulfur-deficient maize, mycorrhizal associations alter key iron homeostasis gene expression, suggesting sulfur's crucial signaling role. Microbes can fortify plants with vitamins, flavonoids, and various micronutrients. These facts indicate that endophytes are at the forefront of microbe-mediated biofortification efforts (Saridis et al., 2019).

1.5 WHEAT BIOFORTIFICATION PROGRAMMES

Numerous global initiatives have been launched to combat micronutrient deficiencies and improve nutritional status, with various organizations and programs playing pivotal roles in this effort.

At the international level, organizations such as the Consultative Group on International Agricultural Research (CGIAR), the International Maize and Wheat Improvement Center (CIMMYT), and the World Health Organization (WHO) are collaborating to address micronutrient malnutrition. One prominent program within CGIAR is "Harvest Plus," which employs conventional breeding techniques to enhance micronutrient levels, particularly focusing on iron, zinc, and provitamin A.

This program operates in countries across Africa, Asia, and Latin America, aiming to reduce micronutrient malnutrition (Ortiz-Monasterio et al., 2007).

In India, central and state governments have initiated several programs and schemes to improve the nutritional health of the population. Some of these programs include the National Iron Plus Initiative (NIPI), the National Iodine Deficiency Disorders Control Programme (NIDDCP), the Integrated Child Development Services (ICDS) scheme, and the Mid-Day Meal Programme. Additionally, in 2018, India launched the flagship program "POSHAN Abhiyaan" under the Ministry of Women and Child Development. This initiative specifically targets the reduction of stunting, undernutrition, and anemia, with a primary focus on children, adolescent girls, and women (NITI Aayog, 2019).

Despite these efforts, there remains a critical need to reinforce innovative, cost-effective, and sustainable agricultural systems that can effectively combat micronutrient deficiencies and improve human health in the future.

1.6 CHALLENGES AND LIMITATIONS

Micronutrient deficiency is a pressing issue worldwide, especially in Asia and Africa, necessitating solutions for its eradication. Genetic biofortification of food crops through advanced techniques offers a cost-effective and highly advantageous approach to addressing the nutritional needs of approximately 40% of the population heavily reliant on food for nourishment.

In rural populations, an initial financial investment in the development of food crop seeds with effective micronutrient absorption can guarantee sufficient intake of vital elements like zinc and iron. It has been shown that genetic biofortification is more economical than other approaches including dietary modification, agronomic fortification, and supplementation programs. According to commercial evaluations, rice biofortification in South Asia can achieve advantageous costbenefit ratios of more than 20% for zinc and between 20% and 30% for iron (Bouis, 2003). Informal assessments also point to high ROIs for fertilization using iodine or selenium.

Biofortification efforts have been hindered by financial constraints and complex plant biology, including the effects of various factors such as CO_2 levels, rhizospheric microbes, and soil conditions on nutrient dynamics (García-Latorre et al., 2021). Issues like heavy metal accumulation, mineral trade-offs, and CO_2 dilution effects pose challenges. Addressing these is vital for future nutrient supply.

Further research is needed on the impact of pesticide use, inorganic fertilizers, and climate change on soil chemistry and microbial diversity. Future studies should focus on understanding and managing plant nutrients in their natural environments, including soil and air. The success of agronomic biofortification varies due to differences in mineral mobility, plant species, and soil compositions across regions. Challenges include antinutrient compounds that hinder mineral bioavailability and the environmental impact of fertilizer accumulation.

Genetic variability for micronutrient quantity within the gene pool of a particular cultivar can be limited. Breeding for specific attributes can be challenging and time-consuming, making it impractical for certain traits, such as selenium concentration in wheat grains (Lyons et al., 2005) and fatty acid content in soybeans

(Oliva et al., 2006). Transgenic crops can overcome genetic limitations but face acceptance issues among end users, along with lengthy and expensive regulatory processes.

Post-harvest processing considerations are essential for optimizing biofortification strategies. For example, many cereal seeds are consumed after polishing or milling despite certain essential minerals being concentrated in the embryo. Antinutrients in crops can reduce mineral bioavailability both within the crops and in the human gut. An integrated strategy and a greater comprehension of the intricate nutrient dynamics in the rhizosphere and inside the plant are needed to solve these issues (Singh et al., 2023).

In India, the prospective effect of biofortification is evaluated in terms of lowering years of life with a handicap. Furthermore, biofortification may significantly reduce the iron deficiency (anemia) and zinc deficiency, resulting in millions of years of life in India with disabilities spared (Stein et al., 2007). Biofortification of key food crops such as maize, wheat, and rice with zinc and iron in African countries is estimated to cost between US$2 and US$20 per saved disability-adjusted life year (DALY). This indicates that biofortification is a cost-effective approach for reducing the impact of micronutrient deficiencies and improving overall nutrition in these regions (Tayade et al., 2023).

1.7 BIOAVAILABILITY OF MICRONUTRIENTS

In addition to the number of micronutrients like zinc and iron in wheat, the availability of these micronutrients for absorption is a critical factor in enhancing their intake. Zinc and iron in wheat grains are often present in forms that are not readily usable by humans, particularly among infants, pregnant, and lactating women who primarily consume cereal-based foods (Al Hasan et al., 2016). To address this, biofortified cereal foods are essential to promote higher micronutrient absorption.

Bioavailability, which quantifies the proportion of micronutrients that the human body can utilize from the total dietary intake, varies in different reports. Kenzhebayeva et al. (2019) reported that zinc has a 25% bioavailability and 5% for iron. Mares and Haase (2020) found that zinc has 16%–50% bioavailability, which is inversely related to oral zinc intake.Finally, mixed or vegetarian diets based on refined cereal grains have a bioavailability of 26%–34% for zinc, while unrefined cereal-based diets have an 18%–26% bioavailability for zinc, as reported by Brown et al. (2004). Depending on the wheat variety, the total amount of accessible zinc in flour might vary from 1.52 to 2.7 mg per 300 g of flour (Rosado et al., 2009).

Numerous internal and external factors affect bioavailability, including

i. dietary micronutrient content,
ii. a specific form of micronutrients in the intestine,
iii. intestinal micronutrient bioaccessibility,
iv. the phytate-to-Zn/Fe molar ratio, and
v. the method used to estimate bioavailability.

For instance, an increase in zinc intake has been shown to reduce bioavailability. Zinc absorption decreases from 21% without phytate to 11%–16% at a phytate-to-Zn molar ratio of 5–15 and even further to 4%–11% at molar ratios exceeding 15. These inhibitory phytate-Zn complexes become stronger in the presence of calcium, although calcium's exacerbating effect has not been consistently observed in human dietary studies. Unlike phytate, fibers like cellulose do not significantly affect zinc absorption. Nevertheless, biofortification has been demonstrated to enhance bioavailability (Rosado et al., 2009).

1.7.1 Phytic Acid and Bioavailability of Micronutrients

About 85% of the total phosphorus in cereals is stored as phytic acid (PA) in the grains and other plant parts (Guttieri et al., 2004; Cominelli et al., 2020). PA, owing to its negative charge, can chelate various cations, including micronutrients like zinc (Zn^{2+}) and iron (Fe^{2+}), forming compounds known as iron or zinc phytate. Due to their highly negative charge across a wide pH range, phytates adversely affect the bioavailability of multiple micronutrients, including Zn^{2+} and $Fe^{2+/3+}$ (Wu et al., 2009).

In human diets, high phytate content is associated with reduced bioavailability of micronutrients such as zinc (Guttieri et al., 2004), iron (Hurrell et al., 2003), and magnesium (Ranhotra, 1983). Among these, the reduced bioavailability of zinc and iron due to elevated PA intake is particularly significant in the context of biofortification efforts (Al Hasan et al., 2016). The bioavailability of iron, when chelated with PA, is also influenced by the oxidation state of iron (Fe^{3+} or Fe^{2+}), which are its common ionic forms involved in chelation. The iron state can be altered by food preparation methods. For example, ferritin can release iron in either the Fe^{3+} or Fe^{2+} state upon heating ($Fe^{3+} \leftrightarrow Fe^{2+}$) (Hoppler et al., 2008). However, in general, Fe^{3+} tends to be chelated and immobilized by PA, thus reducing its bioavailability to plants.

1.8 FUTURE PROSPECTS AND CONCLUSION

Wheat is a vital crop, but traditional breeding has prioritized yield over nutrient content. To address nutritional security, we need to improve both quality and quantity of wheat. Biofortification of wheat involves increasing its nutritional value by leveraging natural genetic diversity, advanced genomic tools, and breeding strategies including molecular breeding, QTL mapping, genomic selection, genome-wide association study (GWAS), and genome engineering.

Studies have identified wild wheat species, *T. dicoccoides*, *Aegilops tauschii*, *T. dicoccoides*, *T. boeoticun*, *T. spelta*, *T. polonicum*, and *T. aestivum* landraces with higher concentrations of minerals like iron, zinc, and manganese. QTL mapping and GWAS have pinpointed specific genetic loci responsible for grain zinc and iron levels. Transgenic methods have also been employed to enhance the content of iron, manganese, and vitamin A in wheat.

However, genome-editing technologies, particularly the CRISPR/Cas system, have gained popularity due to their precision, simplicity, and ability to eliminate foreign DNA from the modified plants. While genome editing has been used in wheat to improve stress tolerance and yield, it has not been extensively explored for biofortification. Therefore, there is a need to utilize genome editing to enhance the content of essential nutrients like Fe, Zn, Se, Ca, and essential amino acids while reducing antinutrients like PA.

Additional genome-editing techniques, such as multiplex gene editing, transient expression of CRISPR/Cas9, base editing, prime editing, and CRISPR/Cas9 ribonucleoproteins, hold promise and should be considered for future research.

Exploring natural genetic diversity through whole-genome sequencing and functional genomics can provide valuable insights for improving wheat traits (Phogat et al., 2020).

To accelerate global efforts to enhance wheat nutritional quality, gene editing in germline cells, CRISPR systems with RNA interference elements, epigenetic genome modifications, and simulation model-based predictions for wheat quality traits under diverse environmental conditions should be explored (Mohan et al., 2022). Combining these techniques with high-yielding traits is crucial for addressing the nutritional needs of the growing population.

REFERENCES

Abdallah, N. A., Prakash, C. S., & McHughen, A. G. (2015). Genome editing for crop improvement: Challenges and opportunities. *GM Crops & Food*, *6*(4), 183–205.

Aciksoz, S. B., Yazici, A., Ozturk, L., & Cakmak, I. (2011). Biofortification of wheat with iron through soil and foliar application of nitrogen and iron fertilizers. *Plant and Soil*, *349*, 215–225.

Al Hasan, S. M., Hassan, M., Saha, S., Islam, M., Billah, M., & Islam, S. (2016). Dietary phytate intake inhibits the bioavailability of iron and calcium in the diets of pregnant women in rural Bangladesh: A cross-sectional study. *BMC Nutrition*, *2*, 1–10.

Alloway, B. J. (2009). Soil factors associated with zinc deficiency in crops and humans. *Environmental Geochemistry and Health*, *31*(5), 537–548.

Bashir, K., & Nishizawa, N. K. (2006). Deoxymugineic acid synthase: A gene important for Fe-acquisition and homeostasis. *Plant Signaling & Behavior*, *1*(6), 290–292.

Bonneau, J., Baumann, U., Beasley, J., Li, Y., & Johnson, A. A. (2016). Identification and molecular characterization of the nicotianamine synthase gene family in bread wheat. *Plant Biotechnology Journal*, *14*(12), 2228–2239.

Borg, S., Brinch-Pedersen, H., Tauris, B., Madsen, L. H., Darbani, B., Noeparvar, S., & Holm, P. B. (2012). Wheat ferritins: Improving the iron content of the wheat grain. *Journal of Cereal Science*, *56*(2), 204–213.

Borrill, P., Connorton, J., Balk, J., Miller, T., Sanders, D., & Uauy, C. (2014). Biofortification of wheat grain with iron and zinc: Integrating novel genomic resources and knowledge from model crops. *Frontiers in Plant Science*, *5*, 73651.

Bouis, H. E. (2003). Micronutrient fortification of plants through plant breeding: Can it improve nutrition in man at low cost? *Proceedings of the Nutrition Society*, *62*(2), 403–411.

Bouis, H. E., & Saltzman, A. (2017). Improving nutrition through biofortification: A review of evidence from HarvestPlus, 2003 through 2016. *Global Food Security*, *12*, 49–58.

Bouis, H. E., & Welch, R. M. (2010). Biofortification—A sustainable agricultural strategy for reducing micronutrient malnutrition in the global south. *Crop science*, *50*, S–20.

Briat, J. F., Rouached, H., Tissot, N., Gaymard, F., & Dubos, C. (2015). Integration of P, S, Fe, and Zn nutrition signals in *Arabidopsis thaliana*: Potential involvement of phosphate starvation response 1 (PHR1). *Frontiers in Plant Science*, *6*, 129608.

Brinch-Pedersen, H., Olesen, A., Rasmussen, S. K., & Holm, P. B. (2000). Generation of transgenic wheat (*Triticum aestivum* L.) for constitutive accumulation of an *Aspergillus phytase*. *Molecular Breeding*, *6*, 195–206.

Brown, K. H., Rivera, J. A., Bhutta, Z., Gibson, R. S., King, J. C., Lönnerdal, B., ... & International Zinc Nutrition Consultative Group (IZiNCG. (2004). International zinc nutrition consultative group (IZiNCG) technical document# 1. Assessment of the risk of zinc deficiency in populations and options for its control. *Food and Nutrition Bulletin*, *25*(2), S99–S203.

Cakmak, I. (2008). Enrichment of cereal grains with zinc: Agronomic or genetic biofortification? *Plant and Soil*, *302*, 1–17.

Cominelli, E., Pilu, R., & Sparvoli, F. (2020). Phytic acid and transporters: What can we learn from low phytic acid mutants? *Plants*, *9*(1), 69.
Commoner, B. (1977). Cost-risk-benefit analysis of nitrogen fertilization: A case history. *Ambio*, *6*, 157–161.
Connorton, J. M., Jones, E. R., Rodríguez-Ramiro, I., Fairweather-Tait, S., Uauy, C., & Balk, J. (2017). Wheat vacuolar iron transporter TaVIT2 transports Fe and Mn and is effective for biofortification. *Plant Physiology*, *174*(4), 2434–2444.
Crespo-Herrera, L. A., Govindan, V., Stangoulis, J., Hao, Y., & Singh, R. P. (2017). QTL mapping of grain Zn and Fe concentrations in two hexaploid wheat RIL populations with ample transgressive segregation. *Frontiers in Plant Science*, *8*, 292565.
De Valença, A. W., Bake, A., Brouwer, I. D., & Giller, K. E. (2017). Agronomic biofortification of crops to fight hidden hunger in sub-Saharan Africa. *Global Food Security*, *12*, 8–14.
Dhaliwal, S. S., Sharma, V., Shukla, A. K., Verma, V., Kaur, M., Shivay, Y. S., … & Hossain, A. (2022). Biofortification: A frontier novel approach to enrich micronutrients in field crops to encounter the nutritional security. *Molecules*, *27*(4), 1340.
Duran, P., Acuña, J. J., Jorquera, M. A., Azcón, R., Borie, F., Cornejo, P., & Mora, M. L. (2013). Enhanced selenium content in wheat grain by co-inoculation of selenobacteria and arbuscular mycorrhizal fungi: A preliminary study as a potential Se biofortification strategy. *Journal of Cereal Science*, *57*(3), 275–280.
El-Saadony, M. T., El-Wafai, N. A., El-Fattah, H. I. A., & Mahgoub, S. A. (2019). Biosynthesis, optimization and characterization of silver nanoparticles using a soil isolate of *Bacillus pseudomycoides* MT32 and their antifungal activity against some pathogenic fungi. *Advances in Animal and Veterinary Sciences*, *7*, 238–249,
Fageria, N. K., & Baligar, V. C. (2008). Ameliorating soil acidity of tropical oxisols by liming for sustainable crop production. *Advances in Agronomy*, *99*, 345–399.
Fageria, N. K., Baligar, V. C., & Clark, R. B. (2002). Micronutrients in crop production. *Advances in Agronomy*, *77*, 185–268.
FAO, IFAD, UNICEF, WFP, WHO. (2021). The state of food security and nutrition in the world. In: *Transforming Food Systems for Food Security, Improved Nutrition and Affordable Healthy Diets for All*, FAO, Québec City, QC, Canada.
Frossard, E., Bucher, M., Mächler, F., Mozafar, A., & Hurrell, R. (2000). Potential for increasing the content and bioavailability of Fe, Zn and Ca in plants for human nutrition. *Journal of the Science of Food and Agriculture*, *80*(7), 861–879.
Garcia-Latorre, C., Rodrigo, S., & Santamaria, O. (2021). Effect of fungal endophytes on plant growth and nutrient uptake in *Trifolium subterraneum* and *Poa pratensis* as affected by plant host specificity. *Mycological Progress*, *20*(9), 1217–1231.
Garg, M., Sharma, N., Sharma, S., Kapoor, P., Kumar, A., Chunduri, V., & Arora, P. (2018). Biofortified crops generated by breeding, agronomy, and transgenic approaches are improving lives of millions of people around the world. *Frontiers in Nutrition*, *5*, 301899.
Gautam, T., Jan, I., Batra, R., Singh, K., Pandey, R., Sharma, P. K., … & Gupta, P. K. (2023). Further studies on nicotianamine aminotransferase (NAAT) genes involved in biofortification in bread wheat (*Triticum aestivum* L.). *Plant Gene*, *33*, 100389.
Grusak, M. A. (2002). Enhancing mineral content in plant food products. *Journal of the American College of Nutrition*, *21*(3), 178S–183S.
Guttieri, M., Bowen, D., Dorsch, J. A., Raboy, V., & Souza, E. (2004). Identification and characterization of a low phytic acid wheat. *Crop Science*, *44*(2), 418–424.
Haydon, M. J., Kawachi, M., Wirtz, M., Hillmer, S., Hell, R., & Krämer, U. (2012). Vacuolar nicotianamine has critical and distinct roles under iron deficiency and for zinc sequestration in Arabidopsis. *The Plant Cell*, *24*(2), 724–737.
Hoppler, M., Schönbächler, A., Meile, L., Hurrell, R. F., & Walczyk, T. (2008). Ferritin-iron is released during boiling and in vitro gastric digestion, 3. *The Journal of Nutrition*, *138*(5), 878–884.

Horton, S., & Mannar, M. V. (2018). Economics of food fortification. In: M. G. Venkatesh Mannar & R. F. Hurrell (Eds.), *Food Fortification in a Globalized World*, Academic Press, London, pp. 299–304.

Hurrell, R. F., Reddy, M. B., Juillerat, M. A., & Cook, J. D. (2003). Degradation of phytic acid in cereal porridges improves iron absorption by human subjects. *The American Journal of Clinical Nutrition*, *77*(5), 1213–1219.

Hurrell, R., Ranum, P., de Pee, S., Biebinger, R., Hulthen, L., Johnson, Q., & Lynch, S. (2010). Revised recommendations for iron fortification of wheat flour and an evaluation of the expected impact of current national wheat flour fortification programs. *Food and Nutrition Bulletin*, *31*(11), S7–S21.

Ibba, M. I., Gupta, O. P., Govindan, V., Johnson, A. A. T., Brinch-Pedersen, H., Nikolic, M., & Taleon, V. (2022). Wheat biofortification to alleviate global malnutrition. *Frontiers in Nutrition*, *9*, 1001443.

Iizumi, T., Furuya, J., Shen, Z., Kim, W., Okada, M., Fujimori, S., … & Nishimori, M. (2017). Responses of crop yield growth to global temperature and socioeconomic changes. *Scientific Reports*, *7*(1), 7800.

Jacoby, R., Peukert, M., Succurro, A., Koprivova, A., & Kopriva, S. (2017). The role of soil microorganisms in plant mineral nutrition—Current knowledge and future directions. *Frontiers in Plant Science*, *8*, 292271.

Jeje, O., Oluwafemi, G., & Omowaye-Taiwo, O. (2023). Mitigation of hidden hunger through biofortification: An appraisal. *International Journal of Food Sciences and Nutrition*, *6*(1), 13–29.

Jhinjer, R. K., Mavi, G. S., Sood, N., Malhotra, A., Kaur, H., Kumar, V., … & Sohu, V. S. (2018). Variation in zinc, iron and quality parameters in wheat lines at different sowing locations. *International Journal of Pure & Applied Bioscience*, *6*(2), 146–155.

Jiang, Y., Wang, W., Xie, Q., Liu, N. A., Liu, L., Wang, D., … & Wang, E. (2017). Plants transfer lipids to sustain colonization by mutualistic mycorrhizal and parasitic fungi. *Science*, *356*(6343), 1172–1175.

Kamble, U., Mishra, C. N., Govindan, V., Sharma, A. K., Pawar, S., Kumar, S., … & Singh, G. (2022). Ensuring nutritional security in India through wheat biofortification: A review. *Genes*, *13*(12), 2298.

Kenzhebayeva, S., Abekova, A., Atabayeva, S., Yernazarova, G., Omirbekova, N., Zhang, G., … & Wang, Y. (2019). Mutant lines of spring wheat with increased iron, zinc, and micronutrients in grains and enhanced bioavailability for human health. *BioMed Research International*, 2019.

Khush, G. S., Lee, S., Cho, J. I., & Jeon, J. S. (2012). Biofortification of crops for reducing malnutrition. *Plant Biotechnology Reports*, *6*, 195–202.

Kumar, A., Kaur, G., Goel, P., Bhati, K. K., Kaur, M., Shukla, V., & Pandey, A. K. (2019). Genome-wide analysis of oligopeptide transporters and detailed characterization of yellow stripe transporter genes in hexaploid wheat. *Functional & Integrative Genomics*, *19*, 75–90.

Lobo, L. L. B., dos Santos, R. M., & Rigobelo, E. C. (2019). Promotion of maize growth using endophytic bacteria under greenhouse and field conditions. *Australian Journal of Crop Science*, *13*(12), 2067–2074.

Ludwig, Y., & Slamet-Loedin, I. H. (2019). Genetic biofortification to enrich rice and wheat grain iron: From genes to product. *Frontiers in Plant Science*, *10*, 450726.

Lyons, G., Ortiz-Monasterio, I., Stangoulis, J., & Graham, R. (2005). Selenium concentration in wheat grain: Is there sufficient genotypic variation to use in breeding? *Plant and Soil*, *269*, 369–380.

Malaguti, M., Dinelli, G., Leoncini, E., Bregola, V., Bosi, S., Cicero, A. F., & Hrelia, S. (2014). Bioactive peptides in cereals and legumes: Agronomical, biochemical and clinical aspects. *International Journal of Molecular Sciences*, *15*(11), 21120–21135.

Mares, M., & Haase, H. (2020). A guide to human zinc absorption: General overview and recent advances of in vitro intestinal models. *Nutrients*, *12*(3), 762.

Mathpal, P., Kumar, U., Kumar, A., Kumar, S., Malik, S., Kumar, N., ... & Kumar, S. (2018). Identification, expression analysis, and molecular modeling of iron-deficiency-specific clone 3 (Ids3)-like gene in hexaploid wheat. *3 Biotech*, *8*, 1–11.

McMullen, M. D., Kresovich, S., Villeda, H. S., Bradbury, P., Li, H., & Sun, Q. (2009). Genetic properties of the maize nested association mapping population. *Science*, *325*,737–740

Meenakshi, J. V. (2009). *Best Practice Paper: Cost-Effectiveness of Biofortification*, Copenhagen Consensus Center, Lowell, MA.

Mohan, D., Sendhil, R., Gupta, O. P., Pandey, V., Krishnappa, G., & Singh, G. P. (2022). Wheat quality index: New holistic approach to identify quality superior genotypes. *Cereal Research Communications*, *50*(4), 1105–1115.

Mulualem, T. (2015). Application of bio-fortification through plant breeding to improve the value of staple crops. *Biomedicine and Biotechnology*, *3*(1), 11–19.

Nestel, P. (1993). *Food Fortification in Developing Countries*, US Agency for International Development, Washington, DC.

NITI Aayog. (2019). India. Retrieved December 11, 2019 from https://www.niti.gov.in/verticals/women-and-child-development.

Oliva, M. L., Shannon, J. G., Sleper, D. A., Ellersieck, M. R., Cardinal, A. J., Paris, R. L., & Lee, J. D. (2006). Stability of fatty acid profile in soybean genotypes with modified seed oil composition. *Crop Science*, *46*(5), 2069–2075.

Ortiz-Monasterio, J. I., Palacios-Rojas, N., Meng, E., Pixley, K., Trethowan, R., & Pena, R. J. (2007). Enhancing the mineral and vitamin content of wheat and maize through plant breeding. *Journal of Cereal Science*, *46*(3), 293–307.

Pfeiffer, W. H., & McClafferty, B. (2007). HarvestPlus: Breeding crops for better nutrition. *Crop Science*, *47*, S–88.

Phogat, B. S., Kumar, S., Kumari, J., Kumar, N., Pandey, A. C., Singh, T. P., ... & Bansal, K. C. (2020). Characterization of wheat germplasm conserved in the Indian National Genebank and establishment of a composite core collection. *Crop Science*, *61*(1), 604–620.

Ramesh, A., Sharma, S. K., Sharma, M. P., Yadav, N., & Joshi, O. P. (2014). Inoculation of zinc solubilizing *Bacillus aryabhattai* strains for improved growth, mobilization and biofortification of zinc in soybean and wheat cultivated in vertisols of central India. *Applied Soil Ecology*, *73*, 87–96.

Ramzani, P. M. A., Khalid, M., Naveed, M., Ahmad, R., & Shahid, M. (2016). Iron biofortification of wheat grains through integrated use of organic and chemical fertilizers in pH affected calcareous soil. *Plant Physiology and Biochemistry*, *104*, 284–293.

Rana, A., Joshi, M., Prasanna, R., Shivay, Y. S., & Nain, L. (2012). Biofortification of wheat through inoculation of plant growth promoting rhizobacteria and cyanobacteria. *European Journal of Soil Biology*, *50*, 118–126.

Ranhotra, G. S. (1983). Bioavailability of magnesium in cereal-based foods. *Cereal Foods World (USA)*, *28*, 6.

Rawat, N., Neelam, K., Tiwari, V. K., & Dhaliwal, H. S. (2013). Biofortification of cereals to overcome hidden hunger. *Plant Breeding*, *132*(5), 437–445.

Rehman, A., Farooq, M., Ozturk, L., Asif, M., & Siddique, K. H. (2018). Zinc nutrition in wheat-based cropping systems. *Plant and Soil*, *422*, 283–315.

Rengel, Z., Batten, G. D., & Crowley, D. D. (1999). Agronomic approaches for improving the micronutrient density in edible portions of field crops. *Field Crops Research*, *60*(1–2), 27–40.

Rosado, J. L., Hambidge, K. M., Miller, L. V., Garcia, O. P., Westcott, J., Gonzalez, K., ... & Krebs, N. F. (2009). The quantity of zinc absorbed from wheat in adult women is enhanced by biofortification. *The Journal of Nutrition*, *139*(10), 1920–1925.

Saltzman, A., Andersson, M. S., Asare-Marfo, D., Lividini, K., De Moura, F. F., Moursi, M., Oparinde, A., & Taleon, V. (2016). Biofortification techniques to improve food security. In *Reference Module in Food Science* (1st ed.), pp. 1–9.

Saridis, G., Chorianopoulou, S. N., Ventouris, Y. E., Sigalas, P. P., & Bouranis, D. L. (2019). An exploration of the roles of ferric iron chelation-strategy components in the leaves and roots of maize plants. *Plants*, *8*(5), 133.

Sharma, D., Ghimire, P., Bhattarai, S., Adhikari, U., Khanal, S., & Poudel, P. B. (2020). Biofortification of wheat: Genetic and agronomic approaches and strategies to combat Iron and Zinc deficiency. *International Journal of Environment, Agriculture and Biotechnology*, *5*, 4.

Sharma, S., Kaur, G., Kumar, A., Meena, V., Kaur, J., & Pandey, A. K. (2019). Overlapping transcriptional expression response of wheat zinc-induced facilitator-like transporters emphasize important role during Fe and Zn stress. *BMC Molecular Biology*, *20*, 1–17.

Shewry, P. R., & Hey, S. J. (2015). The contribution of wheat to human diet and health. *Food and Energy Security*, *4*(3), 178–202.

Singh, S. K., Barman, M., Sil, A., Prasad, J. P., Kundu, S., & Bahuguna, R. N. (2023). Nutrient biofortification in wheat: Opportunities and challenges. *Cereal Research Communications*, *51*(1), 15–28.

Stein, A. J., Nestel, P., Meenakshi, J. V., Qaim, M., Sachdev, H. P. S., & Bhutta, Z. A. (2007). Plant breeding to control zinc deficiency in India: How cost-effective is biofortification? *Public Health Nutrition*, *10*(5), 492–501.

Stewart, R., Van Rooyen, C., Dickson, K., Majoro, M., & De Wet, T. (2010). What is the impact of microfinance on poor people?: A systematic review of evidence from sub-Saharan Africa. *Technical Report*. EPPI-Centre, Social Science Research Unit, University of London, London.

Tan, J., Wang, J., Chai, T., Zhang, Y., Feng, S., Li, Y., … & Chai, X. (2013). Functional analyses of T a HMA 2, a P 1B-type ATP ase in wheat. *Plant Biotechnology Journal*, *11*(4), 420–431.

Tayade, R., Kabange, R., Ali, M. W., Yun, B. W., & Nabi, R. B. S. (2023). Biofortification of wheat using current resources and future challenges. In: R. Deshmukh, A. Nadaf, W. A. Ansari, K. Singh, & H. Sonah (Eds.), *Biofortification in Cereals: Progress and Prospects*, Springer Nature Singapore, Singapore, pp. 173–208.

Vatansever, R., Filiz, E., & Eroglu, S. (2017). Genome-wide exploration of metal tolerance protein (MTP) genes in common wheat (*Triticum aestivum*): Insights into metal homeostasis and biofortification. *Biometals*, *30*, 217–235.

Velu, G., Singh, R. G., Balasubramaniam, A., Mishra, V. K., Chand, R., Tiwari, C., … & Pfeiffer, W. H. (2015). Reaching out to farmers with high zinc wheat varieties through public-private partnerships: An experience from eastern-gangetic plains of India. *Advances in Food Technology and Nutritional Sciences—Open Journal*, *1*(3), 73–75.

Virk, P. S., Andersson, M. S., Arcos, J., Govindaraj, M., & Pfeiffer, W. H. (2021). Transition from targeted breeding to mainstreaming of biofortification traits in crop improvement programs. *Frontiers in Plant Science*, *12*, 703990.

Wang, K. M., Wu, J. G., Li, G., Yang, Z. W., & Shi, C. H. (2011). Distribution of phytic acid and mineral elements in three indica rice (*Oryza sativa* L.) cultivars. *Journal of Cereal Science*, *54*(1), 116–121.

Wang, M., Kawakami, Y., & Bhullar, N. K. (2019). Molecular analysis of iron deficiency response in hexaploid wheat. *Frontiers in Sustainable Food Systems*, *3*, 67.

Wang, Y., Yang, X., Zhang, X., Dong, L., Zhang, J., Wei, Y., … & Lu, L. (2014). Improved plant growth and Zn accumulation in grains of rice (*Oryza sativa* L.) by inoculation of endophytic microbes isolated from a Zn hyperaccumulator, *Sedum alfredii H. Journal of Agricultural and Food Chemistry*, *62*(8), 1783–1791.

White, P. J., & Broadley, M. R. (2005). Biofortifying crops with essential mineral elements. *Trends in Plant Science*, *10*(12), 586–593.

White, P. J., & Broadley, M. R. (2011). Physiological limits to zinc biofortification of edible crops. *Frontiers in Plant Science*, *2*, 15683.

Winkler, J. T. (2011). Biofortification: Improving the nutritional quality of staple crops. In *Access Not Excess by Charles Pasternak.* Smith-Gordon, pp. 100–112.

Wu, H., Yang, H. Y., You, X. L., & Li, Y. H. (2013). Diversity of endophytic fungi from roots of Panax ginseng and their saponin yield capacities. *SpringerPlus*, *2*, 1–9.

Wu, P., Tian, J. C., Walker, C. E., & Wang, F. C. (2009). Determination of phytic acid in cereals–A brief review. *International Journal of Food Science & Technology*, *44*(9), 1671–1676.

Xiaoyan, S., Yan, Z., & Shubin, W. (2012). Improvement Fe content of wheat (*Triticum aestivum*) grain by soybean ferritin expression cassette without vector backbone sequence. *Journal of Agricultural Biotechnology*, 89247109.

Yang, X. W., Tian, X. H., Lu, X. C., Cao, Y. X., & Chen, Z. H. (2011). Impacts of phosphorus and zinc levels on phosphorus and zinc nutrition and phytic acid concentration in wheat (*Triticum aestivum L.*). *Journal of the Science of Food and Agriculture*, *91*(13), 2322–2328.

Zheng, L., Huang, F., Narsai, R., Wu, J., Giraud, E., He, F., … & Shou, H. (2009). Physiological and transcriptome analysis of iron and phosphorus interaction in rice seedlings. *Plant Physiology*, *151*(1), 262–274.

Zou, C., Du, Y., Rashid, A., Ram, H., Savasli, E., Pieterse, P. J., … & Cakmak, I. (2019). Simultaneous biofortification of wheat with zinc, iodine, selenium, and iron through foliar treatment of a micronutrient cocktail in six countries. *Journal of Agricultural and Food Chemistry*, *67*(29), 8096–8106.

2 Molecular Epidemiology of Cotton Leaf Curl Virus Disease

Recent Developments toward Sustainable Fiber Production

Inderjeet Singh, Heena Jain, Pardeep Kumar Sadh, Shiwani Chahal, Satish Kumar Sain, Sayed Sartaj Sohrab, and Priyanka Siwach

2.1 INTRODUCTION

The textile and clothing industry (TCI) is one of the oldest, largest, and most influential business sectors contributing to the economy worldwide by generating about US$1,000 billion each year (Costa et al., 2020). According to a report by the Organisation for Economic Co-operation and Development (OECD)/Food and Agriculture Organization (OECD) (2022), world consumption of raw cotton is expected to increase from 1,709 to 2,340 (kt)[4] in developed countries, while in developing countries, it is anticipated to rise from 23,819 to 27,953 (kt)[4] until 2030. Currently, India, China, Pakistan, and the United States are dominating among cotton-growing nations, producing about two-thirds of the world's cotton; India being the leading producer of cotton occupies 37.56% of the global cotton output. Several biotic and abiotic factors are responsible for this fall-off, but the plant pathogens have been observed to cause major significant damage due to which cotton yield has not been able to cope with the increasing demand. The annual economic impact of viral infections in the agricultural sector is more than US$30 billion worldwide (Jones, 2021). Numerous viruses are devastating to the cotton crop, among them, Begomoviruses of the *Geminiviridae* family have drawn the most attention since they are the most diverse and emerging plant viruses globally. A begomoviral disease, cotton leaf curl disease (CLCuD) is a major havoc that reduces the yield of cotton worldwide (Mahmood et al., 2021). The disease is observed to be caused by a begomoviral complex, namely, the cotton leaf curl virus (CLCuV) which is transmitted through whitefly. In 1988, a severe outbreak of the disease marked its presence in Asia, after which huge devastation in the form of a series of epidemics was observed, causing great loss to the economies of India and

DOI: 10.1201/9781032642567-3

Pakistan (Biswas et al., 2020). During 1997–1998, a sudden increase in CLCuD was reported in Northern India. The variability of CLCuV and betasatellite molecules in Northern India and other regions have been published earlier in various reports. Most of the begomoviruses identified in the Old World are reported as monopartite, while the New World begomoviruses contain bipartite genomes. In the Old World, most of the monopartite begomovirus is associated with betasatellites. Betasatellites have a small (1.4 kb), circular, ssDNA genome with only one gene known as βC1 gene. The βC1 gene is known as a determinant of pathogenicity, post-transcriptional gene-silencing suppressor, and mediates virus movement. Over many years, dedicated efforts like the use of insecticides and the development of resistant breeding lines/varieties have been implemented to control the incidence of diseases. However, due to the high frequency of recombination among various strains of begomoviruses, a satisfactory control strategy is yet to be achieved. CLCuD continues to make devastations with frequent viral outbreaks from time to time. The Indian subcontinent faced a recent major outbreak in 2015–2016, which caused an anticipated loss of US$630–670 million to the Indian economy (Datta et al., 2017). CLCuD continues to cause huge damage to the agricultural sector too.

Epidemiology is the science of investigating the prevalent pathogen of a disease, tracking the history of the disease, analyzing the ecology of the pathogen, determining changes in disease intensity over time, and suggesting efficient tools to tackle the disease (Jeger, 2020). During the past few decades, viral diseases have triggered the application of molecular biology in the area of epidemiology to understand the molecular mechanisms of disease occurrence. Molecular epidemiology, thus, is an advanced science where the epidemiological study is done based on DNA, RNA, or protein structure, which leads to devising advanced molecular methods to predict and control infections, as shown in Figure 2.1. This study is one such attempt for the same. This chapter starts with a detailed background and history of CLCuD followed by

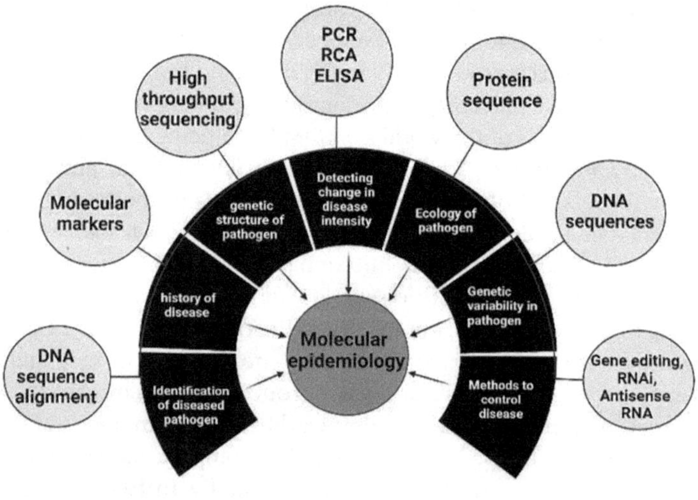

FIGURE 2.1 An overview of molecular epidemiology. Created by Biorender, https://www.biorender.com.

analysis of the genomic structure and complex life cycle of begomoviruses to understand host–vector–virus interactions at the molecular level. Numerous researches have been conducted during the last few years to focus on the mutation and recombination pattern to discover potential targets for controlling the disease, identify gaps, and formulate hypotheses for future steps in the direction of disease management.

2.2 BACKGROUND AND HISTORY

The origin of the disease was observed in Nigeria in 1912 where the symptoms were noticed in the native cultivated species of cotton: *Gossypium vitiform* and *G. peruvianum*. The term "leaf curl" for the disease was devised for the unique and extensive leaf curling symptoms. Downward curling and texture modification in leaves were noticed on the *G. hirsutum* in 1924 in Oyo, Nigeria, and Sudan (Bailey, 1934). Whitefly (*Bemisia tabaci*) was discovered as a vector for the transmission of the leaf curl disease. Studies identified that the geminiviruses act as an agent of disease transmitted by *B. tabaci*. Of 111 virus species transmitted by *B. tabaci*, 90% belong to the begomoviruses (Jones, 2003). CLCuV, a begomoviral species, was revealed as a causative agent of CLCuD in cotton. Some outbreaks of the disease were reported in the cotton fields of Tanzania and Northern America in 1975 (Hussain and Ali, 1975). However, the disease was controlled in Africa and America through vector management and through the cultivation of resistant varieties. However, Asia has suffered highly devastating epidemics in the last four decades. "Multan Epidemic" was the first to be reported in 1988 in Moza Kokhran, Multan, Pakistan, which spread to India during 1990–1999, causing a major loss to cotton production. After this epidemic, the cultivation of resistant varieties of cotton controlled the disease for some time., However, soon after another epidemic known as the Burewala epidemic (2002–2014) was reported in Burewala, Punjab province of Pakistan, and spread to Northern India (Qadir et al., 2019). The same situation was also reported in China and the Philippines for cotton and for many other ornamental species (*Malvaceae*), as well as for some vegetables. During 2019–2020, another peak in yield loss due to this disease was reported in Northern India. The average percent disease index/severity (PDI) was observed as 16.44%, 17.27%, and 11.84% in three major cotton-growing states of India, namely Haryana, Rajasthan, and Punjab, respectively (Blaise and Kranthi, 2019).

2.3 GENOMIC STRUCTURE OF BEGOMOVIRUSES

The genome size of geminiviruses is in the range of 2.7–2.8 kb with circular single-stranded DNA. The *Geminiviridae* family is classified into 14 genera, such as *Becurtovirus*, *Begomovirus*, *Capulavirus*, *Citlodavirus*, *Curtovirus*, *Eragrovirus*, *Grablovirus*, *Maldovirus*, *Mastrevirus*, *Mulcrilevirus*, *Opunvirus*, *Topilevirus*, *Topocuvirus*, and *Turncurtovirus* (Fiallo et al., 2021). All these genera are monopartite except begomoviruses which have both monopartite (DNA-A) and bipartite (DNA-A and DNA-B) genomic organization. Begomoviruses have an approximately equal size of DNA-A and DNA-B, that is, 2.8 kb. These components share a common region containing the nanomeric sequence called "iterons." Some extra single-stranded genomes of the ~1.4 kb were also found to be associated with viral

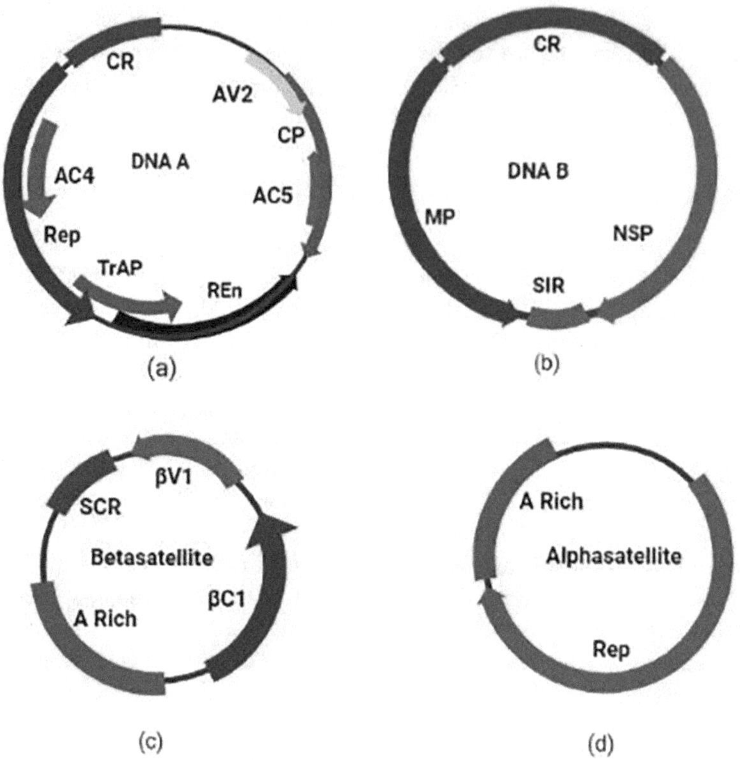

FIGURE 2.2 Genome organization of begomoviruses. Created by Biorender, https://www.biorender.com.

genomes, named α and β satellites, which help with the replication or movement in the host plant (Ben et al., 2021). The viral genome contains open reading frames (ORFs), which code for several proteins required for the infection of the virus to host plants, as shown in Figure 2.2, and the functions of these ORFs are elaborated in Table 2.1. These proteins interact with the vector and host to aid the replication and transmission of the virus which leads to successful infection.

2.4 VIRUS–VECTOR–HOST INTERACTION

Whiteflies act as vectors to transmit the pathogenic virus from one host to another. There are more than 1,000 species of whiteflies. *B. tabaci* (Gennadius; Hemiptera: *Aleyrodidae*) is unique having a broad range of host plants belonging to diverse families—*Solanaceae*, *Euphorbiaceae*, *Malvaceae* (e.g., cotton, okra), *Cucurbitaceae* (e.g., cucumber and squash), as well as some weeds which also cause an economic loss of billions of dollars every year worldwide (Malka et al., 2021). After the acquisition of the virus by *B. tabaci*, it gets transferred to the midgut and during subsequent feeding, the virus is moved to salivary glands to infect the host plant. Intermolecular interactions between viral and vector proteins might trigger the defense system of *B. tabaci* (Liu et al., 2013).

TABLE 2.1
Open Reading Frames (ORFs) of Various Genomic Elements of Begomoviruses along with Their Functions (Ben et al., 2021)

Genomic Elements	ORFs	Proteins Encoded	Functions
DNA-A (complementary strand)	AC1	Rep	Replication of virus, epigenetic regulation, cell cycle reprogramming
	AC2	TrAP	Transactivation, host resistance (HR) suppression, inhibition of phytohormone signaling
	AC3	REn	Replication enhancer
	AC4	C4 protein	Systemic movement, cell cycle regulation
	AC5	C5 protein	HR induction, RNAi suppression
DNA-A (sense-strand)	AV1	CP	Promote vector transmission, nucleo-cytoplasmic trafficking
	AV2	AV2/pre-coat protein	RNAi suppression, HR suppression
	AV3	AV3 protein	RNAi suppression
DNA-B (complementary strand)	BC1	MP	Cell-to-cell movement
DNA-B (sense-strand)	BV1	NSP	Promote vector transmission, pathogenicity determinant, movement
DNA-α	α-rep	Rep	Replication of DNA-α
DNA-β	V1	βV1	HR induction
	C1	βC1	Inhibition of photosynthesis, subvert ubiquitination, movement, induces autophagy

Once inside the host plant cell, the genome of geminiviruses starts replication via rolling circle replication (RCR) using host cell machinery, in which the ssDNA is converted into dsDNA in the nucleus of the host cell, as shown in Figure 2.3. After which, the Rep protein of the virus nicks the newly synthesized viral dsDNA at a specific nonameric sequence of IR and displaces the strand by attaching itself to the 5′' cut.

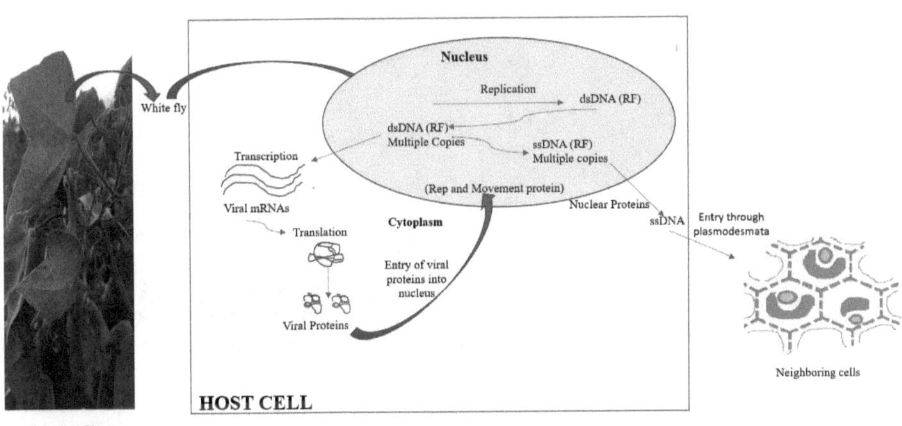

FIGURE 2.3 Mechanism of CLCuD infection.

Rep and REn protein recruit host replication factors at the 3'' end of nick and mediate extension of the strand (Wu et al., 2021).

One RCR cycle produces one viral dsDNA along with one single-stranded (ss) viral DNA. ssDNA is further converted into dsDNA which gets transcribed and translated. Double-stranded DNA (dsDNA) also mediates another cycle of RCR. In this way, several copies of dsDNA and viral ssDNA are made. Both the DNA-A and DNA-B components are replicated using the same approach. After some time, when the replication process stops due to the accumulation of coat proteins, the ssDNA is transported out of the cell with the help of movement protein (MP), coat protein (CP) (in the case of monopartite begomovirus), or NSP (in the case of bipartite begomovirus) through plasmodesmata and spread to neighboring cells (Devendran et al., 2022).

Both biotic and abiotic factors affect the transmission of viruses, that is, host plant, vector behavior, and virus strain. It causes an unremarkable change in the early to the last stage, reducing the yield of the cotton crop and the death of the seedling. Figure 2.4 shows various symptoms of CLCuD-infected cotton plants. Symptoms include cramped and ambiguous leaves with an upward or downward curl at the fringe, thickening of the veins, chlorotic and dark green leaf blotch, and extra leafy outgrowths on the ventral surface of the leaves called leaf enations.

CLCuD symptom depends on the severity of infection in the plant and also on the genotype, as well as the age of the plant at the time of infection. Late-stage infection causes a minor reduction in plant growth and if the plants are infected in the early stage after germination, it leads to tightly rolled leaves and stunted growth (Qazi et al., 2007).

Susceptibility level and symptom severity in the cotton plants show different symptom-rating scales as listed in Table 2.2.

FIGURE 2.4 CLCuD-infected cotton plant showing symptoms of leaf curling, vein thickening, and leaf enation.

TABLE 2.2
Disease Rating Scale of CLCuD (Sain et al., 2020)

Disease Rating Scale/Grade	Symptoms	Disease Response
0	No symptoms of the disease	Immune
1	Vein darkening and thickening of upper leaves	Highly resistant
2	Cupping, curling, and vein thickening in topmost leaves	Resistant
3	25% part of the plants infected	Moderately resistant
4	50% part of the plant is infected	Moderately susceptible
5	75% part of the plant is infected	Susceptible
6	Complete retardation and severely infected plant with stunted growth including	Highly susceptible

2.5 GENETIC VARIABILITY AMONG BEGOMOVIRUSES

Viruses are highly dynamic molecules with the capability of endless variability in the genome. To date, five different strains of CLCuD have been reported, which are cotton leaf curl Gezira virus (CLCuGeV), cotton leaf curl Multan virus (CLCuMuV), cotton leaf curl Kokhran virus (CLCuKoV), cotton leaf curl Alabad virus (CLCuAlV) and cotton leaf curl Bangalore virus (CLCuBaV) from different parts of the world. CLCuGeV was identified to be responsible for the disease in Africa (Ouattara et al., 2020). In the Indian subcontinent, the Multan epidemic was caused by three dominant players: CLCuAlV, CLCuKoV, and CLCuMuV (Zhou, 2013). Following this epidemic, resistant cultivars were developed, which somehow restored cotton productivity, but again CLCuD struck back with the emergence of a new virulent strain—CLCuKoV-Burewala, marking the beginning of the second epidemic. A distinct recombinant strain of CLCuMuV was found to substitute the existing CLCuKoV-Bu strain during the third epidemic (Multan rebound) during 2015–2016. Recently, CLCuMuV was also reported from southeastern China and the Philippines; the spread was attributed mainly to the cross-border movement of traders carrying infected samples (cutting of hibiscus plants, etc.) from Pakistan to China (Datta et al., 2017).

Before and during the Burewala epidemic, the causative agents of CLCuD were having monopartite genome. However, after 2015, many other "non-core" Geminiviruses, with bipartite genomes, were also found in infected cotton plants like Papaya leaf curl virus (PaLCuV), Chickpea chlorotic dwarf virus (CpCDV), Okra enation leaf curl virus (OEnLCuV), Tomato leaf curl virus (ToLCuV), Tomato leaf curl Bangalore virus (ToLCBaV), and Tomato leaf curl New Delhi virus (ToLCNDV).

Multiple strains of the cotton leaf curl virus have been found in different regions of the world. Presently, several recombinant viral strains like "CLCuMuV-Rajasthan (CLCuMuV-Ra), CLCuMuV-Pakistan (CLCuMuV-PK), CLCuKoV-Shadadpur (CLCuKoV-Sha), and CLCuKoV-Burewala (CLCuKoV-Bu)" are predominant in North (Biswas et al., 2020). Isolates from Guanxi, China (GX1), and Guangzhou,

Guangdong (G6), displayed more than 99% and 89% resemblance with the CLCuMuV isolates, respectively. In a recent study, the virus isolates from CLCuD-infected cotton plants from the Manipur region were found to be associated with Tomato leaf curl Patna betasatellite (ToLCPaB); DNA-A of virus presented recombination between "AV1 region" of CLCuMuV-PK and CLCuBuV-IN. In another study, several virus isolates collected from North India (labeled as CLCuMuV-SR13, CLCuMuV-SR14, CLCuMuV-SR15, CLCuMuV-ND14, and CLCuMuV-ND15) exhibited distinct recombination patterns (Qadir et al., 2019).

2.6 DIAGNOSTIC METHODS FOR CLCUD

CLCuD is coupled with diverse species of begomoviruses, and all species display similar sort of symptoms in the host plants. So, it is not feasible to recognize the actual viral species just by looking at visual symptoms. Some earlier attempts were made using techniques like immunoblotting and ELISA, but these showed limited sensitivity and specificity and therefore were not very popular (Verma et al., 2022). With the advancement of molecular techniques, various techniques have been developed for precise detection, and a brief description is given below.

2.6.1 Nucleic Acid Spot Hybridization

Coat protein (CP) and Replication (Rep) genes of whitefly-transmitted geminiviruses, located in DNA-A, are expected to be highly conserved. Nucleic acid spot hybridization (NASH) was found effective in the diagnosis of CLCuRaV in plants and whiteflies. It utilized radio-labeled probes (pCLCA and pCLCβ) for CLCuRaV DNA-A and CLCuRaV DNA-β, respectively. The coat protein gene (~750 bp) and full-length DNA-β (~1.4 kb) were amplified by polymerase chain reaction (PCR) using various primers (CPF and CPR for the CP gene while Beta 1 and Beta 2 primers for DNA-β) (Radhakrishnan et al., 2004).

2.6.2 Basic Polymerase Chain Reaction (PCR)

The PCR-based method had been acquired during 2003–2004 for rapid diagnosis of the Cotton leaf curl virus that infects cotton. Amplification of the coat protein gene (771 bp) was performed for accurate detection of the infected sample. The method gives quick results with amplifications achieved within 1.30–1.45 h (Chakrabarty et al., 2005).

2.6.3 One-Step PCR

A one-step PCR protocol was recognized to diagnose three viruses (CLCuMuV, CLCuKoV-Bur, and CLCuKoV). Primers Beg-01 and Beg-02 were utilized primarily to validate begomoviruses in the genotypes. Afterward, specific diagnostic primers, CLCuKoV-Bur-F, CLCuKoV-Bur-R, CLCuMuV-F, and CLCuKoV-R, were constructed for the diagnosis of associated CLCuD (Saleem et al., 2017).

2.6.4 Quantitative PCR (qPCR)

A quantitative PCR (qPCR) procedure has been developed for proper quantification of CLCuKoV-Ko, CLCuMB, and ClCuMuA. Three primer pairs (BegomoqPCRF1/BegomoqPCRR1, BetaqPCRF2/BetaqPCRR2, and AlphaqPCRF2/AlphaqPCRF2, for the amplification of V2 gene of begomovirus, the βC1 gene of betasatellite and the Rep gene of alphasatellite, respectively) were utilized with variable primer concentrations (1.5, 2.5, 3.5, or 4.5 pM/reaction) at varying annealing temperatures of 54°C, 55°C, 56°C, and 57°C, respectively (Shafiq et al., 2017).

2.6.5 Loop-Mediated Isothermal Amplification (LAMP) Assay

A vigorous method was developed using the loop-mediated isothermal amplification (LAMP) assay to detect the CLCuKoV-Bur strain and CLCuMB. Primers were constructed using CR of CLCuKoV-Bur and CLCuMB-βC1 genes. Primers for the CLCuMB-βC1 gene worked well at 58°C using 100mg of total plant tissue DNA as a template in a 25-mL reaction volume (Rafiq et al., 2021).

2.6.6 Multiplex Polymerase Chain Reaction (mPCR) Assay

A multiplex polymerase chain reaction (mPCR) assay has been established for the concurrent recognition of three viral strains, that is, CLCuMuV, CLCuKoV, and CLCuAlV accompanied by CLCuD of cotton. Three specific primers which together amplify the CP gene (780 nts.) of the begomovirus, the βC1gene (375 nts.) of the betasatellite, and the Rep gene (452 nts.) of the alpha satellite were identified in the mPCR assays (Palchoudhury et al., 2022).

2.7 MITIGATION STRATEGIES

2.7.1 Producing and Cultivating Resistant Varieties

Many CLCuD-resistant/tolerant cultivars such as H-1117, H-1226, and H-1236, F-1861, LH-2076, and LHH-144 (Hybrid), RS-875, RS-810, RS-2013, CSSH 198, CSHH 238 (Hybrids) were produced by the State Agricultural Universities and ICAR-Central Institute for Cotton Research and the private sectors to restore cotton productivity after the first pandemic in India (Monga et al., 2011). Two varieties (CIM-446 & FH-901) and 13 strains (FH-900S, CBVI-473, CM-28,565/98,642/98,726/98, BH-549/99, BH-147, BH-148, VH-137, Z-113, P-6 & P-8) were found resistant among 45 varieties and 40 strains of *G. hirsutum* at Punjab Seed Corporation, Khanewal (Akhtar et al., 2002).

2.7.2 RNA Interference (RNAi)

RNA interference (RNAi) is a potent tool based on transcriptional gene silencing (TGS) and post-transcriptional gene silencing (PTGS). Recently, this technique has been used against viruses such as the African cassava mosaic virus (ACMV) and Mung bean yellow mosaic virus (MYMV) to develop resistant cotton varieties. Recently, the ghr-miR166b of *G. hirsutum* has been reported to show multi-compatible targets in mitochondrial ATP synthase of *B. tabaci*, hence acting as biopesticides for combating the *B. tabaci* population (Wamiq et al., 2018).

2.7.3 Anti-sense RNA

Anti-sense RNA has been effective against numerous viruses including TYLCV and CLCuKoV. Viral AV1 and rep protein were targeted in transgenic cotton by using this technology, which blocked the replication of the virus, viral movement, and encapsidation. Recently, in transgenic cotton (G. *hirsutum* cv. Coker-310), no symptoms were observed after inserting viral gene βC1 in anti-sense orientation under 35S promoter (Sohrab, 2016).

2.7.4 Identification of Plant Host Hormones

Phytohormones initiate specific signaling mechanisms to combat various pathogens. In *Arabidopsis*, plant hormone reduces the resistance against insects and is controlled by miR156-targeted-SPL9; jasmonic acid hormone develops resistance against insects, herbivores, and necrotrophic pathogens (Mao et al., 2017).

2.8 FUTURE PROSPECTS

2.8.1 Cotton Genome Editing Using CRISPR-Cas9

The clustered regularly interspaced palindromic repeat (CRISPR)-CRISPR-CAS is a genome-editing tool that is utilized against pathogens by introducing the Cas9 protein and guiding RNAs into the host cell. Genomes of Bean yellow dwarf virus (BeYDV), Merremia mosaic virus, CLCuKoV, TYLCV, and beet have been edited using CRISPR-Cas (Baltes et al., 2015).

2.8.2 Pathogen-Derived Resistance (PDR)

In pathogen-derived resistance (PDR), a portion of a viral gene is inserted into the host plant, which prevents viral replication. However, PDR was not efficient due to the gene-silencing mechanism of viruses. siRNA strategy is utilized to disturb the structure of the protein. Currently, only the intergenic regions of *Ageratum conyzoides* are transformed against CLCuV (Nagata et al., 2015).

2.8.3 Use of Molecular Markers in Conventional Breeding

Research is under way to recognize the QTLs linked to disease resistance. A random amplified polymorphic DNA (RAPD) assay has been used to detect the markers. A bi-parental F2 mapping population was generated using the highly resistant cultivar *G.hirsutum* LRA-5166 and a vulnerable variety S-12. OPY-21080, OPQ-14325, OPO-19460, OPO-2, OPQ-14, and OPY-19 were found to be correlated with the resistance. Two cotton cultivars CIM-443 and CIM-240 were found resistant to CLCuD in marker-assisted selection (MAS). Two SSR markers, namely CM-43 and PR-91, were the only intolerant genotypes amplified that showed a substantial correlation with CLCuD (Abbas et al., 2015).

2.9 CONCLUSION

Cotton is a major fiber crop globally with an ever-increasing demand due to the rising population and changing lifestyles. Cotton production is constrained by many biotic and abiotic factors among which CLCuD is the dominant one. It is caused by a complex of begomoviruses—having many strains, grouped as core and non-core viruses regarding pathogenicity for CLCuD. The molecular epidemiological study carried out in the present work revealed new updates regarding genetic structure and the pathogen–vector–host relationship. Studies about recombination and evolution in begomoviruses indicate this group as fast-evolving with many novel recombination patterns, and this is quite alarming as it may result in severe outbreaks at times. Although some reliable molecular diagnostic techniques are already available, much remains to be done in effective mitigation strategies. Recent advancements in genome editing and exhaustive virus genotyping seem to offer solutions to this prolonged threat.

REFERENCES

Abbas, A., Iqbal, M. A., Rahman, M. U., & Paterson, A. H. (2015). Estimating genetic diversity among selected cotton genotypes and the identification of DNA markers associated with resistance to cotton leaf curl disease. *Turkish Journal of Botany*, *39*(6), 1033–1041.

Akhtar, K. P., Khan, A. I., Hussain, M., Haq, M. A., & Khan, M. S. I. (2002). Response of cotton varieties to cotton leaf curl virus (CLCuV). *Tropical Agricultural Research and Extension, 5*(1 & 2), 2002.

Bailey, M. A. (1934). Leaf curl disease of cotton in the SudanSudan empire cotton. *Empire Cotton Growing ReviewGr Rev*, *11*, 280–288.

Baltes, N. J., Hummel, A. W., Konecna, E., Cegan, R., Bruns, A. N., Bisaro, D. M., & Voytas, D. F. (2015). Conferring resistance to geminiviruses with the CRISPR–Cas prokaryotic immune system. *Nature Plants*, *1*(10), 1–4.

Ben Chehida, S., Filloux, D., Fernandez, E., Moubset, O., Hoareau, M., Julian, C., ... & Lefeuvre, P. (2021). Nanopore sequencing is a credible alternative to recover complete genomes of geminiviruses. *Microorganisms*, *9*(5), 903.

Biswas, K. K., Bhattacharyya, U. K., Palchoudhury, S., Balram, N., Kumar, A., Arora, R., ... & Mandal, P. K. (2020). Dominance of recombinant cotton leaf curl Multan-Rajasthan virus associated with cotton leaf curl disease outbreak in northwest India. *PLoS One*, *15*(4), e0231886.

Blaise, D., & Kranthi, K. R. (2019). Cotton production in India. In: K. Jabran, & B. Singh Chauhan (Eds.), *Cotton Production*, John Wiley & Sons Ltd., pp. 193–215.

Chakrabarty, P. K., Sable, S., Monga, D., & Mayee, C. D. (2005). Polymerase chain reaction-based detection of *Xanthomonas axonopodis* pv. malvacearum and cotton leaf curl virus. *Indian Journal of Agricultural Sciences*, *75*(8), 524–528.

Costa, C., Azoia, N., Silva, C., & Marques, E. (2020). Textile industry in a changing world: Challenges of sustainable development. *U. Porto Journal of Engineering*, *6*(2), 86–97.

Datta, S., Budhauliya, R., Das, B., Gopalakrishnan, R., Sharma, S., Chatterjee, S., ... & Veer, V. (2017). Rebound of Cotton leaf curl Multan virus and its exclusive detection in cotton leaf curl disease outbreak, Punjab (India). *Scientific Reports*, *7*(1), 17361.

Devendran, R., Namgial, T., Reddy, K. K., Kumar, M., Zarreen, F., & Chakraborty, S. (2022). Insights into the multifunctional roles of geminivirus-encoded proteins in pathogenesis. *Archives of Virology*, *167*(2), 307–326.

Fiallo-Olivé, E., Lett, J. M., Martin, D. P., Roumagnac, P., Varsani, A., Zerbini, F. M., ... & ICTV Report Consortium. (2021). ICTV virus taxonomy profile: Geminiviridae 2021. *Journal of General Virology*, *102*(12), 001696.

Hussain, T., & Ali, M. (1975). A review of cotton diseases of Pakistan. *Pakistan Cottons*, *19*(2), 71–86. Ref. 62.

Jeger, M. J. (2020). The epidemiology of plant virus disease: Towards a new synthesis. *Plants*, *9*(12), 1768.

Jones, D. R. (2003). Plant viruses transmitted by whiteflies. *European Journal of Plant Pathology*, *109*, 195–219.

Jones, R. A. (2021). Global plant virus disease pandemics and epidemics. *Plants*, *10*(2), 233.

Liu, B., Preisser, E. L., Chu, D., Pan, H., Xie, W., Wang, S., ... & Zhang, Y. (2013). Multiple forms of vector manipulation by a plant-infecting virus: *Bemisia tabaci* and tomato yellow leaf curl virus. *Journal of Virology*, *87*(9), 4929–4937.

Mahmood-ur-Rahman, K. H., Khan, M. A., Bakhsh, A., & Rao, A. Q. (2021). 01. An insight of cotton leaf curl virus: A devastating plant pathogenic begomovirus. *Pure and Applied Biology (PAB)*, *1*(3), 52–58.

Malka, O., Feldmesser, E., Van Brunschot, S., Santos-Garcia, D., Han, W. H., Seal, S., ... & Morin, S. (2021). The molecular mechanisms that determine different degrees of polyphagy in the *Bemisia tabaci* species complex. *Evolutionary Applications*, *14*(3), 807–820.

Mao, Y. B., Liu, Y. Q., Chen, D. Y., Chen, F. Y., Fang, X., Hong, G. J., ... & Chen, X. Y. (2017). Jasmonate response decay and defense metabolite accumulation contributes to age-regulated dynamics of plant insect resistance. *Nature Communications*, *8*(1), 1–13.

Monga, D., Chakrabarty, P. K., & Kranthi, K. R. (2011, February). Cotton leaf curl virus disease in India-Recent status and management strategies. In: *Fifth Meeting of Asian Cotton Research and Development Network, Lahore, Pakistan* (pp. 23–25).

Nagata, A. K. K. I., Lima, M. F., & Gilbertson, R. L. (2015). Geminivirus diseases in vegetables and other crops in Brazil: Current status and approaches for management. *Horticultura Brasileira*, *34*, 1.

OECD FAO. (2022). *OECD-FAO Agricultural Outlook 2022–2031*. OECD.

Ouattara, A., Tiendrebeogo, F., Lefeuvre, P., Hoareau, M., Claverie, S., Allibert, A., ... & Lett, J. M. (2020). Diversity, distribution and prevalence of vegetable-infecting geminiviruses in Burkina Faso. *Plant Pathology*, *69*(2), 379–392.

Palchoudhury, S., Khare, V. K., Balram, N., Bhattacharyya, U. K., Das, S., Shukla, P., ... & Biswas, K. K. (2022). A multiplex polymerase chain reaction for the simultaneous detection of the virus and satellite components associated with cotton leaf curl begomovirus disease complex. *Journal of Virological Methods*, *300*, 114369.

Qadir, R., Khan, Z. A., Monga, D., & Khan, J. A. (2019). Diversity and recombination analysis of Cotton leaf curl Multan virus: A highly emerging begomovirus in northern India. *BMC Genomics*, *20*, 1–13.

Qazi, J., Amin, I., Mansoor, S., Iqbal, M. J., & Briddon, R. W. (2007). Contribution of the satellite encoded gene βC1 to cotton leaf curl disease symptoms. *Virus Research*, *128*(1–2), 135–139.

Radhakrishnan, G., Malathi, V. G., & Varma, A. (2004). Detection of DNA A and DNA beta associated with cotton leaf curl and some other plant diseases caused by whitefly transmitted geminiviruses. *Indian Phytopathology*, *57*(1), 53–60.

Rafiq, A., Ali, W. R., Asif, M., Ahmed, N., Khan, W. S., Mansoor, S., ... & Amin, I. (2021). Development of a LAMP assay using a portable device for the real-time detection of cotton leaf curl disease in field conditions. *Biology Methods and Protocols*, *6*(1), bpab010.

Rehman, M., Chakraborty, P., Tanti, B., Mandal, B., & Ghosh, A. (2021). Occurrence of a new cryptic species of *Bemisia tabaci* (Hemiptera: Aleyrodidae): An updated record of cryptic diversity in India. *Phytoparasitica*, *49*(5), 869–882.

Sain, S. K., Monga, D., Mohan, M., Sharma, A., & Beniwal, J. (2020). Reduction in seed cotton yield corresponding with symptom severity grades of cotton leaf curl disease (CLCuD) in upland cotton (*Gossypium hirsutum* L.). *International Journal of Current Microbiology and Applied Sciences*, *9*(11), 3063–3076.

Saleem, H., Khan, A. A., Azhar, M. T., & Nawaz-ul-Rehman, M. S. (2017). Molecular screening of cotton germplasm for cotton leaf curl disease caused by viral strains. *International Journal of Agriculture & Biology*, *19*(1), 125–130.

Shafiq, M., Iqbal, Z., Ali, I., Abbas, Q., Mansoor, S., Briddon, R. W., & Amin, I. (2017). Real-time quantitative PCR assay for the quantification of virus and satellites causing leaf curl disease in cotton in Pakistan. *Journal of Virological Methods*, *248*, 54–60.

Sohrab, S. S., Kamal, M. A., Ilah, A., Husen, A., Bhattacharya, P. S., & Rana, D. (2016). Development of Cotton leaf curl virus resistant transgenic cotton using antisense ßC1 gene. *Saudi Journal of Biological Sciences*, *23*(3), 358–362.

Verma, N. V., Shukla, M., Kulkarni, R., Srivastava, K., Claudic, B., Savara, J., … & Pandya, A. (2022). Emerging extraction and diagnostic tools for detection of plant pathogens: Recent trends, challenges, and future scope. *ACS Agricultural Science & Technology*, *2*(5), 858–881.

Wamiq, G., & Khan, J. A. (2018). Overexpression of ghr-miR166b generates resistance against *Bemisia tabaci* infestation in *Gossypium hirsutum* plants. *Planta*, *247*, 1175–1189.

Wu, M., Wei, H., Tan, H., Pan, S., Liu, Q., Bejarano, E. R., & Lozano-Durán, R. (2021). Plant DNA polymerases α and δ mediate replication of geminiviruses. *Nature Communications*, *12*(1), 2780.

Zhou, X. (2013). Advances in understanding begomovirus satellites. *Annual Review of Phytopathology*, *51*, 357–381.

3 Nanotechnology-Driven Antimicrobial Interventions for Biotic Stress Amelioration in Plants

Pawan Kaur, Chanchal Mony, Anamika Pal, and Rita Choudhary

3.1 INTRODUCTION

Biotic stress caused by pathogenic or non-pathogenic microorganisms is one of the key constraints associated with crop productivity. Crop loss due to diseases and pest infection are among the major threats associated with food security and farming (Pandey et al., 2017). Around 20%–40% of yield losses in crops have been estimated to be caused by biotic stresses (Pestovsky and Martinez-Antonio, 2017), amounting to about US$200 billion per year (Oluwaseun, 2017) worldwide. Accurate plant disease management is a big challenge for scientists in plant sciences, and they have been working on nanotechnology-based tools to manage biotic stresses in plants. The use of agrochemicals to combat infection in field crops often leads to the development of resistant races of microorganisms. It also causes negative impacts on the beneficial organisms that act as a basis for residual food toxicity and create potential hindrances to food security and sustainable agricultural productivity (Duhan et al., 2017), processing and packaging (Mousavi and Rezaei, 2011), which can be employed for overcoming these concerns by developing novel antimicrobial products (Ocsoy et al., 2013). Nanostructured formulations can facilitate enhanced sensitivity, speed, and specificity as an antimicrobial agent, enabling the detection and protection of crops (Khiyami et al., 2014). Management of crop diseases using a siRNA-based nano-delivery system can be explored (Chaudhary et al., 2018) for pest-free crop productivity (Mishra et al., 2017). This chapter highlights the various nanoparticles with different variants, such as antimicrobials, elicitors, and plant growth promoters, used for agricultural application and their mechanism of action in mitigating biotic stresses in plants, as shown in Figure 3.1.

 DOI: 10.1201/9781032642567-4

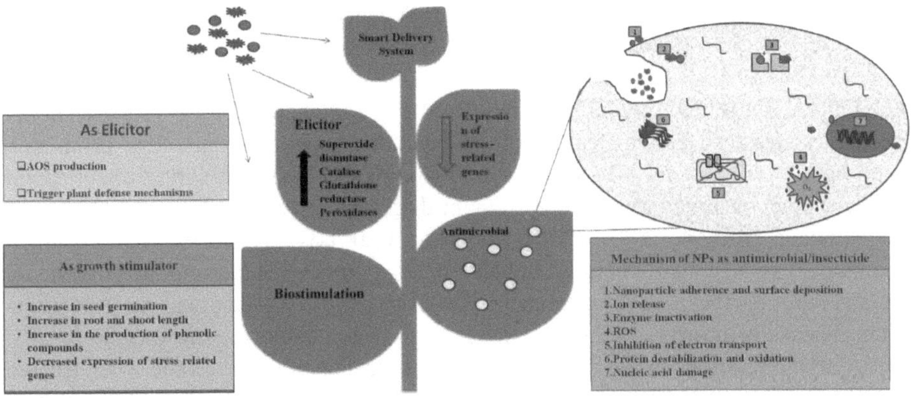

FIGURE 3.1 Described various modes of application of nanoparticles in plants: shows the various applications of nanoparticles in plants as the growth stimulator, as elicitor of plant defense mechanisms, and as antimicrobial agent or insecticide.

3.2 TYPES OF NANOPARTICLES USED IN PLANTS

Nanoparticles are gaining attention in plant defense systems due to their large surface area and easily modifiable properties (Gatoo et al., 2014). The nanoparticle-based antimicrobials in use portray a wide range of altered chemical properties, which can be associated with their diverse modes of action as antimicrobial agents (Beyth et al., 2015). Further, the diversification in the genetic makeup of the bacteria and their physiological state might determine the sensitivity of the bacteria toward the NPs (Nath and Banerjee, 2013). Sometimes, the NP and bacterial ratio plays a crucial role in inducing toxicity (Huh and Kwon, 2011). Generally, the NPs act on the microorganisms disrupting the bacterial membrane potential or inhibiting in protein synthesis or DNA damaging or production of reactive oxygen species (ROS) (Kaur et al., 2011; Pelgrift and Friedman, 2013). In the following section, NPs possessing antimicrobial ability have been categorized.

3.2.1 Metal Nanoparticles

Metals and metal oxide NPs portray highly potent antibacterial effects. Most of the metal and metal oxides show antimicrobial effects by generating ROS, while some of them have high efficiency due to metal ion release and their physical properties. Some of the metal NPs with antimicrobial efficacy have been listed earlier.

3.2.1.1 Silver Nanoparticles

Among nanoparticles, silver NPs are commonly used as an antimicrobial agent due to their widespread strong inhibitory effect and broad-spectrum antimicrobial activity (You et al., 2012; Prameela, 2017). The efficacy of Ag NPs compared with other metal and metal oxide NPs has been reported to be size-dependent (Poulose et al., 2014; Beyth et al., 2015) and showed efficient antimicrobial efficacy *in vitro* and *in vivo* against various pathogens (De Simone et al., 2014), while the mechanism of Ag NP bactericidal effect is not well understood (Majdalawieh et al., 2014). However, these were observed to cause increased membrane permeability by creating pits in the cell wall, thereby causing inactivation of the respiratory chain (Beyth et al., 2010), unwinding of DNA, disruption of cell membrane, and free radical production (Kumari et al., 2017). Despite these debates, Ag NPs have been observed to be one of the promising metal-based antibacterial NPs against huanglongbing disease in *Citrus aurantifolia* and have been observed to be more efficient than the conventional β-lactam antibiotics (Stephano-Hornedo et al., 2020). Mahawar et al. (2020) reported the antifungal behavior of Ag NPs augmented with *Calothrix elenkinii* by increasing the polyphenol oxidase activity, thus assisting in controlling the phytopathogen-causing Alternaria blight disease of tomato. Many researchers have been exploring elevated ROS production and alterations in the properties of the microbial cell membrane when treated with biogenic Ag NPs (Ramalingam et al., 2016) and have managed plant diseases *in vitro* as well as under glasshouse conditions. However, the reactions and responses of plants toward the exposure of NPs in combating diseases are still unknown and need further studies (Hirpara and Gajera, 2020).

3.2.1.2 Silica Nanoparticles

Nowadays, researchers are looking for efficient antimicrobials against phytopathogens with a positive impact on plant growth parameters for plant disease management. Silica nanoparticles fulfill all desired requirements (EI-Mougy, 2013) and showed induced defense mechanisms in plants (Fauteux et al., 2006) and have been observed to increase disease resistance in plants (Brecht et al., 2004; Suriyaprabha et al., 2014). Derbalah et al. (2018) described the antifungal efficacy of SiNPs against *A. solani* phytopathogen causing blight disease in tomatoes and observed higher efficacy as compared with commercial fungicide metalaxyl (recommended fungicide) under *in vitro* and *in vivo* conditions. This study presented SiNPs as an effective and safe alternative to fungicides for plant disease management.

3.2.1.3 Copper Nanoparticles

Copper plays an important role in the prevention of plant diseases. However, their efficacy against bacteria has been reported to be lower than Ag and Zn (Beyth et al., 2015). Copper-based chemicals have been used as effective antimicrobials in agriculture against bacteria and fungal phytopathogens (Kaur et al., 2016). Thus, it can be stated that CuO NPs act as fungicides and compete with other commercially available copper-based fungicides in the market.

3.2.1.4 Zinc Oxide Nanoparticles

ZnO NPs have been reported to show better antimicrobial properties and can be used for plant disease management (Kaur et al., 2011; Chaudhary et al., 2019). Helaly et al. (2014) conducted studies for biotic stress tolerance in plants by exposing them to ZnO NP and observed a positive impact on plant growth and also possess broad-spectrum antimicrobial activity against phytopathogens (Palanikumar et al., 2014; Siddiqui et al., 2018). ZnO NPs showed low toxicity toward human cells (Reddy et al., 2007) and size dependency against pathogens such as *Streptococcus mutans*, *Lactobacillus* (Kasraei et al., 2014), *E. coli* (Kaur et al., 2011), and *Klebsiella pneumonia*

(Reddy et al., 2014). They also affect the bacterial cells by binding to cell membranes, causing disruption and inducing the production of ROS (Pati et al., 2014).

3.2.1.5 Titanium Dioxide Nanoparticles

Titanium dioxide (TiO_2) NPs have also been reported to possess bactericidal properties and showed their efficiency against various viruses and parasites (Allahverdiyev et al., 2011). These are also photocatalytic and visible light, or UV, causes toxicity (Pelgrift and Friedman, 2013), induced ROS production eventually damaging the biomolecules and bacterial cell functioning (Blecher et al., 2011) and showed effectiveness against *Xanthomonas perforansa* to control Bacterial Spot of Tomato (Siddiqui et al., 2018). It has been observed to be highly efficient against spores of *Bacillus* (Hamal et al., 2010), known to be the most resistant organism.

3.2.1.2 Other Nanoparticles

Iron oxide (Fe_3O_4) has been reported to show anti-adherent properties, which significantly affect gram-negative and gram-positive bacterial growth (Anghel et al., 2014). Bactericidal activity has also been reported by magnesium oxide (MgO) NPs against both gram-positive and gram-negative bacteria, spores, and even viruses. Magnesium-containing NP has been observed to inhibit the essential enzymes present in bacteria (Blecher et al., 2011). The formation of biofilm of *E. coli* and *S. aureus* was found to be prevented by nitric oxide. This low-risk antibacterial compound has been reported to be associated with multiple antimicrobial mechanisms (Carpenter and Schoenfisch, 2012). Nitrogen oxide is highly reactive, making its clinical value limited. However, its encapsulation, targeted delivery, and controlled release can be used to exploit its antimicrobial potential (Kutner and Friedman, 2013).

3.2.2 Polymeric Nanoparticles

Polymeric NPs have been used as a carrier for many agrochemicals (Grillo et al., 2016; Mishra et al., 2017) that can be efficiently used for encapsulating agrochemi cals to protect them from degradation and helping in controlled release (Seabra and Duran, 2017). Polymeric polysaccharides, a low-cost, biocompatible, biodegradable green reagent, have been widely used in biological applications (Kollarigowda, 2017). As far as agriculture is concerned, natural polymers such as chitosan and pectin are widely used in the development of NPs (Li et al., 2017). These polymers also act as antimicrobials for the protection of crops, and cationic groups of these polymers disrupt the bacterial cell wall. At the same time, some of them require hydrophobic side chains to penetrate and burst the membrane (Beyth et al., 2015).

3.2.2.1 Chitosan-Based Nanoparticles

Among polymers, chitosan has been used for agrochemicals delivery with improved efficacy of the loaded active ingredients (Kashyap et al., 2015) and the ability to act as a chelating agent for inorganic compounds for assisting in controlled delivery in plants (Saharan et al., 2015; Kumar et al., 2022). It has been shown to have broad-spectrum antibacterial, antiviral, and antifungal activity and is recognized as an immuno-modulatory and plant growth-promoting agent with a stimulating plant defense system (Malerba and Cerana, 2016). Chitosan also plays a significant role in the plant defense system, which includes protein inhibitor callose formation in response to pathogen attack (El-Hadrami et al., 2010) and modulating positive immune responses in plants (Chandra et al., 2015; Syafiq et al., 2020).

3.2.2.2 Polymer Metal Nanocomposites

Polymer metal complexes and metal-organic-framework are other compounds that have also been in the pipeline for use in agriculture areas. Chitosan-metal-based nanocomposites have been shown to have very high antibacterial and antifungal activities against phytopathogens (Kaur et al., 2018). The utilization of metal oxide framework (MOF) is also at a nascent stage in agriculture. It might help maintain nutrient efficiency in soil to enhance fertility (Anstoetz et al., 2016).

3.3 ROLE OF NANOPARTICLES IN PLANT DEFENSE SYSTEM

3.3.1 Antimicrobial Mechanism against Target Microbes

Nanoparticles affect the microbes by directly attaching themselves to their surface and hindering their physiological processes via DNA damage, penetration in cell membranes, or interfering in the protein synthesis process of target microorganisms (Abkhoo and Panjehkeh, 2017). The ability of NPs to bind to the bacteria is dependent on the surface area available for interaction. Many researchers reported the role of nanoparticles such as magnesium oxide NPs and carbon nanomaterials in the suppression of *Ralstonia solanacearum* cells (Cai et al., 2018) and *F. graminearum* spores, respectively (Wang et al., 2014). However, the mechanism of NPs inhibiting spore germination is not yet understood and needs more exploration of research in this area (Hajong et al., 2019).

3.3.2 ROS Production

Nanoparticles have also been reported to interfere with the oxidative mechanism of target pathogens and cause ROS production for antimicrobial activity (Cvjetko et al., 2017). Nanoparticles in plants affect the plant's hormonal balance, thereby affecting the plant's metabolism. The toxic effect of NPs in plants was primarily due to ROS production.

In case the antioxidants are not able to regulate the increased ROS production, then the cellular macromolecules are oxidized by ROS and ultimately cause apoptosis or necrosis in plants, leading to their death (Sharma et al., 2012; Faisal et al., 2013).

3.3.3 Other Mechanisms

Nanoparticles have been observed to cause membrane disruption, hindrance of ATPase activity, inactivation of thiol groups in the fungal cell wall, and DNA abrasion (Velmurugan et al., 2009). Disruption of cellular mechanisms has been observed to severely affect the ATP synthesis, the permeability of the membrane, induction of oxidative stress, and ROS formation in NP-exposed nematodes (Lim et al., 2012), and expression of stress response genes were also observed in nanoparticles exposed to target phytopathogens.

In summary, it can be stated that further research about NPs' mode of action is required, and several killing pathways are still unexplored. Further, the impacts of combinations of NPs-based treatment are not well understood, and the dependent microcidal effect needs further investigation.

3.4 MARKET SCENARIO OF NANO-BASED PLANTS DEFENSE SYSTEM

According to the European Commission, nanotechnology-based translation research has the potential to revolutionize global agriculture and food systems (European Commission, 2013). Nano-revolution in the Agri-Food sector is based on three aspects: first is the development of nanoproducts; second is an increase in agricultural productivity and food security, and the third one is a reduction in environmental pollution. Nano-agri input products (NAIP) and nano-agri products (NAP) are used in agriculture. Many copper, sulfur, and gold-based nanoparticles are used in India for managing plant diseases. Some of the commercially available products in the market, among which copper- and sulfur-based fungicides are prominent, have been used extensively for plant defense as shown in Table 3.1

TABLE 3.1
Some Important Nano-based Fungicides Available in the Market for Crop Protection

S. No.	Fungicide	Formulation Types	Mode of Action	Company Name
1.	NanoFungi	Liquid Concentrate	To control all types of fungus such as *rhizoctonia*, root rot, leaf spots, powdery mildew on plants. Immediate knock-down action.	VM Crop Care
2.	Thymox Control	Nano-emulsion	Broad-spectrum, contact fungicide and cause cell membrane disruption of target fungus.	Kemin Crop Technologies
3.	NanoCu	Copper nanoparticles	Acts as bactericide and fungicide.	Bio Nano Technology
4.	Shield Plant	Copper nanoparticles (Cu: 3%; K silicate: 5%)	Acts as bactericide and fungicide.	Bio Nano Technology
5.	AEGIS Herbal Nanotechnology	Metallic nanoparticle-embedded herbal extract	Highly antifungal and interacts with proteins, entering through membrane and cause structural deformation of cell membrane.	Swaroop Agro Chemical Industries Private Limited
6.	Allosperse	Polymer nanoparticle	Combination of Azteroid and Bifender is used as a broad-spectrum control.	Vive Crop Protection
7.	Nano-Sulfur	S nanoparticle	Acts as a fungicide.	IPI, Singapore
8.	Nanosulf Drenching	S nanoparticle	Acts as a fungicide.	Alert Biotech, India

(*Continued*)

TABLE 3.1 (*Continued*)
Some Important Nano-based Fungicides Available in the Market for Crop Protection

S. No.	Fungicide	Formulation Types	Mode of Action	Company Name
9.	Acurosil Nano	Nanosilver and hydrogen peroxide-based disinfectant	Silver is a main component with its oligo-dynamic and bactericidal effect.	Kimaya Trading Corporation
10.	NanoGold	Gold nanoparticle	Prevention of disease like *Rhizectonia* spp, Black spot, *Botrytis cirerea, Alternaria alternata*, Red Rot, white fungus, and fusarium wilt.	Sanjeevany Chemicals
12.	Knock Out Nano	4% chelated copper along with 10% yeast extract and 500 ppm Nano copper	Works as a Systemic acquired resistance system having a complex mode of action to control fungi in a preventive and curative manner by acting directly on fungus, reducing growth with a rapid accumulation of acetylate. Effective against late blight Downy mildew, powdery mildew, early Blight canker.	Zeal Biologicals
13.	NanoBlitz	Nanotechnical organic extract	Effective against brinjal fruit shoot borer.	Annadata Organics
14.	Bonide 142-Sulfur	Dust	Prevents fungal spores from germinating, so it is most effective before the disease develops.	BONIDE
15.	Dusting S	Dust	M2 Fungicide used to delay fungicide resistance.	Universal Crop Protection (Pty) Ltd.
16.	Vine Dusting S	Dusting powder	Control of powdery mildew and thrips.	Nulandis
17.	Liquid copper fungicide	Liquid copper solution	These fungicides are prepared from the liquid copper and possess extreme fungicidal properties.	Royal Agrochem Private Limited
18.	Coplus	Liquid formulation used as sprayer	Effective for soil-borne diseases like downy mildew, anthracnose, blights and many more.	Krishidhoot Bio-herbals
19.	Fungicide green blue powder copper oxychloride	Wettable powder (WP), water-dispersible granule (WDG)	Effective fungicide.	Falcon International
20.	Copper oxy chloride	Wettable powder	Possess fungicidal properties.	Aroxa Crop Science Private Limited

(*Continued*)

TABLE 3.1 (*Continued*)
Some Important Nano-based Fungicides Available in the Market for Crop Protection

S. No.	Fungicide	Formulation Types	Mode of Action	Company Name
21.	Copper fungicide	Liquid copper solution	Contact mode fungicide.	Bonide
22.	Liquid copper fungicide	Liquid formulation used as sprayer	Effective against rust of many fruits, vegetables.	Southern Ag
23.	Cuprofix Ultra 40 Disperse	Wettable powder (WP)	Effective fungicide against blight.	UPL Ltd
24.	Cuproxat, Cuproxat FL	Liquid formulation	Engineered with proprietary chemistry to deliver uniform, smaller, precisely sized particles for better plant coverage and protection against fungal and bacterial diseases.	Nufarm Americas Inc.
25.	Nordox 75 WG	Wettable powder (WP)	Superior wettability, suspensibility, and high retention.	Nordox

3.5 CONCLUSION AND FUTURE PROSPECTS

The use of nano-agrochemicals is an attractive advancement in the field of plant protection that focuses on a healthy environment. Target-specific delivery of agrochemicals possessing high solubility, better shelf-life, lower evaporation rate, and increased ability to withstand pesticide-resistant insects and pests has been one of the key advantages of nano-agrochemicals. However, commercial deployment of nano-agrochemicals in the area of plant protection has been limited. The lack of knowledge about the permissible limit and uptake capacity of nanomaterials is a key concern, along with their bioavailability and ecotoxicity aspects. Further, field-scale data on the nano-agrochemicals is scarce. In addition to this, limited data are available on the impacts of nano-agrochemicals on the beneficial soil microorganisms and their bioaccumulation and biotransformation upon entry into plants. Therefore, further research on the behavior and fate of agricultural applications of nanomaterials and their interaction with the ecosystem is required.

REFERENCES

Abkhoo, J., & Panjehkeh, N. (2017). Evaluation of antifungal activity of silver nanoparticles on *Fusarium oxysporum*. *International Journal of Infection*, *4*, 2.

Allahverdiyev, A. M., Abamor, E. S., Bagirova, M., & Rafailovich, M. (2011). Antimicrobial effects of TiO_2 and Ag_2O nanoparticles against drug-resistant bacteria and Leishmania parasites. *Future Microbiology*, *6*(8), 933–940.

Anghel, A. G., Grumezescu, A. M., Chirea, M., Grumezescu, V., Socol, G., Iordache, F., … & Holban, A. M. (2014). MAPLE fabricated Fe_3O_4@ Cinnamomum verum antimicrobial surfaces for improved gastrostomy tubes. *Molecules*, *19*(7), 8981–8994.

Anstoetz, M., Sharma, N., Clark, M., & Yee, L. H. (2016). Characterization of an oxalate-phosphate-amine metal–organic framework (OPA-MOF) exhibiting properties suited for innovative applications in agriculture. *Journal of Materials Science*, *51*, 9239–9252.

Beyth, N., Houri-Haddad, Y., Domb, A., Khan, W., & Hazan, R. (2015). Alternative antimicrobial approach: Nano-antimicrobial materials. *Evidence-Based Complementary and Alternative Medicine,* 24601, 16 p. .

Beyth, N., Yudovin-Farber, I., Perez-Davidi, M., Domb, A. J., & Weiss, E. I. (2010). Polyethyleneimine nanoparticles incorporated into resin composite cause cell death and trigger biofilm stress in vivo. *Proceedings of the National Academy of Sciences*, *107*(51), 22038–22043.

Blecher, K., Nasir, A., & Friedman, A. (2011). The growing role of nanotechnology in combating infectious disease. *Virulence*, *2*(5), 395–401.

Cai, L., Chen, J., Liu, Z., Wang, H., Yang, H., & Ding, W. (2018). Magnesium oxide nanoparticles: Effective agricultural antibacterial agent against *Ralstonia solanacearum. Frontiers in Microbiology*, *9*, 335574.

Carpenter, A. W., & Schoenfisch, M. H. (2012). Nitric oxide release: Part II. Therapeutic applications. *Chemical Society Reviews*, *41*(10), 3742–3752.

Chandra, S., Chakraborty, N., Dasgupta, A., Sarkar, J., Panda, K., & Acharya, K. (2015). Chitosan nanoparticles: A positive modulator of innate immune responses in plants. *Scientific Reports*, *5*(1), 15195.

Chaudhary, A., Kumar, N., Kumar, R., & Salar, R. K. (2019). Antimicrobial activity of zinc oxide nanoparticles synthesized from Aloe vera peel extract. *SN Applied Sciences*, *1*, 1–9.

Chaudhary, V., Jangra, S., & Yadav, N. R. (2018). Nanotechnology based approaches for detection and delivery of microRNA in healthcare and crop protection. *Journal of Nanobiotechnology*, *16*, 1–18.

Cvjetko, P., Milošić, A., Domijan, A. M., Vrček, I. V., Tolić, S., Štefanić, P. P., … & Balen, B. (2017). Toxicity of silver ions and differently coated silver nanoparticles in Allium cepa roots. *Ecotoxicology and Environmental Safety*, *137*, 18–28.

Derbalah, A., Shenashen, M., Hamza, A., Mohamed, A., & El Safty, S. (2018). Antifungal activity of fabricated mesoporous silica nanoparticles against early blight of tomato. *Egyptian Journal of Basic and Applied Sciences*, *5*(2), 145–150.

Duhan, J. S., Kumar, R., Kumar, N., Kaur, P., Nehra, K., & Duhan, S. (2017). Nanotechnology: The new perspective in precision agriculture. *Biotechnology Reports*, *15*, 11–23.

El-Hadrami, A., Adam, L. R., El Hadrami, I., & Daayf, F. (2010). Chitosan in plant protection. *Marine Drugs*, *8*(4), 968–987.

European Commission (DG SANCO). (2013). Assessment of nanomaterials in food, health and consumer products. In: *Consumers: Workshop on the Second Regulatory Review on Nanomaterials*, DHA, Brussels.

Faisal, M., Saquib, Q., Alatar, A. A., Al-Khedhairy, A. A., Hegazy, A. K., & Musarrat, J. (2013). Phytotoxic hazards of NiO-nanoparticles in tomato: A study on mechanism of cell death. *Journal of Hazardous Materials*, *250*, 318–332.

Gatoo, M. A., Naseem, S., Arfat, M. Y., Mahmood Dar, A., Qasim, K., & Zubair, S. (2014). Physicochemical properties of nanomaterials: Implication in associated toxic manifestations. *BioMed Research International*, 498420, 8 p.

Grillo, R., Abhilash, P. C., & Fraceto, L. F. (2016). Nanotechnology applied to bio-encapsulation of pesticides. *Journal of Nanoscience and Nanotechnology*, *16*(1), 1231–1234.

Hajong, M., Devi, N. O., Debbarma, M., & Majumder, D. (2019). Nanotechnology: An emerging tool for management of biotic stresses in plants. *Plant Nanobionics: Approaches in Nanoparticles, Biosynthesis, and Toxicity*, *2*, 299–335.

Hamal, D. B., Haggstrom, J. A., Marchin, G. L., Ikenberry, M. A., Hohn, K., & Klabunde, K. J. (2010). A multifunctional biocide/sporocide and photocatalyst based on titanium dioxide (TiO_2) codoped with silver, carbon, and sulfur. *Langmuir*, *26*(4), 2805–2810.

Helaly, M. N., El-Metwally, M. A., El-Hoseiny, H., Omar, S. A., & El-Sheery, N. I. (2014). Effect of nanoparticles on biological contamination of "in vitro' cultures and organogenic regeneration of banana. *Australian Journal of Crop Science*, *8*(4), 612–624.

Huh, A. J., & Kwon, Y. J. (2011). "Nanoantibiotics": A new paradigm for treating infectious diseases using nanomaterials in the antibiotics resistant era. *Journal of Controlled Release*, *156*(2), 128–145.

Kashyap, P. L., Xiang, X., & Heiden, P. (2015). Chitosan nanoparticle based delivery systems for sustainable agriculture. *International Journal of Biological Macromolecules*, *77*, 36–51.

Kasraei, S., Sami, L., Hendi, S., AliKhani, M. Y., Rezaei-Soufi, L., & Khamverdi, Z. (2014). Antibacterial properties of composite resins incorporating silver and zinc oxide nanoparticles on Streptococcus mutans and Lactobacillus. *Restorative Dentistry & Endodontics*, *39*(2), 109.

Kaur, P., Duhan, J. S., & Thakur, R. (2018). Comparative pot studies of chitosan and chitosan-metal nanocomposites as nano-agrochemicals against Fusarium wilt of chickpea (*Cicer arietinum* L.). *Biocatalysis and Agricultural Biotechnology*, *14*, 466–471.

Kaur, P., Thakur, R., & Chaudhury, A. (2016). Biogenesis of copper nanoparticles using peel extract of *Punica granatum* and their antimicrobial activity against opportunistic pathogens. *Green Chemistry Letters and Reviews*, *9*(1), 33–38.

Kaur, P., Thakur, R., Kumar, S., & Dilbaghi, N. (2011, December). Interaction of ZnO nanoparticles with food borne pathogens *Escherichia coli* DH5α and *Staphylococcus aureus* 5021 & their bactericidal efficacy. *AIP Conference Proceedings*, *1393*(1), 153–154.

Khiyami, M. A., Almoammar, H., Awad, Y. M., Alghuthaymi, M. A., & Abd-Elsalam, K. A. (2014). Plant pathogen nanodiagnostic techniques: Forthcoming changes? *Biotechnology & Biotechnological Equipment*, *28*(5), 775–785.

Kumar, R., Duhan, J. S., Manuja, A., Kaur, P., Kumar, B., & Sadh, P. K. (2022). Toxicity assessment and control of early blight and stem rot of *Solanum tuberosum* L. by mancozeb-loaded chitosan–gum acacia nanocomposites. *Journal of Xenobiotics*, *12*(2), 74–90.

Kumari, M., Shukla, S., Pandey, S., Giri, V. P., Bhatia, A., Tripathi, T., … & Mishra, A. (2017). Enhanced cellular internalization: A bactericidal mechanism more relative to biogenic nanoparticles than chemical counterparts. *ACS Applied Materials & Interfaces*, *9*(5), 4519–4533.

Kutner, A. J., & Friedman, A. J. (2013). Use of nitric oxide nanoparticulate platform for the treatment of skin and soft tissue infections. *Wiley Interdisciplinary Reviews: Nanomedicine and Nanobiotechnology*, *5*(5), 502–514.

Li, J., Sang, H., Guo, H., Popko, J. T., He, L., White, J. C., … & Xing, B. (2017). Antifungal mechanisms of ZnO and Ag nanoparticles to *Sclerotinia homoeocarpa*. *Nanotechnology*, *28*(15), 155101.

Lim, D., Roh, J. Y., Eom, H. J., Choi, J. Y., Hyun, J., & Choi, J. (2012). Oxidative stress-related PMK-1 P38 MAPK activation as a mechanism for toxicity of silver nanoparticles to reproduction in the nematode *Caenorhabditis elegans*. *Environmental Toxicology and Chemistry*, *31*(3), 585–592.

Malerba, M., & Cerana, R. (2016). Chitosan effects on plant systems. *International Journal of Molecular Sciences*, *17*(7), 996.

Mishra, S., Keswani, C., Abhilash, P. C., Fraceto, L. F., & Singh, H. B. (2017). Integrated approach of agri-nanotechnology: Challenges and future trends. *Frontiers in Plant Science*, *8*, 254477.

Mousavi, S. R., & Rezaei, M. (2011). Nanotechnology in agriculture and food production. *Journal of Applied Environmental and Biological Sciences*, *1*(10), 414–419.

Nath, D., & Banerjee, P. (2013). Green nanotechnology–A new hope for medical biology. *Environmental Toxicology and Pharmacology*, *36*(3), 997–1014.

Ocsoy, I., Paret, M. L., Ocsoy, M. A., Kunwar, S., Chen, T., You, M., & Tan, W. (2013). Nanotechnology in plant disease management: DNA-directed silver nanoparticles on graphene oxide as an antibacterial against *Xanthomonas perforans*. *ACS Nano*, *7*(10), 8972–8980.

Oluwaseun, A. C., & Sarin, N. B. (2017). Impacts of biogenic nanoparticle on the biological control of plant pathogens. *Pollution Research*, *24*(15), 13700–13709.

Palanikumar, L., Ramasamy, S. N., & Balachandran, C. (2014). Size-dependent antimicrobial response of zinc oxide nanoparticles. *IET Nanobiotechnology*, *8*(2), 111–117.

Pandey, P., Irulappan, V., Bagavathiannan, M. V., & Senthil-Kumar, M. (2017). Impact of combined abiotic and biotic stresses on plant growth and avenues for crop improvement by exploiting physio-morphological traits. *Frontiers in Plant Science*, *8*, 237767.

Pati, R., Mehta, R. K., Mohanty, S., Padhi, A., Sengupta, M., Vaseeharan, B., … & Sonawane, A. (2014). Topical application of zinc oxide nanoparticles reduces bacterial skin infection in mice and exhibits antibacterial activity by inducing oxidative stress response and cell membrane disintegration in macrophages. *Nanomedicine: Nanotechnology, Biology and Medicine*, *10*(6), 1195–1208.

Pelgrift, R. Y., & Friedman, A. J. (2013). Nanotechnology as a therapeutic tool to combat microbial resistance. *Advanced Drug Delivery Reviews*, *65*(13–14), 1803–1815.

Pestovsky, Y. S., & Martínez-Antonio, A. (2017). The use of nanoparticles and nanoformulations in agriculture. *Journal of Nanoscience and Nanotechnology*, *17*(12), 8699–8730.

Poulose, S., Panda, T., Nair, P. P., & Theodore, T. (2014). Biosynthesis of silver nanoparticles. *Journal of Nanoscience and Nanotechnology*, *14*(2), 2038–2049.

Prameela, K. L. (2017). Nanomaterial's applications in agriculture. *Journal of Chemical and Pharmaceutical Sciences*, *10*(1), 593–596.

Reddy, K. M., Feris, K., Bell, J., Wingett, D. G., Hanley, C., & Punnoose, A. (2007). Selective toxicity of zinc oxide nanoparticles to prokaryotic and eukaryotic systems. *Applied Physics Letters*, *90*, 21.

Reddy, L. S., Nisha, M. M., Joice, M., & Shilpa, P. N. (2014). Antimicrobial activity of zinc oxide (ZnO) nanoparticle against *Klebsiella pneumoniae*. *Pharmaceutical Biology*, *52*(11), 1388–1397.

Saharan, V., Sharma, G., Yadav, M., Choudhary, M. K., Sharma, S. S., Pal, A., … & Biswas, P. (2015). Synthesis and in vitro antifungal efficacy of Cu–chitosan nanoparticles against pathogenic fungi of tomato. *International Journal of Biological Macromolecules*, *75*, 346–353.

Seabra, B. A., & Duran, N. (2017). Nanoparticulated nitric oxide donors and their biomedical applications. *Mini reviews in Medicinal Chemistry*, *17*(3), 216–223.

Sharma, P., Jha, A. B., Dubey, R. S., & Pessarakli, M. (2012). Reactive oxygen species, oxidative damage, and antioxidative defense mechanism in plants under stressful conditions. *Journal of Botany*, 217037, 26 p.

Siddiqui, Z. A., Khan, A., Khan, M. R., & Abd-Allah, E. F. (2018). Effects of zinc oxide nanoparticles (ZnO NPs) and some plant pathogens on the growth and nodulation of lentil (*Lens culinaris* Medik.). *Acta Phytopathologica et Entomologica Hungarica*, *53*(2), 195–211.

Suriyaprabha, R., Karunakaran, G., Kavitha, K., Yuvakkumar, R., Rajendran, V., & Kannan, N. (2014). Application of silica nanoparticles in maize to enhance fungal resistance. *IET Nanobiotechnology*, *8*(3), 133–137.

Syafiq, R., Sapuan, S. M., Zuhri, M. Y. M., Ilyas, R. A., Nazrin, A., Sherwani, S. F. K., & Khalina, A. (2020). Antimicrobial activities of starch-based biopolymers and biocomposites incorporated with plant essential oils: A review. *Polymers*, *12*(10), 2403.

Velmurugan, N., Kumar, G. G., Han, S. S., Nahm, K. S., & Lee, Y. S. (2009). Synthesis and characterization of potential fungicidal silver nano-sized particles and chitosan membrane containing silver particles. *Iranian Polymer Journal*, *18*(5), 383–392.

Wang, X., Liu, X., Chen, J., Han, H., & Yuan, Z. (2014). Evaluation and mechanism of antifungal effects of carbon nanomaterials in controlling plant fungal pathogen. *Carbon*, *68*, 798–806.

You, C., Han, C., Wang, X., Zheng, Y., Li, Q., Hu, X., & Sun, H. (2012). The progress of silver nanoparticles in the antibacterial mechanism, clinical application and cytotoxicity. *Molecular Biology Reports*, *39*, 9193–9201.

4 Exome Sequencing
Genomic Revolution of Unlocking Nutritional Potential of Food Crops

Neeru S. Redhu, Bharat Garg, Jyoti Taunk, Vikram Singh, Jayanti Tokas, and Shikha Yashveer

4.1 INTRODUCTION

The human body requires a balanced intake of nutrients—carbohydrates, proteins, fats, vitamins, and minerals—to function optimally. A healthy diet not only fuels our energy needs but also supports vital body functions, including growth and repair. Failure to obtain sufficient nutrients causes malnutrition, leading to various physical and mental illnesses including abnormalities, hindered growth, reduced cognitive function, and heightened vulnerability to diseases. Although both macro and micronutrients are vital for every individual, still across the globe the provision of balanced and nutritious food, especially in poor regions, experiences a significant setback due to micronutrients leading to their deficiency. Mineral deficiency is one of the most significant global concerns, affecting both physical and mental health. According to the World Health Organization (WHO), humans require at least 22 mineral elements for optimal health, and a low nutritional diet would stimulate detrimental effects on the health, especially of pregnant ladies, mothers, and children living in low-income and middle-income countries. Diverse interventions such as supplementation, dietary diversity, and fortification have been implemented to effectively mitigate malnutrition in specific contexts. Of all, dietary supplementation either in the form of supplements or pills is a simple, low-cost solution that improves human health. They can be used to tackle target-specific micronutrient deficiency or health-related conditions, such as turmeric supplements can help in reducing inflammation-related conditions or folic acid supplements during pregnancy can reduce the risk of birth defects.

Currently, no single food source can provide complete nutrition; therefore, the concept of a balanced diet is emphasized. Moreover, individual dietary needs can vary based on factors like age, sex, physical activity level, and underlying health conditions, highlighting the importance of tailored nutritional strategies for different individuals. Thus, the consumption of a suitable dietary regimen is necessary to fulfill an entire nutrient requirement. Dietary diversity mainly aims to provide a balanced diet through a combination of food choices. Another strategy where the premeditated increase or addition of a specific micronutrient concentration during crop/

DOI: 10.1201/9781032642567-5

food processing is done is known as food fortification. The addition of micronutrients such as iron and zinc to flour or the addition of iodine to table salt are examples of food fortification (Yashveer et al., 2010). Other strategy includes bio-fortification, wherein the increase or addition of a specific micronutrient is achieved at the DNA level through methods of agronomic practices like conventional plant breeding or modern biotechnology. Some examples of biofortification are an increase in iron concentration in rice, beans, legumes, and sweet potato or Provitamin A and carotenoid concentration in maize, and sweet potato or amino acid and protein composition in cassava and sorghum (Redhu et al., 2023).

Given the above facts, agricultural advancements can play a crucial role in addressing nutritional challenges on a global scale as (i) agriculture provides a foundation for a diverse range of food products, including grains, fruits, vegetables, and livestock, for meeting all the nutritional requirements of a diverse population. (ii) It can lead to increased yields and enhanced crop quality. (iii) It can make nutrient-rich foods more affordable and accessible to a wider range of people, especially for those living in food-insecure regions or marginalized communities. (iv) It can facilitate the fortification of staple foods with essential nutrients to address common deficiencies. (v) It can support the cultivation of traditional and local foods, which often have unique nutritional value and cultural significance. To improve the nutritional composition of crops, it is imperative to possess a comprehensive understanding of the genetic foundation of nutrient formation and accumulation, as well as the complex interplay among genes, environmental elements, and crop physiology (Nkhata et al., 2020).

Recent advances in molecular strategies and sequencing technologies aid in developing an extensive understanding of the complex interplay among genes related to nutrient formation, accumulation, and availability. Broadly, this understanding can be achieved and improved through "omics" studies like genomics, transcriptomics, proteomics, and metabolomics. Additionally, the development of specialized studies like exome sequencing wherein the focus is explicitly on the protein-coding regions (exons) [thereby also known as whole exome sequencing (WES)] is a cost-effective and highly informative approach for studying the genetic basis. Exome typically constitutes approximately 1%–2% of the entire genome, depending on the species concerned. For instance, in humans with a total of ~233,785 exons, it represents 1.1% of the total genome. Even though the exome constitutes a relatively minor portion of the complete genome, it encompasses the bulk of functional genetic variants that play a crucial role in protein production and function. For instance, 85% of the mutation observed in the human genome is found to be present in the exonic regions.

Typically, WES encompasses a hybrid approach for studying exonic regions, that is, exome capture or target-specific amplification of exome, followed by high-throughput sequencing. Currently, several exome capture assays are being offered by Agilent, Twist, Illumina, IDT, and NimbleGen that are compatible with the Illumina next-generation sequencing (NGS) platform. The detection of genetic alterations, such as insertions, deletions, single-nucleotide variants (SNVs), and copy number variations (CNVs), specifically inside the coding regions of genes responsible for protein synthesis, can be straightforwardly done through WES. It offers an economical option compared with other techniques to identify induced mutations within the genome that not only have genotypic change but also phenotypic impact. This approach has been employed to detect undesirable intracultivar genetic heterogeneity in the plant exome that further affected its phenotype (Haun et al., 2011). Latest interventions also include the expansion in the exome sequencing strategy to encompass functional non-protein-coding components such as microRNA, long intergenic noncoding RNA, and other similar entities, along with specific candidate loci.

Exome sequencing has garnered substantial significance across many disciplines, encompassing the agricultural sector, owing to its ability to discern genetic variations impacting desirable characteristics in crops. The technique also enables the identification and validation of genes linked to nutrient-related traits and characteristics, offering a valuable understanding of the underlying molecular mechanisms that regulate nutrient accumulation, transport, and storage in crops. This information can be further utilized for biofortification, reducing the antinutrient components, or crop improvement through targeted genetic modification, genetic engineering, or precision breeding. Figure 4.1 depicts a general overview of exome sequencing and its involvement in various crop improvement strategies for enhancing nutritional components.

4.2 FUNDAMENTALS OF DNA SEQUENCING: WHOLE GENOME SEQUENCING

DNA sequencing is the scientific procedure used to ascertain the precise arrangement of nucleotides, namely, adenine, cytosine, guanine, and thymine, within a DNA molecule. The sequencing of the entire genome including coding and noncoding regions is generally known as genome or whole genome sequencing (WGS). With the aid of DNA-sequencing techniques, the genome sequencing of various crops, livestock, and microbes has been achieved which helped in gaining

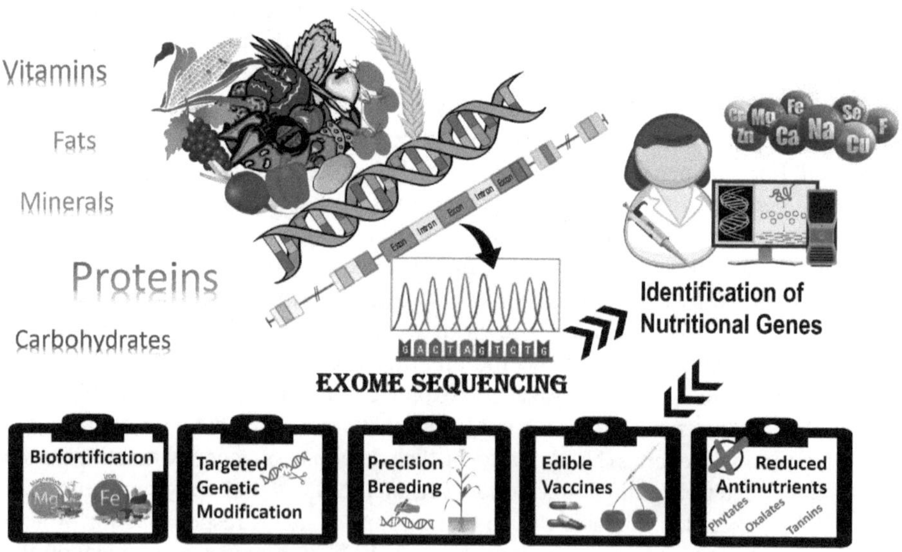

FIGURE 4.1 Exome sequencing and its involvement in various crop improvement strategies for enhancing nutritional components.

valuable insights into the genetic makeup of those organisms. These insights can be also utilized to improve productivity, disease resistance, and nutritional values of food crops.

Currently, numerous sequencing techniques and technologies can be employed for DNA and genome sequencing. The first DNA-sequencing method was proposed by Allan Maxam and Walter Gilbert wherein a chemical-modification-dependent cleavage of DNA was performed followed by gel electrophoresis. Another method was Sanger sequencing, also known as dideoxy sequencing or chain-termination method, which was developed by Frederick Sanger and his colleagues in the 1970s. This technique entails DNA amplification in the presence of chain-terminating dideoxynucleotides that are labeled with fluorescent or radioactive markers followed by gel electrophoresis. Out of the two, Sanger sequencing was more popular and later acted as a basis for the automation of sequencing technologies. The first automated DNA sequencer was invented by Lloyd M. Smith (later introduced by Applied Biosystems in 1987), which enabled the completion of the Human Genome Project in 2003.

The automation of DNA sequencers was a milestone for genome sequencing and other omics studies that ushered the scientific world into the development of numerous sequencing technologies, collectively called NGS technologies. NGS uses highly parallel methods for sequencing, enabling efficient and affordable large-scale sequencing/re-sequencing. The development of NGS technologies removed the bottleneck of WGS. However, the high availability of WGS has shifted this bottleneck to the annotation and interpretation of the sequencing data. Some common NGS platforms include Illumina, Ion Torrent, and PacBio. They use various sequencing-by-synthesis methods, nanopore sequencing, or single-molecule real-time sequencing to determine the DNA sequence. NGS data analysis involves aligning short DNA fragments to a reference genome and assembling the sequence to reconstruct the whole genome. Using NGS technologies and performing WGS, not only laid the foundation to uncover various genetic potential of plants and to understand the domestication process that further can be utilized to improve food's nutritional value and to eradicate malnutrition through crop improvement.

4.3 WHOLE GENOME SEQUENCING FOR NUTRITIONAL IMPROVEMENT OF FOOD CROPS

Once nutritional genes are identified through WGS, they can be used for molecular breeding. WGS can also guide genome-editing techniques like CRISPR-Cas9 to precisely modify genes responsible for nutritional attributes. Whole genome sequencing can also enable the comparison of the genome of different varieties or species of food crops and the identification of genes associated with superior nutritional profiles that can be transferred through breeding or genetic engineering in desired crops/varieties. WGS enabled the identification and characterization of genes responsible for the synthesis of beta carotene, a precursor of vitamin A in the rice genome. This knowledge enabled the development of the Golden rice variety with rice grains rich in beta-carotene content. Similarly, WGS was used to identify genes responsible for lysine biosynthesis and metabolism. This information aided in the development of maize varieties with increased lysine content and addressing protein malnutrition.

Quinoa is known for its high protein content but lacks some essential amino acids. WGS of quinoa has allowed the identification of genes involved in amino acid biosynthesis. The information has been used to select and breed quinoa varieties with improved protein quality. Similarly, WGS has been utilized to identify genes responsible for fatty acid biosynthesis and metabolism, which are further used in the development of soybean varieties with altered fatty acid profiles, including those with increased omega-3 fatty acids (Harris, 2012). However, Cassava is a staple crop in many parts of the world, but it contains toxic cyanogenic glucosides. Through WGS the genes involved in the biosynthesis of these compounds were identified, which further helped in the development of Cassava varieties with reduced cyanogenic glycoside content, making them safe for consumption. Various food crops like beans, wheat, and rice have been biofortified to increase the iron and zinc content. WGS has been instrumental in identifying genes related to iron and zinc uptake, transport, and storage in these crops for their betterment.

4.4 WHOLE EXOME SEQUENCING (WES)

Broadly, exome sequencing or WES can be treated as a subset of genome sequencing as targeted sequencing for only protein-coding regions of the genome is performed. Even though the whole exome only represents 1%–2% of an entire genome, >85% of mutations that cause diseases are found in it. This significantly contributed to the emergence of exome sequencing as a crucial genetic tool for unveiling the molecular basis of diseases and related phenotypic traits. WES has become a more cost-effective alternative to WGS due to its ability to balance cost, genome coverage, diagnostic success, and interpretability. Targeted exome sequencing aided in the identification of specific SNP markers for genes associated with sodicity tolerance, powdery mildew resistance, leaf color, and defective kernel phenotypes in bread wheat (Zhao et al., 2023). Similarly, exome sequencing detected a pivotal role of the *HvCSLC1* gene in the elongation of root hairs in barley. Another study identified 11 quantitative trait loci (QTLs) linked to genes for resistance against leaf rust in barley (Matros et al., 2023). Exome sequencing also identified four unique genetic groups within *Pinus densata* that had a potential role in maintaining species boundaries between newly evolved taxa and their progenitors (*Pinus tabuliformis* and *Pinus yunnanensis*) (Guo et al., 2023). These genetic groups reflected migration history and major gene-flow barriers across the landscape. Exome sequencing is a significant tool for studying various plant traits, and host–pathogen interactions and discovering natural evolution in plants, as the interpretation of allelic variation concerning their phenotype depends on the exon. Integration of molecular techniques, such as QTL mapping, marker-assisted selection, sequencing with developing reference genome, *de novo* genome assembly, pan-, core- and/or super-pan-genomes through bioinformatics tools and analysis, will boost breeding programs.

WGS and WES are two prominent methodologies used in genomics, enabling comprehensive examination of an individual's genetic composition. Both methods offer unique benefits and can be applied in many areas of biological science. Table 4.1 lists the differences between WGS and WES techniques.

TABLE 4.1
Differences between Whole-Genome Sequencing and Whole Exome Sequencing

	Attribute(s)	Whole Genome Sequencing	Whole Exome Sequencing
1	Scope	Sequencing of an entire genome including coding and noncoding regions. It also included mitochondrial and chloroplast genomes.	Selectively targets and sequences only the protein-coding regions of the genome, known as exons. These regions constitute roughly 1%–2% of the entire genome.
2	Mutation detection	A comprehensive view of the genome, making it suitable for identifying a wide range of genctic variants, including insertions, deletions, single nucleotide polymorphisms (SNPs), and structural variations.	Primarily focused on exonic regions. It is well-suited for identifying mutations within protein-coding genes. It may miss significant variants in noncoding regions.
3	Coverage depth	Generally, provides lower coverage depth for each base since it covers the entire genome. Typical coverage depth can range from 30× to 50×.	Offers higher coverage depth for exonic regions because it focuses on a smaller portion of the genome. It can achieve much deeper coverage, often exceeding 100×.
4	Cost	Typically, more expensive than WES due to the broader coverage of the genome. Costs have decreased but still can be a limiting factor for large-scale studies.	Generally, more cost-effective as it targets specific regions, making it suitable for large-scale studies and clinical applications.
5	Data analysis and interpretation	The larger volume of data generated requires more extensive computational resources and data storage. Analyzing noncoding regions can be challenging.	Simplifies data analysis, focusing on known coding regions, which are more readily interpretable.
6	Applications	Ideal for the identification of specific genes and genetic variations that are responsible for important crop traits such as disease resistance, yield, domestication, and drought tolerance.	Particularly useful in studying host-pathogen interactions, natural evolution in plants, and betterment of crop production.

The choice between WGS and WES depends on the goals of the study, the budget, and the available computing resources. WGS is more comprehensive, whereas WES is less expensive and works well for targeted genetic research.

4.5 WORKFLOW OF EXOME SEQUENCING

The process of exome sequencing can be broadly divided into two steps wherein the first step involves capturing of exomic region of the genome, and the second step involves the sequencing of the captured exomic region through any of the NGS approaches. Typically, the complete workflow of exome sequencing includes collection, preparation, and isolation of genomic samples followed by target enrichment for

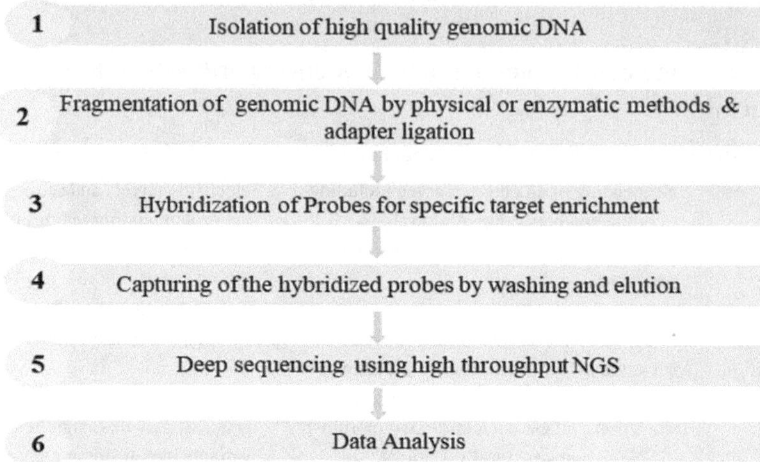

FIGURE 4.2 A general workflow of exome sequencing.

isolation of exome and their subsequent washing and elution of the desired product. This is followed by sequencing and bioinformatics data analysis. Figure 4.2 depicts a general workflow of WES.

4.5.1 Collection, Preparation, and Isolation of Genomic Samples: Library Preparation

Sample collection refers to the collection of target tissues such as leaves, roots, stems, and so on in the case of plants. This step is followed by DNA isolation. As the size of DNA is usually too long, DNA isolation is usually followed by DNA fragmentation either through physical or enzymatic methods. The physical methods of DNA fragmentation include acoustic shearing, sonication, and hydrodynamic shear. On the other hand, enzymatic approaches involve using nuclease or transposase enzymes for DNA fragmentation. Nucleases through hydrolysis of the phosphodiester linkages, whereas transposases through cleaving and inserting a short double-stranded oligonucleotide perform DNA degradation or fragmentation. The use of restriction endonuclease is preferred over non-specific nucleases. The fragmented DNA samples are also known as fragment libraries. DNA fragmentation step is followed by the adapters ligation that would help in the subsequent target enrichment procedure of this fragmented library.

4.5.2 Target-Enrichment Methods for Isolation of Exome

Target enrichment is referred to as the process of isolation/identification of the genomic regions of interest. In terms of exome sequencing, this step is also known as exome capturing. The DNA fragments ligated with adapters are targeted for target enrichment using complementary primers or probes against adapter sequences.

Exome capture technology can be broadly classified into two primary groups, solution-based and array/chip-based. Both technologies capture exome, utilizing the principle of hybridization using specially designed probes or baits to selectively enrich target sequences from a sequencing library. Array-based exome enrichment uses high-density microarrays (a two-dimensional array constructed on a glass slide or a silicon thin film) with bound oligonucleotide probes for capturing the exome. These oligonucleotide probes are specifically designed to complement and hybridize with specific regions of the target genome, that is, exome or protein coding region. The solution-based method is also known as the in-solute or in-solution capturing method. This method involves exome capturing using magnetic beads that have bound oligonucleotide probes or functional chemical components for binding with a specific target genome, that is, exome. Similar to the array-based technique, wherein the exome is selectively hybridized with the designed probes; the solution-based technique attracts and bonds to magnetic beads while the remaining portions of the genome remain dissociated. In-solute-based capture requires a higher concentration of baits relative to the DNA library, while array-based capture requires an excess library compared to the probes to achieve enrichment. One advantage of the in-solute capture method is that using magnetic beads enhances the reaction's efficiency through mechanisms like agitation or thermal stimulation.

Next-generation sequencing primarily uses PCR-based amplification for target enrichment. Popular packages and kits for exome sequencing includes Agilent SureSelect XT2 V6 Exome, the IDT xGEN Exome Panel, the Roche NimblegenSeqCap EZ Exome v3.0, and the Illumina Nextera Rapid Capture Expanded Exome. The washing step is one of the necessary steps of target enrichment after the addition of complementary primers or probes to achieve purity by eliminating undesirable substances, such as other genome segments, proteins, and electrolytes. Washing is mainly done using either distilled water or a specific eluent tailored to the kit to facilitate the elution of the desired product. The eluent serves as the reagent to remove the exomic region from microarray or magnetic beads by disturbing the interaction among the exome and the binding compounds, that is, probes or baits. The washing and elution processes can be repeated many times to get higher purity in exomes. The target enrichment allows the capturing of exome for subsequent steps of exome sequencing or exome library generation.

4.5.3 Exome Sequencing Using NGS

The exome sequencing or exome library generation was made easier with the advent of NGS. Currently, numerous NGS platforms such as Illumina HiSeq flowcell and Roche-454 have been developed, each using unique methods to read and sequence DNA in large volumes. Illumina, Ion Torrent, PacBio, and Nanopore are some of the most popular sequencing technologies. Illumina uses sequencing by synthesis wherein fluorescently labeled enzymes grow a lawn of short single-stranded DNA molecules into double-stranded DNA on a flow cell. Sequencing machines read the fluorescent signal from each newly incorporated base in real-time. Ion Torrent uses a semiconductor to read the change in acidity due to the release of a proton on incorporation of new bases as short single-stranded DNA is turning into double-stranded DNA. PacBio uses DNA polymerase which is fixed on the bottom of the well and fluorescently tagged nucleotides to turn short single-stranded

DNA molecules into double-stranded DNA molecules. Similar to Illumina, it also reads the fluorescent signal from each newly incorporated base in real time. The nanopore sequencing uses nanometer-wide pores in a membrane wherein DNA is threaded/moved through, which results in fluctuation (specific to each base) in the ionic current. This fluctuation in ionic current is monitored by the sequencing machine to know the DNA sequence. Among these, Illumina and Ion Torrent are PCR amplification-based and collectively treated as the second generation of NGS technology while PacBio and Nanopore are based on single-molecule sequencing (the PCR amplification step is skipped) and collectively treated as the third generation of NGS technology.

The fundamental concept of NGS involves the utilization of specific sequencing platforms to capture and amplify or sequence exome samples, wherein a signal is generated during elongation and subsequently detected. Ultimately, the comprehensive and large-scale data of sequence is incorporated using a biological information algorithm. The generation of high-throughput sequence data, also known as reads, is followed by the analysis of data.

4.5.4 Data Analysis

The high throughput data obtained from sequencing experiments can be challenging to comprehend and analyze without bioinformatics tools. This is primarily because many sequencing techniques generate tiny pieces of sequence, also known as sequence reads, necessitating the process of sequence assembly to arrive at a conclusive outcome. The data analysis includes the quality check of sequence reads and their filtering of high-quality reads and removal of low-quality reads followed by their assembly. The sequence assembly can be performed either through reference-based assembly or *de novo*-based assembly to obtain longer lengths of sequences. The reference-based assembly utilizes previously known sequence information of target samples (i.e., plant from which leaf or root samples are collected), whereas *de novo*-based assembly is developed from scratch by aligning the reads on themselves and using various sequence algorithms. Typically, the generation of sequence assembly is followed by the annotation and classification of sequences. Annotation broadly describes the gene ontology, that is, the biological process, molecular function, and cellular compartment. The classification of sequences can provide insight into the evolution within the genome. In the case of exome sequencing, the sequence assembly will provide the sequence of the protein-coding part, while analysis will provide information related to gene ontology and the evolutionary history of the protein-coding regions within the genome. Further analysis also provides greater insight into the potential transcripts of the genome as well as the identification and detection of SNPs or other similar variations.

4.6 APPLICATIONS OF EXOME SEQUENCING FOR CROP IMPROVEMENT

The applications of exome sequencing are widespread. Exome sequencing or WES exhibits extensive utility in the realm of crop improvement and enhancement.

Majorly, WES can aid in targeting various plant traits such as plant yield, nutritional composition, and nutrient bioavailability among others using the following means:

i. Exploring the molecular underpinnings of natural variation and induced mutation;
ii. Detecting novel genetic markers and rare variants;
iii. Serving as a viable alternative for genomics research on non-model and closely related wild species;
iv. Investigating alternative splicing events leading to a change in protein or nutritional composition; and
v. Identifying coding variants with a role in plant population genetics.

Numerous investigations have been performed in plants by specifically studying the exomic region of the genome which aids in crop improvement. For instance, natural variation and genetic diversity among wild and cultivated durum wheat accessions and hexaploid wheat varieties were studied using exome capture and sequencing. The study reported novel co-dominant SNP markers and monitored alien introgressions to mitigate linkage drag (Kaur and Gaikwad, 2017). In another study, hexaploid wheat exome was studied and utilized to map the candidate genes associated with specific phenotypes, such as a yellow–green leaf mutant. The study also demonstrated the potential impact of mutations in gibberellic acid biosynthesis and signaling genes on plant height over the wheat genome. Additionally, a publicly accessible database of wheat mutation coupled with the corresponding germplasm repository was developed to facilitate further research (Xiong et al., 2023). Exome sequencing was also used to screen the coding region mutations in EMS-induced TILLING populations in wheat, barley, and tomato (Singh et al., 2023). In another study, a recombinant inbreed population developed from a cross between wheat varieties Yang16G216 and Yang16M6393 was genotyped using exome capture sequencing, which ultimately helped in mapping and characterization of a novel adult plant leaf rust resistance gene (Zhao et al., 2023). Earlier, Jiang et al. (2022) and Mu et al. (2022) identified a gene for all-stage resistance to stripe rust and powdery mildew resistance gene, respectively, using bulked segregant analysis after exome sequencing and reviewed the application and limitations of the exome capture in hexaploid wheat genome. Self-pollination causes severe inbreeding depression in the case of alfalfa. Using bulked target capture sequencing based on exome sequencing, Lin et al. (2023) identified numerous genetic loci associated with vigorous plant growth in this crop. By analyzing exomic regions, key genetic variations associated with the ability of certain rice varieties to withstand temporary submergence, a critical trait for flood-prone regions, were pinpointed. Targeted exome sequencing of an unselected heavy iron beam irradiated population revealed less-biased mutation characteristics in the rice genome (Ichida et al., 2019). A study of the soybean exome confirmed the existence of undesired intracultivar heterogeneity, which was found to impact the observed variance in the Williams 82 cultivar (Haun et al., 2011). In the case of barley, the investigation targeting exome detected natural variations and mutations that are associated with early maturation. Also, the genes responsible for the nodding dwarf phenotype were identified along with the identification of markers to distinguish between *Hordeum vulgare* and *H. bulbosum* and to identify alien introgression (Wendler et al., 2014). In another study, the whole exome of barley established the link between the glucan synthase gene (*HvCSLC1*) and root hair phenotype. The study reported the splice-junction mutation within the *HvCSLC1* gene, which resulted in truncated protein and subsequently exhibited a short root hair phenotype. Similar linkage has also been unequivocally affirmed by identifying three independent allelic mutations in rice and maize. The study imparted HvCSLC1 enzyme as a key enzyme responsible for synthesizing the XyG backbone during the elongation of root hair tubes, thereby aiding in the structural integrity and flexibility necessary for the dynamic process of tip-growing root hair cells. Exome sequencing has been used to identify the genetic variations related to fruit size and yields in tomatoes (Song et al., 2022). Moreover, exome sequencing has also been significant in examining genetic diversity in essential perennial plants such as miscanthus, black cottonwood, and pine. Exome sequencing was employed to study the genetic basis of resistance to maize lethal necrosis disease (MLND) to develop MLND-resistant maize varieties (Gowda et al., 2015). Targeting exome has facilitated the identification of stronger connections between their genotypes and phenotypes, thereby contributing to advancements in agricultural enhancement.

4.7 CURRENT STATUS OF EXOME SEQUENCING FOR NUTRITIONAL IMPROVEMENT OF FOOD CROPS

Exome sequencing has been gaining traction in crop genomics and research, including efforts to improve the nutritional quality of food crops. Pootakham et al. (2014) utilized this approach to genotype Cassava F1 progeny to segregate the population for starch viscosity phenotypes. Similarly, exome capture was utilized to target and characterize variation in genomic/QTL regions of Brassica genotypes that are associated with agronomical or nutritionally important traits (Schiessl et al., 2014). The approach has been utilized to identify genes responsible for grain quality traits such as protein composition in the case of rice. Exome sequencing has helped discover genes responsible for enhanced protein quality and oil content in soybean varieties. Analysis of the soybean thiol esterase gene family revealed the role of *GmFAT* in improving fatty acid composition. Likewise in potatoes, the approach has been used to study genes associated with tuber quality, such as starch content and cooking properties, which revealed a very complex genetic architecture for tuber sprouting and tuber growth (Sharma et al., 2021). Dixon et al. (2022) introduced an approach to improve the nutritional quality of wheat grain by way of genes that regulate wheat inflorescence development. Using exome sequencing, semi-dominant alleles for HOMOBOX domain 2 on wheat A and D sub-genomes which are responsible for generating more flower-bearing spikelets and enhanced grain protein content were identified. These alleles increased HB-2 expression which in turn increased amino acid supply to inflorescence during grain development through leaf modification and vascular tissue development. Till now very little work has been done for targeting nutrient synthesis, composition, accumulation transport and storage using exome sequencing. Figure 4.3 depicts various potential pathways and plants that can be targeted for achieving a better overall nutrient composition of plants for a healthy diet.

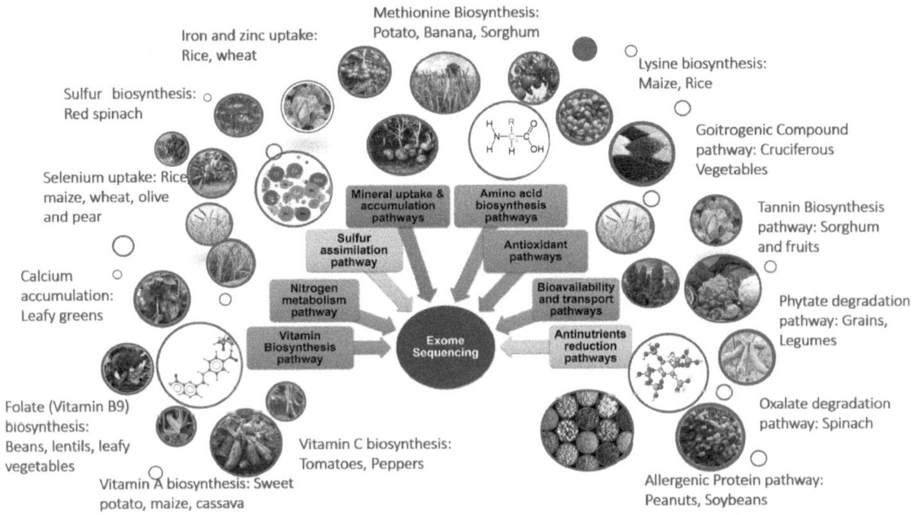

FIGURE 4.3 Applications of exome sequencing through potential pathways that can be targeted for achieving better overall nutritional composition of plants for a healthy diet.

4.8 CONCLUSION AND FUTURE PERSPECTIVES

Exome sequencing has become an invaluable tool and has assisted in crop management and improvement by understanding its genetic makeup. NGS technologies in plant exomics have accelerated crop improvement, especially in wheat, barley, rice, maize, and cotton. As the tools for WES are increasingly developed, various aspects of crop improvement can be targeted including mechanisms for host–pathogen interaction, natural evolution of crops, and biodiversity. Even though not much has been done to target the nutrient component yet. Still, exome sequencing has provided the platform that can help in the identification of key genes or protein-coding regions and subsequently help in achieving better nutrients in a fast-track manner. By determining potential genes, mutations, and variants linked to nutrient-related traits, WES can serve a valuable role in understanding the underlying molecular mechanisms that regulate nutrient accumulation, transport, and storage in crops. This can provide valuable insights for the development of focused breeding approaches for enhancing nutrient content, increasing nutrients' bioavailability, and optimizing the nutritional profiles of crops. By leveraging this technology, the development of crops that are not only more nutritious but also resilient and tailored to meet the dietary needs of diverse populations can be accelerated. Advances in sequencing technology have increased the throughput of exome sequencing, allowing for the analysis of multiple crop varieties or lines simultaneously. This is beneficial for large-scale breeding programs aimed at enhancing nutritional traits. Integration of exome sequencing data with genome-wide association analysis enables the determination of specific genomic regions linked to desirable nutrient profiles. Further, exome sequencing can be coupled with functional annotation to understand the roles of specific genes in nutrient biosynthesis and accumulation. The use of exome sequencing is likely to continue to grow as sequencing technologies advance, making it even more cost-effective and efficient. This will further support efforts to enhance the nutritional quality of food crops. However, presently there is a necessity to build large biological databases precisely for exome to improve the interpretation of various traits related to crop improvement. Also, novel analysis techniques need to be developed that can deal with all of the variants within a genome. Moreover, not all sequence variants are present in the exonic portion of genes that are necessary for controlling transcriptional regulation. Hence, combining and integrating exome information with other genomic regions is also necessary for interpreting the genome's structural variation. Overall, the incorporation of exome sequencing for improving the nutritional quality of food crops represents a significant step towards addressing global nutritional challenges and enhancing the sustainability of agriculture.

REFERENCES

Dixon, L. E., Pasquariello, M., Badgami, R., Levin, K. A., Poschet, G., Ng, P. Q., … & Boden, S. A. (2022). MicroRNA-resistant alleles of HOMEOBOX DOMAIN-2 modify inflorescence branching and increase grain protein content of wheat. *Science Advances*, *8*(19), eabn5907.

Gowda, M., Das, B., Makumbi, D., Babu, R., Semagn, K., Mahuku, G., … & Prasanna, B. M. (2015). Genome-wide association and genomic prediction of resistance to maize lethal necrosis disease in tropical maize germplasm. *Theoretical and Applied Genetics*, *128*, 1957–1968.

Guo, J. F., Zhao, W., Andersson, B., Mao, J. F., & Wang, X. R. (2023). Genomic clines across the species boundary between a hybrid pine and its progenitor in the eastern Tibetan Plateau. *Plant Communications*, *4*, 4.

Harris, W. S. (2012). Stearidonic acid–enhanced soybean oil: A plant-based source of (n-3) fatty acids for foods. *The Journal of Nutrition*, *142*(3), 600S–604S.

Haun, W. J., Hyten, D. L., Xu, W. W., Gerhardt, D. J., Albert, T. J., Richmond, T., … & Stupar, R. M. (2011). The composition and origins of genomic variation among individuals of the soybean reference cultivar Williams 82. *Plant Physiology*, *155*(2), 645–655.

Ichida, H., Morita, R., Shirakawa, Y., Hayashi, Y., & Abe, T. (2019). Targeted exome sequencing of unselected heavy-ion beam-irradiated populations reveals less-biased mutation characteristics in the rice genome. *The Plant Journal*, *98*(2), 301–314.

Jiang, Y., Duan, L., Guan, F., Yao, F., Long, L., Wang, Y., … & Chen, G. (2022). Exome sequencing from bulked segregant analysis identifies a gene for all-stage resistance to stripe rust on chromosome 1AL in Chinese wheat landrace 'Xiaohemai'. *Plant Disease*, *106*(4), 1209–1215.

Kaur, P., & Gaikwad, K. (2017). From genomes to GENE-omes: Exome sequencing concept and applications in crop improvement. *Frontiers in Plant Science*, *8*, 304784.

Lin, S., Medina, C. A., Wang, G., Combs, D., Shewmaker, G., Fransen, S., … & Yu, L. X. (2023). Identification of genetic loci associated with five agronomic traits in alfalfa using multi-environment trials. *Theoretical and Applied Genetics*, *136*(5), 121.

Matros, A., Schikora, A., Ordon, F., & Wehner, G. (2023). QTL for induced resistance against leaf rust in barley. *Frontiers in Plant Science*, *13*, 1069087.

Mu, Y., Gong, W., Guo, J., Han, R., Han, G., & Ma, P. (2022). Identification of the powdery mildew resistance gene in wheat breeding line Yannong 99102-06188 via bulked segregant exome capture sequencing. *Frontiers in Plant Science*, *13*, 1005627.

Nkhata, S. G., Chilungo, S., Memba, A., & Mponela, P. (2020). Biofortification of maize and sweetpotatoes with provitamin A carotenoids and implication on eradicating vitamin A deficiency in developing countries. *Journal of Agriculture and Food Research*, *2*, 100068.

Pootakham, W., Shearman, J. R., Ruang-Areerate, P., Sonthirod, C., Sangsrakru, D., Jomchai, N., … & Tangphatsornruang, S. (2014). Large-scale SNP discovery through RNA sequencing and SNP genotyping by targeted enrichment sequencing in cassava (*Manihot esculenta* Crantz). *PloS ONE*, *9*(12), e116028.

Redhu, N. S., Yashveer, S., Taunk, J., Banga, P., Singh, V., Tokas, J., … & Arora, I. (2023). Plant microRNAs: Unexplored biogenesis, prediction tools and applications. *Gene Reports*, *32*, 101799.

Schiessl, S., Samans, B., Hüttel, B., Reinhard, R., & Snowdon, R. J. (2014). Capturing sequence variation among flowering-time regulatory gene homologs in the allopolyploid crop species *Brassica napus*. *Frontiers in Plant Science*, *5*, 102709.

Sharma, S. K., McLean, K., Colgan, R. J., Rees, D., Young, S., Sønderkær, M., … & Bryan, G. J. (2021). Combining conventional QTL analysis and whole-exome capture-based bulk-segregant analysis provides new genetic insights into tuber sprout elongation and dormancy release in a diploid potato population. *Heredity*, *127*(3), 253–265.

Singh, D., Chaudhary, P., Taunk, J., Singh, C. K., Chinnusamy, V., Sevanthi, A. M., … & Pal, M. (2023). Targeting Induced Local Lesions in Genomes (TILLING): Advances and opportunities for fast tracking crop breeding. *Critical Reviews in Biotechnology*, *44*, 1–20.

Song, J., Shang, L., Li, C., Wang, W., Wang, X., Zhang, C., … & Zhang, J. (2022). Variation in the fruit development gene POINTED TIP regulates protuberance of tomato fruit tip. *Nature Communications*, *13*(1), 5940.

Wendler, N., Mascher, M., Nöh, C., Himmelbach, A., Scholz, U., Ruge-Wehling, B., & Stein, N. (2014). Unlocking the secondary gene-pool of barley with next-generation sequencing. *Plant Biotechnology Journal*, *12*(8), 1122–1131.

Xiong, H., Guo, H., Fu, M., Xie, Y., Zhao, L., Gu, J., … & Liu, L. (2023). A large-scale whole-exome sequencing mutant resource for functional genomics in wheat. *Plant Biotechnology Journal*, *21*(10), 2047–2056.

Yashveer, S., Singh, V., Kaswan, V., Kaushik, A., & Tokas, J. (2010). Green biotechnology, nanotechnology and bio-fortification: Perspectives on novel environment-friendly crop improvement strategies. *Biotechnology and Genetic Engineering Reviews*, *30*(2), 113–126.

Zhao, R., Liu, B., Wan, W., Jiang, Z., Chen, T., Wang, L., & Bie, T. (2023). Mapping and characterization of a novel adult-plant leaf rust resistance gene LrYang16G216 via bulked segregant analysis and conventional linkage method. *Theoretical and Applied Genetics*, *136*(1), 1.

5 Millets or Miracles
Biotechnological Interventions and The Future Strategies

Deepika and Raj Kumar Salar

5.1 INTRODUCTION

Cereals are major food crops all over the world. These grains not only contribute to the nutritional needs of humans and animals, but they also have a significant impact on enhancing socio-economic aspects. Cereals are used as raw materials in many industries, especially in food industries (Nayik et al., 2023). Production of a higher amount of cereals has become a necessity to fulfill the demand of the increasing population. Also, cereals can play a good role in substituting non-vegetarian diets and save the lives of many innocent animals used as food. Saving animals can have positive effects on our ecosystems and the planet. Owing to human activities, many ecosystems have suffered negatively. Soil, water, and air have been polluted, which has resulted in climate change and shift in seasons. Droughts, low rainfalls, and floods are common. The land has been overexploited for producing agricultural products and has become unfit for producing few crops. Poverty and malnutrition are prominent in many countries, especially in arid regions. In such a scenario, we require cereal crops which can help us to overcome all these problems. Millets are cereal crops that have great nutritional value (Mishra et al., 2022). These are used in producing various bio-industrial products also. Further, these crops are better adapted to biotic and abiotic stress especially drought and high temperatures. This makes these crops a suitable option to address malnutrition and poverty concerns, especially from the arid regions of the world. Also being highly nutritive, these crops can have a positive impact on the food plates of people living in developed countries. They can become major food in developing countries with low rainfall and arid lands. Further, they may help in improving the nutritional and economic conditions of people living in such areas. This chapter is designed to highlight the advantages and future strategies of cultivating and consuming millets.

5.2 DIFFERENT TYPES OF MILLETS

Millets refer to a collection of monocotyledonous flowering plants that produce numerous little seeds. These plants are commonly cultivated as cereals or used as pasture in various regions across the globe. These plants are classified in the Poaceae family and are typically categorized into the subfamilies Chloridoideae and Panicoideae.

 DOI: 10.1201/9781032642567-6

Figure 5.1 shows the classifications of major and minor millets. Millets are known by various names in vernacular languages. Table 5.1 provides a comprehensive list of millets, including their scientific names, popular names, subfamilies, and families.

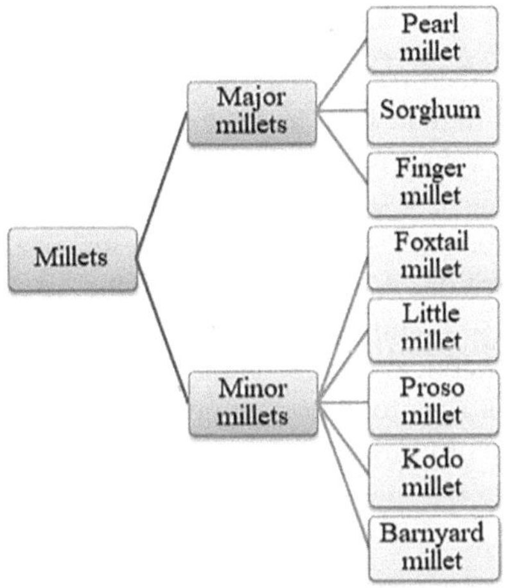

FIGURE 5.1 Major and minor millets.

TABLE 5.1
List of a Few Millet Species with Common Names, Subfamily, and Family

S. No.	Scientific Name of the Millet	Common Names	Subfamily	Family
1.	*Brachiaria deflexa*	Guinea millet	*Panicoideae*	*Poaceae*
	Brachiaria ramosa	Browntop millet	*Panicoideae*	*Poaceae*
2.	*Coix lacryma-jobi*	Adlay millet, Job's tears	*Panicoideae*	*Poaceae*
3.	*Digitaria compacta*	Raishan	*Panicoideae*	*Poaceae*
4.	*Digitaria exilis*	Fonio millet, White fonio, hungry rice	*Panicoideae*	*Poaceae*
5.	*Digitaria iburua*	Black fonio	*Panicoideae*	*Poaceae*
6.	*Digitaria sanguinalis*	Polish millet	*Panicoideae*	*Poaceae*
7.	*Echinochloa crus-galli*	Barnyard grass, Cockspur grass	*Panicoideae*	*Poaceae*
8.	*Echinochloa esculenta*	Japanese barnyard millet	*Panicoideae*	*Poaceae*
10.	*Echinochloa stagnina*	Burgu millet	*Panicoideae*	*Poaceae*
11.	*Eleusine coracana*	Finger millet	*Chloridoideae*	*Poaceae*
12.	*Eragrostis tef*	Teff	*Chloridoideae*	*Poaceae*
15.	*Panicum miliaceum*	Proso millet, Common millet, Broomcorn millet, Hog millet, White millet,	*Panicoideae*	*Poaceae*
16.	*Panicum sonorum*	Sonoran millet	*Panicoideae*	*Poaceae*
17.	*Panicum sumatrense*	Little millet	*Panicoideae*	*Poaceae*
18.	*Paspalum scrobiculatum*	Kodo millet	*Panicoideae*	*Poaceae*
19.	*Pennisetum glaucum*	Pearl millet	*Panicoideae*	*Poaceae*
20.	*Setaria italica*	Foxtail millet, Italian millet	*Panicoideae*	*Poaceae*
21.	*Sorghum bicolor*	Great millet, sorghum	*Panicoideae*	*Poaceae*
22.	*Spodiopogon formosanus*	Taiwan oil millet	*Panicoideae*	*Poaceae*

The most commonly cultivated species of millets are described in the followed sections.

5.2.1 Sorghum

Sorghum is a versatile cereal crop that serves as a food source for humans and as animal feed. Its grain is used for human food and the production of various industrial products such as bioethanol. Sorghum varieties are usually drought-resistant (Borrell et al., 2006), heat-tolerant, and nitrogen-efficient. It contributes to the global economy also as it is part of international trade. Sorghum is incredibly important for health and nutrition from the viewpoint of nutrition. When it comes to minerals, lipids, proteins, vitamins, and phenolic compounds, sorghum is an excellent source of all of them. The fact that it possesses a diverse phytochemical content, notably in the bran layer and other beneficial components. This makes it a more important crop that has the potential to be utilized as a functional meal and to minimize the threat of chronic noncommunicable diseases.

5.2.2 Pearl Millet

Pearl millet is the most extensively grown type of millet. It is also called Indian millet and Bajra in Hindi. It is well adapted to growing in droughts, high-temperature stress, and soil with low fertility. It has the potential to be cultivated in regions that are inaccessible for other cereal crops, as it can also grow in high salinity and low pH. It is used as human food and is rich in nutrition. Being a gluten-free grain, pearl millet consumption is beneficial in the treatment of celiac illnesses as well as constipation. Pearl s contains significant amounts of iron and zinc, which can potentially increase levels of hemoglobin.

5.2.3 Finger Millet

The scientific name of finger millet is *Eleusine coracana* (L.) Gaertn. It is called Ragi in Hindi. It performs well as a famine cereal crop. Finger millet may be grown under harsh conditions like heat stress and drought stress, and this can be done without affecting the overall productivity. It is commonly eaten food in many countries throughout Africa and South Asia. The finger millet contains a rich supply of vitamins, minerals, proteins, and carbohydrates. The amino acid profile of this grain is abundant in thiamine, iron, riboflavin, methionine, leucine, isoleucine, and phenylalanine (Nakarani et al., 2021). Additionally, it possesses a substantial quantity of phenylalanine. Its intake significantly improves one's ability to fight against a wide variety of diseases. Finger millet possesses not only anti-inflammatory and antidiabetic properties but also exhibits antidiarrheal properties. It also exhibits antibacterial and antioxidant characteristics.

5.2.4 Foxtail Millet

Setaria italica (L.) is commonly known as foxtail millet. Apart from being used for human food, it is also used as feed for birds. It has a vertical leafy stem. It has very small seeds which possess health-promoting properties. Foxtail millets have a significant concentration of essential amino acids. Foxtail millets contain a substantial quantity of flavonoids. In addition to being an essential source of food, foxtail millet also serves vital functional reasons in the diets of many different people. It is often regarded as one of the most easily digestive and hypoallergenic grains. This plant has been effectively adapted to dry and semi-arid environments in Asia, North Africa, South America, and North America during its extended cultivation history. Most foxtail millet accessions show substantial tolerance to abiotic stressors, particularly drought and abnormal concentrations of salt.

5.2.5 Proso Millet

Proso millet is a low-maintenance crop. The scientific name for this millet is *Panicum miliaceum* (L.). Proso millet is mostly noted for its exceptionally short lifecycles. Only a limited number of cultivars are capable of yielding mature grains within a mere 60-day period after being sown. This crop has a minimal water need. Apart from African nations, it is also cultivated in America, Canada, Russia, etc., to provide feed for cattle. Birdseed industries significantly contribute to the primary market share of proso millet. Even though proso millet has several health advantages, such as being gluten-free, being high in proteins, and having a low glycemic index, Proso millet is facing difficulties in penetrating the human food market. The recent surge in gluten-free trends and the growing emphasis on health awareness among individuals have sparked a renewed fascination with ancient grains including proso millet.

5.2.6 Little Millet

Panicum sumatrense is a millet that has very small seeds. It is also called little millet. It is a plant that can withstand both drought and waterlogging. Like most of the millets, it serves both purposes, that is, human food and animal feed. Little millet has high cholesterol content, yet its ingestion can increase the levels of good kind of cholesterol. The complex carbohydrate in question undergoes sluggish digestion, which proves highly beneficial for individuals with diabetes (Gayathri, 2015). Minerals such as potassium, iron, phosphorus, magnesium, and zinc are abundant in little millet, which is also a good source of vitamin B. Therefore, it can address the field of nutrition-sensitive agriculture, which aims to address the prevailing problem of micronutrient deficiencies by enhancing nutrition (Arunachalam et al., 2005; Kundgol et al., 2014; Selvi et al., 2015).

5.3 BENEFITS OF MILLETS

Millets are ancient grains that are packed with various micronutrients and macronutrients. Being rich in various biotic compounds and nutrients makes them very useful for various purposes. Many new food products are being developed using millet as the raw material. Millets can survive in semi-dry regions and that makes them a crop that can provide food security as well as economic security. Apart from providing nutritional security millets exhibit many nutraceutical properties too. Figure 5.2 describes the advantages of millets.

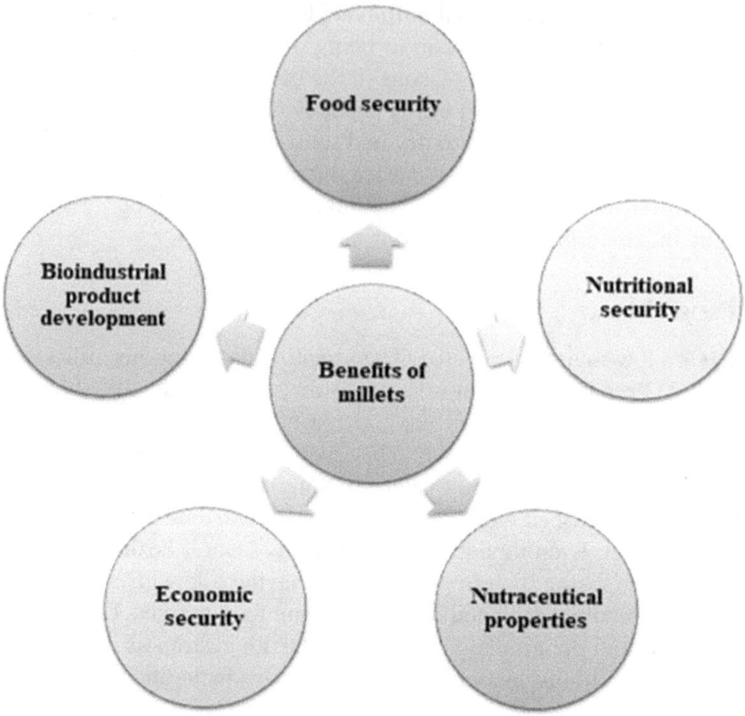

FIGURE 5.2 Benefits of millets.

5.4 FOOD SECURITY IN ARID REGIONS

Food security refers to a situation of consistently having enough affordable, wholesome food available to people. A very large section of the population fails to get proper food even in today's world. Millets can tolerate biotic stress as well as abiotic stress. There are many countries where land is less fertile. Water scarcity, high temperature, high salinity, and low pH are some common issues faced by farmers of arid regions. In such areas, millets can serve as a cereal crop which can provide food security. Millets have a high level of resistance to the severe weather and soil conditions that are typically encountered in semi-arid locations. Every type of millet can thrive in challenging environmental conditions, especially in areas with poor soil quality and scarce rainfall, which are not ideal for most of the major crops cultivated globally. There have been several studies that have demonstrated that sorghum, fox millet, and finger millet show notable resilience to the impacts of heat stress and drought. Proso millet's ability to withstand drought allows it to evade the terminal phase of drought, typically experienced toward the end growing season. The explanation for this is that proso millet reaches maturity within a span of about three months. Consequently, proso millet is classified as a variety of millet that necessitates minimal water usage (Goron and Raizada, 2015). Studies suggesting that millets have the potential to tolerate abiotic stress are listed in Table 5.2.

TABLE 5.2
Studies Suggesting Abiotic Stress Tolerance of Millets

S.No.	Name of Millet	Study	References
1	Pearl millet	The examination of gene expression profiles revealed that PRLT2/89-33 is naturally able to detect drought at the developmental stage and flowering stage.	Shivhare et al. (2020)
2	Foxtail millet	The drought resistance of foxtail millet during the germination phase under PEG stress is highly connected with phenylpropanoid-related pathways and this was revealed by transcriptome and metabolite analysis.	Yu et al. (2020)
3	Sorghum	The transcriptome of sorghum bicolor was functionally annotated in response to osmotic stress and abscisic acid.	Dugas et al. (2011)
4	Finger millet	Functional annotation and transcription factor mining have revealed that the finger millet genome has a substantial number of genes associated with drought tolerance.	Hittalmani et al. (2017)
5	Barnyard millet, pearl millet, common millet, foxtail millet, finger millet, Job's tears	Pearl millet exhibited the greatest resilience to drought conditions and the lowest resistance to waterlogging conditions, as determined by the ratio of water use efficiency (WUE) to the control condition. On the other hand, barnyard millet had the highest level of stability under both drought and waterlogging conditions.	Zegada-Lizarazu and Iijima (2005)

5.5 NUTRITIONAL SECURITY

We understand that a well-balanced diet is essential for the optimal growth and development of human beings. Many diseases occur only due to the deficiency of important micronutrients and vitamins or due to an unbalanced diet. A vast population on this Earth is suffering from malnutrition and is undernourished, hence nutritional security is the need of the hour. Millets are rich in nutrition which makes them a very good food that can provide nutritional security to the human population (Tripathi et al., 2023). Due to its ability to thrive in arid regions, this crop serves a crucial role in ensuring nutritional security for populations lacking access to popular food crops found in other parts of the world. Millets can provide essential micronutrients such as zinc, iodine, calcium iron, etc. Millets are high in bioactive substances and can have a very good amino acid profile. In terms of fat content, millets have a higher fat composition compared with wheat and rice. Proteins serve as a crucial reservoir of amino acids, and humans must obtain the required amino acids from their diet. Millets are a good source of proteins also in fact they provide a comparable amount of protein as provided by wheat, rice, and maize. The proximate composition and dietary fiber per 100 g of main millets are presented in Table 5.3.

TABLE 5.3
Proximate Composition of Millet Crops (Per 100 g)

Millet	Moisture	Ash	Protein	Fat	Carbohydrate	Crude Fiber	References
Pearl millet	4.50 ± 0.49	2.00 ± 0.02	10.57 ± 0.26	7.69 ± 0.24	74.41 ± 0.64	1.07 ± 0.39	Owheruo et al. (2019)
Sorghum	8.33 ± 0.02	1.94 ± 0.04	14.80 ± 0.34	4.33 ± 0.08	68.34 ± 0.41	2.26 ± 0.05	Gassem et al. (2003)
Foxtail millet	10.8 ± 0.02	2.3 ± 0.06	11.2 ± 0.05	4.5 ± 0.01	65 ±0.02	6.2 ±0.01	Patangare Suwarna et al. (2019)
Finger millet	12.3	2.7	7.2	1.5	75.0	1.3	Singh et al. (2012)
Barnyard millet	5.0	4.1	8.6	4.6	65.5	12.2	Singh et al. (2012)
Proso millet	11.2	3.6	12.5	1.1	70.1	1.5	Singh et al. (2012)
Little millet	10.6	20.51	6.26	2.04	47.85	13.08	Dangeti et al. (2013)

5.6 PHARMACEUTICAL PROPERTIES

Pharmaceutical properties refer to the properties of the food which can be used to treat some disorder or disease. The concept of nutraceuticals is gaining popularity these days. Millets are rich in various bioactive components which make them a food that can be used to help in various diseases and disorders.

5.6.1 Antioxidative Properties

Antioxidative properties mean the capacity to reduce oxidative stress, which is a state that occurs when there is an excessive generation of reactive species, leading to the oxidative destruction of lipids, proteins, and DNA molecules. Oxidative stress may lead to various neurodegenerative, cardiovascular, and respiratory diseases. Our bodies have a natural defense system for fighting oxidative stress but usually, it is not enough. We need to take antioxidants from food sources also. Millets, such as barnyard, kodo, proso, foxtail, finger, and little millet, contain flavonoids and polyphenols, which are natural antioxidants. Consuming millets can therefore help reduce oxidative stress. Finger millet is a highly beneficial grain that contains abundant natural antioxidants and is rich in nutrients (Xiang et al., 2019). As oxidative stress itself induces various diseases, hence it can be inferred that the consumption of millet protects us from the instigations of several diseases caused by oxidative stress. After researching 13 different types of barnyard millet (*Echinochloa* spp.), Kim et al. (2011) concluded that barnyard millet possesses antioxidative properties and has the potential to be used as a functional diet.

5.6.2 Diabetes

Diabetes mellitus is a metabolic condition that causes a person to experience hyperglycemia and changes in the metabolism of proteins, carbohydrates, and lipids throughout their body. It is a chronic endocrine condition that is extremely common. Synthetic inhibitors of alpha-glucosidase and pancreatic amylase are available these days but natural inhibitors are considered safer. Many studies have revealed that consumption of millets helps in protection from diabetes. According to Rao et al. (2017), the consumption of millets regularly can be beneficial in the prevention of diabetes. This is due to the increased magnesium content of millets, which in turn enhances the efficiency of insulin and glucose receptors. In diabetic rats, pearl millet causes a hypoglycemic impact and enhances lipidemic control (Sukar et al., 2020).

5.6.3 Cardiovascular Disease

Cardiovascular diseases are those that are related to heart or blood vessels such as cardiomyopathy, heart failure, rheumatic heart disease, venous thrombosis, and so on. Major causes of cardiovascular diseases are unhealthy diet, stress, physical inactivity, obesity, and smoking. It can be prevented by a healthy diet. Considering that millets are abundant in magnesium, lignans, potassium, fiber, and antioxidants, they have the potential to reduce the incidence of cardiovascular diseases. Saleh et al. (2013) reported that phenolic extracts of finger millets, tiny, proso, pearl, kodo, and foxtail exhibited inhibitory effects on lipid peroxidation. The researchers discovered that these extracts blocked lipid peroxidation by 1%–41%, and kodo millet exhibited the highest level of inhibition. Millets such as foxtail, proso, barnyard, finger, foxtail, and pearl millet can drastically reduce triglycerides and hyperlipidemia, which in turn leads to a reduction in the risk of cardiovascular diseases. The findings of a study carried out by Alzahrani and colleagues (2022) demonstrated that the consumption of pearl millet by obese rats resulted in antihyperlipidemic and hypoglycemic physiological effects.

5.6.4 Anticancer Properties

In cancer, the patient shows abnormal cell growth, which ultimately leads to tumor formation. Cancer can spread to other parts of the body too, after originating from one specific site in the body. Millets can minimize cancer risk due to their high presence of phenolic acids, phytate, tannins, and other phytonutrients. An investigation was carrried out by Shen et al. (2018) on the antioxidant and antiproliferative properties of Proso millet that involved the examination of five waxy and four non-waxy kinds of grain. It was found that J7, J3, and J1 varieties had better anticancer activity compared to other varieties. In addition, Shi et al. (2019) conducted a separate investigation and found that polyphenols linked to foxtail millet bran had antiproliferative effects on HT-29 cells by modifying miR-149-mediated aerobic glycolysis.

5.6.5 Celiac Disease

Celiac disease is a condition characterized by intolerance of gluten and primarily affects the small intestine. This is an autoimmune disease. Gluten is a kind of protein found in commonly used cereal like wheat, barley, and rye but not in millets. Therefore, millets are considered a safe diet for celiac patients. There is a significant amount of potential for millets to be used in the production of a variety of beverages and packaged foods for individuals who are afflicted with this ailment. Gluten-free upma using pearl millet and rice and suggested that upma made of equal proportions of millet and rice can help to treat diseases related to deficiency in nutrients.

5.6.6 Antihypertensive Properties

Hypertension is a pathological state characterized by elevated blood pressure. Prolonged hypertension is a primary contributor to the development of cardiovascular disease. Chen et al. (2017) observed that the bioactive protein from foxtail millet show angiotensin-converting enzyme (ACE) inhibitory activity. He extracted protein from the foxtail millet and then hydrolyzed those using digestive proteases to conduct his study. Protein hydrolysates significantly lowered the blood pressure in hypersensitive rats. A separate investigation carried out by Hou et al. (2018) demonstrated that the ingestion of foxtail millet effectively reduced blood pressure levels in individuals with moderate hypertension.

5.6.7 Antimicrobial Properties

Many studies have been conducted that show that millets have antimicrobial properties. Polyphenols of finger millet and pearl millet exhibited antimicrobial activity against *Escherichia coli*, *Staphylococcus aureus*, *Proteus mirabilis*, *Klebsiella pneumonia*, *Enterococcus* sp., *Pseudomonas aeruginosa*, *Serratia marcescens*, *Salmonella* sp., and *Shigella dysenteriae*. According to research conducted by Singh et al. (2015), the antimicrobial properties of phenolic acids are derived from finger millet and pearl millet. A novel antifungal peptide with a molecular mass of 26.9 kDa has been extracted from the seeds of foxtail millet. This peptide has demonstrated antifungal properties against *Alternaria alternate*, *Trichoderma viride*, *Fusarium oxysporum*, and *Botrytis cinerea* (Xu et al., 2011).

5.7 ECONOMIC SECURITY

Economic security refers to a condition where commoners can earn enough money for their survival. Growing uses of millets and their awareness can generate various occupations in society which can allow people to enhance their economic status. Millets are climate-resilient crops, and their production requires low investment. Due to the low cost of production farmers can generate a sustainable amount of income by growing millets. Value addition in these crops can lead to further economic gains. According to the Agricultural and Processed Food Products Export Development Authority (APEDA), despite having a roughly 40% share of the world's millet output,

India only exported 1% of it in 2021–2022, earning $64.28 million while in 2020–2021 India earned $59.75 million. The United States of America, Canada, Russia, and Ukraine import millets and export value-added goods but India exports only millets according to APEDA. As per a report by the *Economic Times*, 200 new millets entrepreneurs have been found by the Indian Institute of Millet Research. A few of the start-ups are beginning to export small quantities of the value-added commodities that they are generating. APEDA is advertising them at global trade shows.

5.8 BIOTECHNOLOGICAL INTERVENTIONS

Millets contain a high concentration of nutrients and exhibit nutraceutical qualities. Many biotechnological interventions have been done to date focusing on various properties and uses of these crops. These interventions include crop improvement using morden day biotechnology techniques, food and nutraceutical properties enhancement, and the development of various industrial products using these crops. Figure 5.3 highlights three major biotechnological interventions done by scientists using millet crops.

5.8.1 Crop Improvement

Wide research has been conducted on various millets to improve the quality of grain and to improve other traits that help in coping with biotic and abiotic stress. Biotechnological interventions including crop improvement using traditional breeding, marker-assisted breeding, gene editing, transgenic technology, and bioinformatics have strengthened crop improvement. The genotyping-by-sequencing method was used to identify several single nucleotide polymorphisms (SNPs) in finger millet and pearl millet, which was reported by Hu et al. (2015) and Gimode et al. (2016). A recent study by Yu et al. (2020) revealed that the drought tolerance of *Setaria italica* (foxtail millet) is closely linked to phenylpropanoid pathways during the germination phase of seedlings under polyethylene glycol (PEG) stress. This finding was based on transcriptome and metabolite studies. This information can be used to enhance the crop even further. Ceasar et al. (2017) documented the use of agrobacterium-mediated transformation to suppress the activity of phosphate transporters in foxtail millet. Four finger-millet cultivars have also been reproduced using the newly developed agrobacterium transformation protocol (Satish et al., 2017).

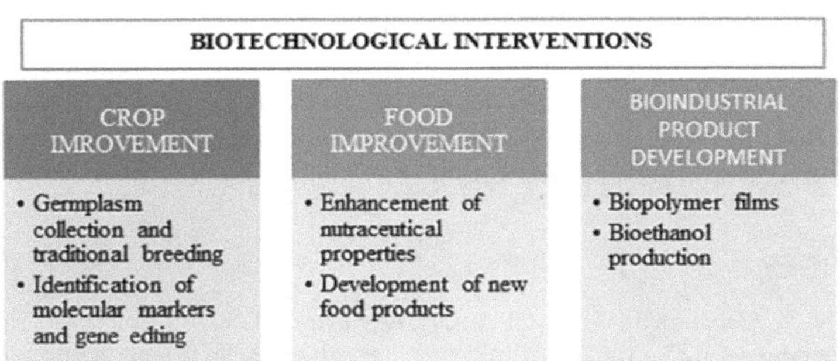

FIGURE 5.3 Biotechnological interventions.

5.8.2 Food Improvement

Millets-based food has various benefits but still, there is scope for further value addition. Different types of approaches have been applied to increase the nutritional qualities and digestibility of millet-based foods such as fermentation, enzymatic hydrolysis, designing complementary foods using millets and other cereals, and development of different kinds of food products. According to the findings of a study carried out by Purewal et al. in 2019, it was found that the process of solid-state fermentation of different types of pearl millet using *Aspergillus oryzae* and *Rhizopus azygosporus* resulted in an improvement in the antioxidative characteristics and the ability to protect against DNA damage. Salar et al. (2017) demonstrated similar findings when they fermented pearl millet with *Aspergillus sojae.*

5.8.3 Bio-Industrial Product Development

We use synthetic polymers such as polyethylene and polypropylene for the packaging in food industries and other sectors of life too. Millets can be used to make edible biopolymers and coatings. Apart from being biodegradable and nontoxic, these are rich in polysaccharides, proteins, and bioactive compounds. Baek and Song (2019) studied the biofilm prepared using Proso millet with various concentrations of curcumin and found that the biofilm showed higher antioxidant activity. They also suggested that this film can be used as active material for packing. Bioethanol is another bio-industrial product that can be obtained from millets. In place of sugarcane molasses, sweet sorghum juice might be a useful alternative raw material for ethanol synthesis (Ratnavathi et al., 2010).

5.9 FUTURE ASPECTS AND STRATEGIES

Significant efforts have been made to genetically enhance millets but these cereals are still underutilized globally despite their numerous advantages. It is required that we develop varieties according to specific regions these days. New product development may help in increasing the popularity of these cereal grains. Increased commercialization can have a substantial impact on motivating farmers to cultivate a greater quantity of millets. It will also help in popularizing the crop among commoners. Blending millets with other protein sources might assist in compensating for the lack of specific amino acids like lysine. Enhancing these qualities would be a vital factor in broadening the range of millet grain applications. Significant preclinical evidence has become available in recent years, indicating that millets have a beneficial impact on a range of ailments, including diabetes, cardiovascular diseases, and gastrointestinal disorders. The promotion and popularization of millet consumption should be prioritized for both the food and industrial sectors, considering the multitude of benefits they offer.

REFERENCES

Alzahrani, N. S., Alshammari, G. M., El-Ansary, A., Yagoub, A. E. A., Amina, M., Saleh, A., & Yahya, M. A. (2022). Anti-hyperlipidemia, hypoglycemic, and hepatoprotective impacts of pearl millet (*Pennisetum glaucum* L.) grains and their ethanol extract on rats fed a high-fat diet. *Nutrients*, *14*(9), 1791.

Arunachalam, V., Rengalakshmi, R., & Kubera Raj, M. S. (2005). Ecological stability of genetic diversity among landraces of little millet (*Panicum sumatrense*) in south India. *Genetic Resources and Crop Evolution*, *52*, 15–19.

Baek, S. K., & Song, K. B. (2019). Characterization of active biodegradable films based on proso millet starch and curcumin. *Starch-Stärke*, *71*(3–4), 1800174.

Borrell, A., Jordan, D., Mullet, J., Henzell, B., & Hammer, G. (2006). Drought adaptation in sorghum. In J. M. Ribaut (Ed.), *Drought Adaptation in Cereals*, Food Products Press, Binghamton, pp. 335–399.

Ceasar, S. A., Baker, A., & Ignacimuthu, S. (2017). Functional characterization of the PHT1 family transporters of foxtail millet with development of a novel Agrobacterium-mediated transformation procedure. *Scientific Reports*, *7*(1), 14064.

Chen, J., Duan, W., Ren, X., Wang, C., Pan, Z., Diao, X., & Shen, Q. (2017). Effect of foxtail millet protein hydrolysates on lowering blood pressure in spontaneously hypertensive rats. *European Journal of Nutrition*, *56*, 2129–2138.

Dangeti, S. R., Karthikeyan, S., Kumar, G. R., & Desai, S. (2013). Proximate and phytochemical analysis of seed coat from *P. sumantranse* (Little Millet). *Biochem Anal Biochem*, *2*(134), 2161–1009.

Dugas, D. V., Monaco, M. K., Olson, A., Klein, R. R., Kumari, S., Ware, D., & Klein, P. E. (2011). Functional annotation of the transcriptome of *Sorghum bicolor* in response to osmotic stress and abscisic acid. *BMC Genomics*, *12*, 1–21.

Gassem, M. A., & Osman, M. A. (2003). Proximate composition and the content of sugars, amino acids and anti-nutritional factors of three sorghum varieties. *Agricultural Research Center, King Saud University, Research Bulletin*, *125*, 5–19.

Gayathri, A. (2015). Little millet nutrition, health benefits and facts you must know.

Gimode, D., Odeny, D. A., de Villiers, E. P., Wanyonyi, S., Dida, M. M., Mneney, E. E., ... & De Villiers, S. M. (2016). Identification of SNP and SSR markers in finger millet using next generation sequencing technologies. *PLoS One*, *11*(7), e0159437.

Goron, T. L., & Raizada, M. N. (2015). Genetic diversity and genomic resources available for the small millet crops to accelerate a New Green Revolution. *Frontiers in Plant Science*, *6*, 132042.

Hittalmani, S., Mahesh, H. B., Shirke, M. D., Biradar, H., Uday, G., Aruna, Y. R., ... & Mohanrao, A. (2017). Genome and transcriptome sequence of finger millet (*Eleusine coracana* (L.) Gaertn.) provides insights into drought tolerance and nutraceutical properties. *BMC Genomics*, *18*, 1–16.

Hou, D., Chen, J., Ren, X., Wang, C., Diao, X., Hu, X., ... & Shen, Q. (2018). A whole foxtail millet diet reduces blood pressure in subjects with mild hypertension. *Journal of Cereal Science*, *84*, 13–19.

Hu, Z., Mbacké, B., Perumal, R., Guèye, M. C., Sy, O., Bouchet, S., ... & Morris, G. P. (2015). Population genomics of pearl millet (*Pennisetum glaucum* (L.) R. Br.): Comparative analysis of global accessions and Senegalese landraces. *BMC Genomics*, *16*, 1–12.

Kim, J. Y., Jang, K. C., Park, B. R., Han, S. I., Choi, K. J., Kim, S. Y., ... & Seo, W. D. (2011). Physicochemical and antioxidative properties of selected barnyard millet (*Echinochloa utilis*) species in Korea. *Food Science and Biotechnology*, *20*, 461–469.

Kundgol, N. G., Kasturiba, B., Math, K. K., & Kamatar, M. Y. (2014). Screening of little millet landraces for chemical composition. *International Journal of Farm Sciences*, *4*(2), 33–38.

Mishra, A., Dutta, T., Baitharu, I. J. J. N., & Baitharu, I. (2022). Nutritional values and potential health benefits of millets-A. *Journal of Nutrition*, *8*(1), 9–26.

Nakarani, U. M., Singh, D., Suthar, K. P., Karmakar, N., Faldu, P., & Patil, H. E. (2021). Nutritional and phytochemical profiling of nutracereal finger millet (Eleusine coracana L.) genotypes. *Food Chemistry*, *341*, 128271.

Nayik, G. A., Tufail, T., Anjum, F. M., & Ansari, M. J. (Eds.). (2023). *Cereal Grains: Composition, Nutritional Attributes, and Potential Applications*. CRC Press, Boca Raton, FL.

Owheruo, J. O., Ifesan, B. O., & Kolawole, A. O. (2019). Physicochemical properties of malted finger millet (*Eleusine coracana*) and pearl millet (*Pennisetum glaucum*). *Food Science & Nutrition*, *7*(2), 476–482.

Patangare Suwarna, S., Pawar, V. S., Syed, H. M., & Shinde, S. T. (2019). Studies on physical properties and nutritional profile of foxtail millet. *The Pharma Innovation Journal*, *8*(3), 286–288.

Purewal, S. S., Salar, R. K., Bhatti, M. S., Sandhu, K. S., Singh, S. K., & Kaur, P. (2019). Solid-state fermentation of pearl millet with *Aspergillus oryzae* and *Rhizopus azygosporus*: Effects on bioactive profile and DNA damage protection activity. *Journal of Food Measurement and Characterization*, *14*, 150–162.

Rao Dayakar, B., Bhaskarachary, K., Arlene Christina, G. D., Sudha Devi, G., Vilas, A. T., & Tonapi, A. (2017). *Nutritional and Health Benefits of Millets*. ICAR_Indian Institute of Millets Research (IIMR), Rajendranagar, Hyderabad.

Ratnavathi, C. V., Suresh, K., Kumar, B. V., Pallavi, M., Komala, V. V., & Seetharama, N. (2010). Study on genotypic variation for ethanol production from sweet sorghum juice. *Biomass and Bioenergy*, *34*(7), 947–952.

Salar, R. K., Purewal, S. S., & Sandhu, K. S. (2017). Fermented pearl millet (*Pennisetum glaucum*) with in vitro DNA damage protection activity, bioactive compounds and antioxidant potential. *Food Research International*, *100*, 204–210.

Saleh, A. S., Zhang, Q., Chen, J., & Shen, Q. (2013). Millet grains: Nutritional quality, processing, and potential health benefits. *Comprehensive Reviews in Food Science and Food Safety*, *12*(3), 281–295.

Satish, L., Ceasar, S. A., & Ramesh, M. (2017). Improved Agrobacterium-mediated transformation and direct plant regeneration in four cultivars of finger millet (*Eleusine coracana* (L.) Gaertn.). *Plant Cell, Tissue and Organ Culture (PCTOC)*, *131*, 547–565.

Selvi Manimozhi, V., Nirmalakumari, A., & Senthil, N. (2015). Genetic diversity for zinc, calcium and iron content of selected little Millet Genotypes. *Journal of Nutrition & Food Sciences*, *5*(417), 2.

Shen, R., Ma, Y., Jiang, L., Dong, J., Zhu, Y., & Ren, G. (2018). Chemical composition, antioxidant, and antiproliferative activities of nine Chinese proso millet varieties. *Food and Agricultural Immunology*, *29*(1), 625–637.

Shi, J., Shan, S., Zhou, G., Li, H., Song, G., Li, Z., & Yang, D. (2019). Bound polyphenol from foxtail millet bran exhibits an antiproliferative activity in HT-29 cells by reprogramming miR-149-mediated aerobic glycolysis. *Journal of Functional Foods*, *56*, 246–254.

Shivhare, R., Asif, M. H., & Lata, C. (2020). Comparative transcriptome analysis reveals the genes and pathways involved in terminal drought tolerance in pearl millet. *Plant Molecular Biology*, *103*(6), 639–652.

Singh, K. P., Mishra, A., & Mishra, H. N. (2012). Fuzzy analysis of sensory attributes of bread prepared from millet-based composite flours. *LWT-Food Science and Technology*, *48*(2), 276–282.

Singh, N., Meenu, G., Sekhar, A., & Jayanthi, A. (2015). Evaluation of antimicrobial and anticancer properties of finger millet (*Eleusine coracana*) and pearl millet (*Pennisetum glaucum*) extracts. *The Pharma Innovation*, *3*(11), 82.

Sukar, K. A. O., Abdalla, R. I., Humeda, H. S., Alameen, A. O., & Mubarak, E. I. (2020). Effect of pearl millet on glycaemic control and lipid profile in streptozocin induced diabetic wistar rat model. *Asian Journal of Medicine and Health*, *18*, 40–51.

Tripathi, G., Jitendrakumar, P. H., Borah, A., Nath, D., Das, H., Bansal, S., … & Singh, B. V. (2023). A review on nutritional and health benefits of millets. *International Journal of Plant & Soil Science*, *35*(19), 1736–1743.

Welcome to APEDA. (n.d.). Welcome to APEDA. https://apeda.gov.in/apedawebsite/#

Xiang, J., Apea-Bah, F. B., Ndolo, V. U., Katundu, M. C., & Beta, T. (2019). Profile of phenolic compounds and antioxidant activity of finger millet varieties. *Food Chemistry*, *275*, 361–368.

Xu, W., Wei, L., Qu, W., Liang, Z., Wang, J., Peng, X., … & Huang, K. (2011). A novel antifungal peptide from foxtail millet seeds. *Journal of the Science of Food and Agriculture*, *91*(9), 1630–1637.

Yu, A., Zhao, J., Wang, Z., Cheng, K., Zhang, P., Tian, G., … & Wang, Y. (2020). Transcriptome and metabolite analysis reveal the drought tolerance of foxtail millet significantly correlated with phenylpropanoids-related pathways during germination process under PEG stress. *BMC Plant Biology*, *20*, 1–17.

Zegada-Lizarazu, W., & Iijima, M. (2005). Deep root water uptake ability and water use efficiency of pearl millet in comparison to other millet species. *Plant Production Science*, *8*(4), 454–460.

Part II

Healthcare Biotechnology

6 RNAi
A Promising Next-Generation Therapeutic Approach

Aman Kumar, Anubha Sharma, Kanisht Batra, Savita Budania, Deepika Chaudhary, Sunil Mor, and Sushila Maan

6.1 INTRODUCTION

In the ever-evolving landscape of medicine, the quest for innovative therapeutic modalities has led to the exploration of RNAi as a promising frontier. RNAi, a natural mechanism of gene regulation, has garnered significant attention for its potential to revolutionize the treatment of various diseases, ranging from genetic disorders to cancer and infectious diseases. This introduction aims to provide a foundational understanding of RNAi, delineating its mechanisms, applications, and transformative impact, supported by relevant citations from the scientific literature.

RNAi functions at the molecular level, orchestrating the silencing of specific genes through the degradation of messenger RNA (mRNA) molecules. This process is mediated by small RNA molecules, including small interfering RNAs (siRNAs) and microRNAs (miRNAs), which guide the RNA-induced silencing complex (RISC) to the target mRNA. Through sequence-specific recognition, RISC initiates cleavage or translational repression of the target mRNA, modulating gene expression with remarkable precision (Fire et al., 1998; Elbashir et al., 2001).

The exceptional specificity of RNAi, coupled with its ability to target previously inaccessible disease-causing genes, underscores its potential as a therapeutic modality. Unlike conventional small-molecule drugs, RNAi offers a tailored approach to treatment, enabling the selective inhibition of pathogenic genes while sparing healthy tissues (Hannon, 2002). Furthermore, the modular nature of RNAi allows for the development of customizable therapies tailored to the unique genetic profiles of individual patients, heralding a new era of personalized medicine (McManus and Sharp, 2002).

Despite its immense promise, the clinical translation of RNAi therapeutics has encountered challenges, particularly in the realms of delivery and off-target effects. Efficient delivery of RNAi molecules to target tissues remains a formidable challenge, necessitating the development of innovative delivery vehicles capable of traversing biological barriers and achieving therapeutic concentrations at the desired site of action (Wittrup and Lieberman, 2015). Additionally, concerns regarding off-target

DOI: 10.1201/9781032642567-8

effects and immune activation necessitate meticulous optimization of RNAi triggers to ensure safety and efficacy in clinical settings (Jackson and Linsley, 2010).

Nonetheless, recent advancements in RNAi technology, coupled with a deeper understanding of its biological mechanisms, have propelled the field forward, bringing RNAi-based therapeutics closer to reality. Through collaborative efforts between academia, industry, and regulatory bodies, the journey toward unlocking the full therapeutic potential of RNAi is well underway, offering renewed hope for patients and clinicians alike.

6.2 HISTORICAL BACKGROUND

The discovery of RNAi stands as a landmark in molecular biology, tracing its origins to fundamental research on gene regulation and post-transcriptional control. RNAi's journey from a perplexing phenomenon in *Caenorhabditis elegans* to a promising next-generation therapeutic approach reflects a convergence of scientific inquiry, technological innovation, and clinical translation. This historical background provides a narrative of RNAi's discovery and evolution, supported by key citations from seminal studies and foundational research.

The roots of RNAi can be traced back to the late 20th century when scientists Andrew Fire and Craig Mello made a groundbreaking discovery in the nematode *C. elegans*. In 1998, Fire and Mello demonstrated that double-stranded RNA (dsRNA) could trigger potent and specific gene silencing, leading to the phenomenon known as RNA interference (Fire et al., 1998). This seminal study unveiled a previously unrecognized mechanism for gene regulation, challenging conventional views of how genetic information is expressed and controlled.

Following the pioneering work of Fire and Mello, subsequent studies elucidated the molecular mechanisms underlying RNAi and its widespread occurrence across diverse organisms. In 2001, Elbashir et al. demonstrated the efficacy of short interfering RNAs (siRNAs), synthetic RNA molecules designed to mimic the dsRNA triggers of RNAi, in mediating gene silencing in mammalian cells (Elbashir et al., 2001). This milestone study laid the foundation for harnessing RNAi as a powerful tool for experimental gene knockdown and therapeutic intervention.

The elucidation of RNAi's mechanisms and its potential applications sparked a surge of research activity, propelling RNAi into the forefront of biomedical research. Subsequent studies uncovered the role of microRNAs (miRNAs), endogenous small RNA molecules involved in regulating gene expression, further expanding the scope of RNAi-mediated gene regulation (Lee et al., 1993; Lagos-Quintana et al., 2001).

As RNAi continued to captivate the scientific community, its therapeutic potential became increasingly evident. RNAi-based therapeutics emerged as a promising strategy for selectively targeting disease-causing genes, offering a novel approach to treat a wide range of disorders, including cancer, genetic diseases, viral infections, and neurodegenerative diseases (Davidson and McCray, 2011).

Despite early enthusiasm, translating RNAi from bench to bedside has posed challenges, particularly in the realms of delivery, stability, and off-target effects. However, ongoing research efforts and technological advancements continue to address these challenges, paving the way for the clinical realization of RNAi-based therapies (Wittrup and Lieberman, 2015).

6.3 MECHANISM OF ACTION

RNAi exerts its therapeutic potential through a highly orchestrated molecular mechanism, leveraging small RNA molecules to selectively silence target genes at the post-transcriptional level. This mechanism of action, elucidated through seminal studies, underscores RNAi's precision and versatility in modulating gene expression. This section provides an overview of the molecular intricacies underlying RNAi, supported by key citations from foundational research.

The RNAi pathway is initiated by the introduction of double-stranded RNA (dsRNA) molecules into the cell, either exogenously or through endogenous processing. Once inside the cell, dsRNA is recognized and processed by the enzyme Dicer, an RNase III endonuclease, into short double-stranded fragments known as small interfering RNAs (siRNAs) (Bernstein et al., 2001). These siRNAs typically consist of 21–23 nucleotide duplexes with characteristic 2-nucleotide overhangs at the 3' ends. Following Dicer-mediated processing, one of the siRNA strands, known as the guide strand, is incorporated into the RNA-induced silencing complex (RISC), while the passenger strand is degraded. The guide strand serves as a template for target recognition, guiding RISC to complementary sequences within the target mRNA molecule (Hannon, 2002). Upon binding to the target mRNA, RISC facilitates mRNA cleavage or translational repression, depending on the degree of complementarity between the siRNA guide strand and the target sequence shown in Figure 6.1 (Liu et al., 2004). Perfect or near-perfect complementarity between the siRNA and target mRNA leads to mRNA cleavage mediated by the endonuclease activity of Argonaute proteins within RISC, resulting in the degradation of the target mRNA (Hutvagner and Zamore, 2002). Alternatively, imperfect base pairing may result in translational repression without mRNA cleavage, thereby inhibiting protein synthesis from the target mRNA (Filipowicz et al., 2005). The exquisite specificity of RNAi stems from the sequence-dependent recognition between the siRNA guide strand and the target mRNA, enabling the selective inhibition of disease-associated genes while sparing non-targeted genes. Furthermore, the modular nature of RNAi allows for the design of custom siRNAs tailored to specific target sequences, offering a versatile platform for therapeutic intervention (Davidson and McCray Jr, 2011).

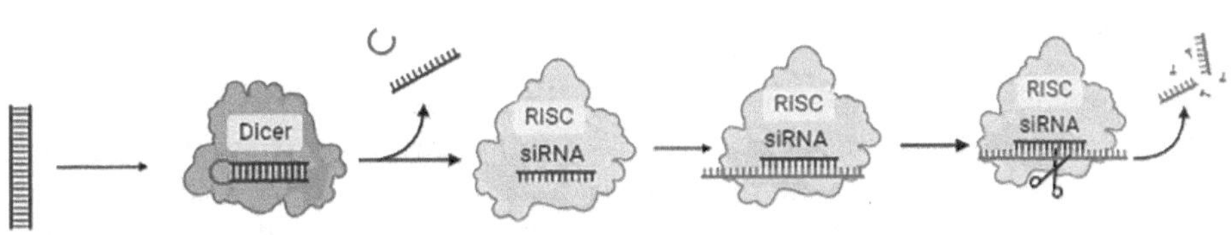

FIGURE 6.1 Mechanism of RNAi.

6.4 PRESENT STATUS OF RNAI THERAPEUTICS

RNAi continues to hold immense promise as a next-generation therapeutic approach, with ongoing research and clinical trials advancing its applications across diverse disease areas. The present status of RNAi-based therapeutics reflects both progress and challenges in translating this technology from the laboratory to the clinic. This section provides an overview of the current landscape of RNAi therapeutics, supported by recent citations from scientific literature and clinical trials.

6.4.1 Clinical Trials and Therapeutic Applications

Over the past decade, RNAi-based therapeutics have progressed from preclinical studies to clinical trials across a spectrum of diseases. Clinical trials have evaluated the safety, efficacy, and pharmacokinetics of RNAi-based drugs in indications such as rare genetic disorders, viral infections, cancer, and ocular diseases (Nair et al., 2014; Dhillon and Winstanley, 2020). Notable examples include the approval of Patisiran and Givosiran for the treatment of hereditary transthyretin-mediated amyloidosis (hATTR) and acute hepatic porphyria, respectively, marking significant milestones in the clinical translation of RNAi (Adams et al., 2018; Coelho et al., 2013).

6.4.2 Delivery Challenges and Innovations

Despite clinical successes, delivery remains a major hurdle for RNAi therapeutics, particularly for systemic administration. Efforts to overcome delivery challenges have spurred the development of innovative delivery platforms, including lipid nanoparticles, polymer-based carriers, and viral vectors (Wittrup and Lieberman, 2015). These delivery systems aim to enhance the stability, bioavailability, and tissue-specific targeting of RNAi molecules, thereby improving therapeutic outcomes and minimizing off-target effects.

6.4.3 Optimization of RNAi Triggers

Advances in RNAi technology have led to the optimization of RNAi triggers, including siRNAs, miRNAs, and short hairpin RNAs (shRNAs), to improve efficacy and reduce immunogenicity (Jackson and Linsley, 2010). Chemical modifications, such as 2'-O-methyl and locked nucleic acid (LNA) modifications, enhance the stability and specificity of RNAi triggers, enabling their clinical application (Kumar et al., 2007).

6.4.4 Expanded Therapeutic Targets

RNAi-based therapeutics are being explored for a widening array of therapeutic targets, including previously "undruggable" targets such as oncogenic drivers and disease-associated mutations (Buchholz et al., 2021). The advent of CRISPR-Cas9-mediated RNAi screening has facilitated the discovery of novel therapeutic targets and synergistic combinations for RNAi-based therapies (Wang et al., 2014).

6.4.5 Regulatory and Commercial Landscape

Regulatory agencies such as the U.S. Food and Drug Administration (FDA) and the European Medicines Agency (EMA) continue to refine guidelines for the development and approval of RNAi therapeutics, providing a regulatory framework to support their clinical translation (Liang et al., 2017). Moreover, collaborations between academic institutions, biopharmaceutical companies, and government agencies have accelerated the development and commercialization of RNAi-based drugs, fostering a vibrant ecosystem for innovation and investment in this field.

6.4.6 FDA-Approved RNAi Therapeutics

6.4.6.1 Patisiran (Onpattro)

Indication: Patisiran is indicated for the treatment of polyneuropathy in adults with hereditary transthyretin-mediated amyloidosis (hATTR) (Adams et al., 2018).

Mechanism of Action: Patisiran is an RNAi therapeutic that targets and silences the transthyretin (TTR) gene, thereby reducing the production of abnormal TTR protein implicated in hATTR amyloidosis (Adams et al., 2018).

6.4.6.2 Givosiran (Givlaari)

Indication: Givosiran is indicated for the treatment of adults with acute hepatic porphyria (AHP) (Coelho et al., 2013).

Mechanism of Action: Givosiran utilizes RNAi to target and silence aminolevulinic acid synthase 1 (ALAS1), an enzyme involved in the heme biosynthesis pathway. By reducing ALAS1 expression, givosiran decreases the accumulation of neurotoxic intermediates responsible for symptoms of AHP (Coelho et al., 2013).

6.4.6.3 Lumasiran (Oxlumo)

Indication: Lumasiran is indicated for the treatment of primary hyperoxaluria type 1 (PH1) to lower urinary oxalate levels in pediatric and adult patients (Garrelfs et al., 2021).

Mechanism of Action: Lumasiran is a small interfering RNA (siRNA) that targets hydroxyacid oxidase 1 (HAO1), an enzyme involved in the glyoxylate metabolism pathway. By inhibiting HAO1 expression, lumasiran reduces the hepatic production of oxalate, thereby lowering urinary oxalate levels and mitigating the progression of PH1 (Garrelfs et al., 2021).

6.4.6.4 Inclisiran (Leqvio)

Indication: Inclisiran is indicated for the treatment of adults with heterozygous familial hypercholesterolemia (HeFH) or established atherosclerotic cardiovascular disease (ASCVD) who require additional lowering of low-density lipoprotein cholesterol (LDL-C) (Ray et al., 2017).

Mechanism of Action: Inclisiran is a small interfering RNA (siRNA) that targets proprotein convertase subtilisin/kexin type 9 (PCSK9), a protein involved in regulating LDL receptor degradation. By reducing PCSK9 expression, inclisiran enhances LDL receptor recycling and clearance, leading to sustained reductions in LDL-C levels (Ray et al., 2017).

The FDA-approved RNAi therapeutics represent significant milestones in the clinical translation of RNAi technology and offer new treatment options for patients with hATTR amyloidosis, acute hepatic porphyria, primary hyperoxaluria type 1 and dyslipidemia offering new treatment options for patients with these conditions. Please note that the status of FDA approvals may have changed since my last update, so it's a good idea to consult the most recent sources or official FDA announcements for the latest information on RNAi-based therapeutics.

6.4.7 RNAi Therapeutics under Clinical Trials

Several RNAi therapeutics are under clinical investigation in various stages of clinical trials. A few examples are as follows.

6.4.7.1 Fitusiran

a. Phase: Fitusiran is being evaluated in multiple phase 3 clinical trials for the treatment of hemophilia A and B with or without inhibitors.
b. Indication: Fitusiran is a small interfering RNA (siRNA) therapeutic targeting antithrombin (AT) for the treatment of hemophilia, aiming to rebalance hemostasis by reducing AT levels and increasing thrombin generation (Machin and Ragni, 2018).

6.4.7.2 Givosiran (ALN-AS1)

a. Phase: Givosiran has completed phase 3 clinical trials for the treatment of acute hepatic porphyria (AHP) and has been approved by the FDA (Puy et al., 2020).
b. Indication: Givosiran is an RNAi therapeutic targeting aminolevulinic acid synthase 1 (ALAS1), aiming to reduce the accumulation of neurotoxic heme intermediates in patients with AHP (Puy et al., 2020).

6.4.7.3 Vutrisiran (ALN-TTRsc02)

a. Phase: Vutrisiran is currently being evaluated in phase 3 clinical trials for the treatment of hereditary transthyretin-mediated amyloidosis (hATTR) (Benson et al., 2018).
b. Indication: Vutrisiran is an RNAi therapeutic targeting transthyretin (TTR) for the treatment of hATTR amyloidosis, aiming to reduce TTR protein levels and slow disease progression (Benson et al., 2018).

6.4.7.4 Glembatumumab Vedotin (CDX-011)

a. Phase: Glembatumumab vedotin is undergoing phase 2 clinical trials for the treatment of advanced melanoma and breast cancer.
b. Indication: Glembatumumab vedotin is an RNAi therapeutic targeting glycoprotein NMB (gpNMB) conjugated to monomethyl auristatin E (MMAE), aiming to deliver cytotoxic payload selectively to gpNMB-expressing tumor cells (Yardley et al., 2020).

6.4.7.5 Nedosiran (*N*-Acetyl-D-Mannosamine, *N*-Acetyl-D-Galactosamine Conjugate) (Formerly Known as DCR-PHXC)

a. Phase: Nedosiran has been evaluated in phase 3 clinical trials for the treatment of primary hyperoxaluria type 1 (PH1) (Garrelfs et al., 2021).
b. Indication: Nedosiran is an RNAi therapeutic targeting hydroxy acid oxidase 1 (HAO1) for the treatment of PH1, aiming to reduce hepatic oxalate production and lower urinary oxalate levels (Garrelfs et al., 2021).

6.4.7.6 Teprasiran (Formerly Known as QPI-1002)

a. Phase: Teprasiran has been evaluated in phase 2 clinical trials for the prevention of acute kidney injury (AKI) in patients undergoing cardiac surgery (Asgari et al., 2018).
b. Indication: Teprasiran is an RNAi therapeutic targeting p53 mRNA for the prevention of AKI associated with cardiac surgery, aiming to reduce renal cell death and inflammation (Asgari et al., 2018).

6.4.7.7 Tivanisiran

a. Phase: Tivanisiran has been evaluated in phase 3 clinical trials for the treatment of familial amyloid polyneuropathy (FAP) (Adams et al., 2018).
b. Indication: Tivanisiran is an RNAi therapeutic targeting transthyretin (TTR) for the treatment of FAP, aiming to reduce TTR protein levels and slow disease progression (Adams et al., 2018).

These examples illustrate the ongoing clinical investigation of RNAi therapeutics for various indications. For the most up-to-date information on clinical trials involving Nedosiran, Teprasiran, and Tivanisiran, it is recommended to consult clinical trial registries or databases such as ClinicalTrials.gov.

6.5 CHALLENGES

While RNAi holds immense promise as a next-generation therapeutic approach, several challenges must be addressed to fully realize its potential in clinical applications. These challenges encompass various aspects of RNAi-based therapeutics, including delivery, off-target effects, immunogenicity, and specificity. This section highlights key challenges facing RNAi therapeutics, supported by relevant citations from scientific literature and clinical studies.

6.5.1 Delivery Challenges

One of the major hurdles in RNAi therapeutics is efficient and targeted delivery of RNAi molecules to specific tissues and cells. Naked RNA molecules are susceptible to degradation by nucleases and face barriers such as renal clearance and cellular uptake limitations (Wittrup and Lieberman, 2015). Overcoming these delivery challenges requires the development of innovative delivery systems, including lipid nanoparticles, polymer-based carriers, and viral vectors, to enhance stability, biodistribution, and intracellular delivery of RNAi payloads (Akinc et al., 2008; Wittrup and Lieberman, 2015).

6.5.2 Off-Target Effects

Off-target effects, unintended silencing of genes with partial sequence complementarity to the RNAi trigger, pose a significant challenge to RNAi therapeutics (Jackson and Linsley, 2010). Off-target effects may lead to unintended physiological consequences and undermine the specificity of RNAi-mediated gene silencing. Addressing off-target effects requires careful design and optimization of RNAi triggers, as well as the use of bioinformatics tools and experimental validation to minimize off-target silencing (Jackson and Linsley, 2010).

6.5.3 Immunogenicity and Immune Responses

RNAi triggers, particularly exogenous dsRNA molecules, can elicit innate immune responses and activation of the interferon pathway, leading to systemic inflammation and immunogenicity (Judge et al., 2005). Immune responses to RNAi therapeutics may result in dose-limiting toxicity and compromise therapeutic efficacy. Strategies to mitigate immunogenicity include the use of chemically modified RNAi triggers, immunomodulatory agents, and delivery systems that minimize immune recognition and activation (Judge et al., 2005; Wittrup and Lieberman, 2015).

6.5.4 Specificity and Selectivity

Achieving specificity and selectivity in RNAi-mediated gene silencing is essential to avoid unintended off-target effects and ensure therapeutic efficacy. However, designing RNAi triggers with high specificity remains a challenge, particularly for targets with homologous sequences or functional redundancy (Jackson and Linsley, 2010). Improving specificity requires comprehensive target validation, rational design of RNAi triggers, and experimental validation using appropriate controls and assays (Jackson and Linsley, 2010).

6.5.5 Clinical Translation and Regulatory Approval

Despite significant progress in preclinical studies and clinical trials, the clinical translation and regulatory approval of RNAi therapeutics pose challenges related to safety, efficacy, and manufacturing scalability (Amarzguioui and Rossi, 2008). Regulatory agencies such as the U.S. Food and Drug Administration (FDA) and the European Medicines Agency (EMA) require extensive preclinical data on pharmacokinetics, biodistribution, and toxicity, as well as rigorous clinical trial evidence to support the approval of RNAi-based drugs (Amarzguioui and Rossi, 2008).

In summary, while RNAi holds great promise as a next-generation therapeutic approach, several challenges must be overcome to realize its full potential in clinical applications. Addressing these challenges requires interdisciplinary collaboration, innovative technologies, and ongoing research efforts to advance the field of RNAi therapeutics.

6.6 FUTURE PROSPECTS OF RNAI-BASED THERAPEUTICS

The future prospects of RNAi-based therapeutics are promising and encompass a wide range of applications and advancements. The following are some key areas where RNAi holds considerable potential.

6.6.1 Precision Medicine

RNAi offers the ability to precisely target specific genes or pathways implicated in diseases, making it well-suited for personalized medicine approaches. By tailoring RNAi therapeutics to individual genetic profiles or disease signatures, treatments can be optimized for greater efficacy and reduced off-target effects.

6.6.2 Diverse Disease Targets

RNAi-based therapeutics have the potential to target a wide range of diseases, including genetic disorders, viral infections, cancer, neurological diseases, metabolic disorders, and autoimmune conditions. As our understanding of disease mechanisms improves and new therapeutic targets are identified, RNAi provides a versatile platform for developing targeted therapies.

6.6.3 Combination Therapies

RNAi can be combined with other treatment modalities, such as small molecules, antibodies, or gene editing technologies, to achieve synergistic effects or overcome resistance mechanisms. Combinatorial approaches may enhance therapeutic outcomes and enable the development of more effective treatment regimens for complex diseases.

6.6.4 Advancements in Delivery Systems

Ongoing research is focused on developing improved delivery systems for RNAi therapeutics, including lipid nanoparticles, polymer-based carriers, and viral vectors. These advancements aim to enhance the stability, specificity, and tissue-targeting capabilities of RNAi molecules, ultimately improving their clinical efficacy and safety.

6.6.5 Noninvasive Administration Routes

Efforts are underway to explore noninvasive routes of administration for RNAi therapeutics, such as oral delivery, inhalation, or topical application. Noninvasive delivery methods could improve patient compliance, reduce treatment burden, and expand the therapeutic potential of RNAi across different disease indications.

6.6.6 Regulatory Approval and Commercialization

With the successful approval of RNAi-based drugs by regulatory agencies such as the FDA, there is growing confidence in the clinical translation and commercialization of RNAi therapeutics. Continued investment and collaboration between academia, industry, and regulatory bodies will support the development of RNAi-based drugs and their transition into clinical practice.

6.6.7 Emerging Technologies and Applications

Advances in RNAi technology, including novel RNAi triggers, chemical modifications, and delivery strategies, are driving innovation in the field. Additionally, emerging applications such as RNAi-based vaccines, gene editing, and regenerative medicine hold promise for addressing unmet medical needs and advancing the field of RNAi therapeutics.

Overall, the future of RNAi-based therapeutics is characterized by continued innovation, expanded therapeutic applications, and the potential to transform the treatment landscape across a wide range of diseases. As research progresses and technologies evolve, RNAi is poised to play an increasingly prominent role in personalized medicine and the development of targeted therapies tailored to individual patients.

6.7 CONCLUSIONS

RNAi stands as a promising next-generation therapeutic approach with the potential to revolutionize the treatment of various diseases. Over the years, RNAi has emerged as a powerful tool for precise gene modulation, offering advantages such as specificity, versatility, and the ability to target previously "undruggable" genes. The clinical success of FDA-approved RNAi therapeutics such as patisiran, givosiran, and inclisiran underscores the transformative impact of RNAi technology in addressing unmet medical needs and improving patient outcomes. Ongoing research and technological advancements in RNAi delivery systems, chemical modifications, and target identification continue to drive innovation in the field. Emerging technologies, such as RNAi-based vaccines, gene editing, and regenerative medicine, further expand the therapeutic potential of RNAi. Despite its promise, RNAi-based therapeutics face challenges such as delivery optimization, off-target effects, immunogenicity, and regulatory hurdles. Addressing these challenges requires interdisciplinary collaboration, innovative technologies, and rigorous preclinical and clinical evaluation. Overcoming these obstacles presents opportunities for advancing the field and unlocking the full potential of RNAi as a therapeutic approach. With continued research, investment, and collaboration, RNAi is poised to make significant contributions to the future of medicine, improving patient outcomes and quality of life.

REFERENCES

Adams, D., Gonzalez-Duarte, A., O'Riordan, W. D., Yang, C. C., Ueda, M., Kristen, A. V., ... & Suhr, O. B. (2018). Patisiran, an RNAi therapeutic, for hereditary transthyretin amyloidosis. *New England Journal of Medicine, 379*(1), 11–21.

Akinc, A., Zumbuehl, A., Goldberg, M., Leshchiner, E. S., Busini, V., Hossain, N., ... & Anderson, D. G. (2008). A combinatorial library of lipid-like materials for delivery of RNAi therapeutics. *Nature Biotechnology, 26*(5), 561–569.

Amarzguioui, M., & Rossi, J. J. (2008). Principles of dicer substrate (D-siRNA) design and function. In S. Barik (Ed.), *RNAi: Design and Application*, Humana, Totowa, NJ, 3–10.

Asgari, E., Leow, R., & Moradi, S. (2018). QPI-1002 (an siRNA targeting p53) for prevention of acute kidney injury: Efficacy and safety in phase II clinical trial. *Journal of Investigative Medicine, 66*, 1077–1081.

Benson, M. D., Waddington-Cruz, M., Berk, J. L., Polydefkis, M., Dyck, P. J., Wang, A. K., ... & Coelho, T. (2018). Inotersen treatment for patients with hereditary transthyretin amyloidosis. *New England Journal of Medicine, 379*(1), 22–31.

Bernstein, E., Caudy, A. A., Hammond, S. M., & Hannon, G. J. (2001). Role for a bidentate ribonuclease in the initiation step of RNA interference. *Nature, 409*(6818), 363–366.

Buchholz, K., Goff, L. A., & Trapnell, C. (2021). Intergenic and repeat transcription in human mRNAseq samples that are depleted of ribosomal RNA. *Nature Communications, 12*, 2094.

Coelho, T., Adams, D., Silva, A., Lozeron, P., Hawkins, P. N., Mant, T., ... & Suhr, O. B. (2013). Safety and efficacy of RNAi therapy for transthyretin amyloidosis. *New England Journal of Medicine, 369*(9), 819–829.

Davidson, B. L., & McCray Jr, P. B. (2011). Current prospects for RNA interference-based therapies. *Nature Reviews Genetics, 12*(5), 329–340.

Dhillon, A. P., & Winstanley, A. (2020). RNA Interference (RNAi) Therapeutics Market. *Nature Reviews Drug Discovery, 19*, 245.

Elbashir, S. M., Harborth, J., Lendeckel, W., Yalcin, A., Weber, K., & Tuschl, T. (2001). Duplexes of 21-nucleotide RNAs mediate RNA interference in cultured mammalian cells. *Nature, 411*(6836), 494–498.

Filipowicz, W., Jaskiewicz, L., Kolb, F. A., & Pillai, R. S. (2005). Post-transcriptional gene silencing by siRNAs and miRNAs. *Current Opinion in Structural Biology, 15*(3), 331–341.

Fire, A., Xu, S., Montgomery, M. K., Kostas, S. A., Driver, S. E., & Mello, C. C. (1998). Potent and specific genetic interference by double-stranded RNA in Caenorhabditis elegans. *Nature, 391*(6669), 806–811.

Garrelfs, S. F., Frishberg, Y., Hulton, S. A., Koren, M. J., O'Riordan, W. D., Cochat, P., ... & Lieske, J. C. (2021). Lumasiran, an RNAi therapeutic for primary hyperoxaluria type 1. *New England Journal of Medicine, 384*(13), 1216–1226.

Hannon, G. J. (2002). RNA interference. *Nature, 418*(6894), 244–251.

Hutvagner, G., & Zamore, P. D. (2002). A microRNA in a multiple-turnover RNAi enzyme complex. *Science, 297*(5589), 2056–2060.

Jackson, A. L., & Linsley, P. S. (2010). Recognizing and avoiding siRNA off-target effects for target identification and therapeutic application. *Nature Reviews Drug Discovery, 9*(1), 57–67.

Judge, A. D., Sood, V., Shaw, J. R., Fang, D., McClintock, K., & MacLachlan, I. (2005). Sequence-dependent stimulation of the mammalian innate immune response by synthetic siRNA. *Nature Biotechnology, 23*(4), 457–462.

Kumar, P., Wu, H., McBride, J. L., Jung, K. E., Hee Kim, M., Davidson, B. L., ... & Manjunath, N. (2007). Transvascular delivery of small interfering RNA to the central nervous system. *Nature, 448*(7149), 39–43.

Lagos-Quintana, M., Rauhut, R., Lendeckel, W., & Tuschl, T. (2001). Identification of novel genes coding for small expressed RNAs. *Science*, *294*(5543), 853–858.

Lee, R. C., Feinbaum, R. L., & Ambros, V. (1993). The C. elegans heterochronic gene lin-4 encodes small RNAs with antisense complementarity to lin-14. *Cell*, *75*(5), 843–854.

Liang, X. H., Sun, H., Nichols, J. G., & Crooke, S. T. (2017). RNase H1-dependent antisense oligonucleotides are robustly active in directing RNA cleavage in both the cytoplasm and the nucleus. *Molecular Therapy*, *25*(9), 2075–2092.

Liu, J., Carmell, M. A., Rivas, F. V., Marsden, C. G., Thomson, J. M., Song, J. J., ... & Hannon, G. J. (2004). Argonaute2 is the catalytic engine of mammalian RNAi. *Science*, *305*(5689), 1437–1441.

Machin, N., & Ragni, M. V. (2018). An investigational RNAi therapeutic targeting antithrombin for the treatment of hemophilia A and B. *Journal of Blood Medicine*, *9*, 135–140.

McManus, M. T., & Sharp, P. A. (2002). Gene silencing in mammals by small interfering RNAs. *Nature Reviews Genetics*, *3*(10), 737–747.

Nair, J. K., Willoughby, J. L., Chan, A., Charisse, K., Alam, M. R., Wang, Q., ... & Manoharan, M. (2014). Multivalent N-acetylgalactosamine-conjugated siRNA localizes in hepatocytes and elicits robust RNAi-mediated gene silencing. *Journal of the American Chemical Society*, *136*(49), 16958–16961.

Puy, H., Gouya, L., & Deybach, J. C. (2020). ALAS1 reduction by givosiran: A new era in acute hepatic porphyria treatment. *Molecular Genetics and Metabolism*, *131*, 1–2.

Ray, K. K., Landmesser, U., Leiter, L. A., Kallend, D., Dufour, R., Karakas, M., ... & Kastelein, J. J. (2017). Inclisiran in patients at high cardiovascular risk with elevated LDL cholesterol. *New England Journal of Medicine*, *376*(15), 1430–1440.

Wang, T., Wei, J. J., Sabatini, D. M., & Lander, E. S. (2014). Genetic screens in human cells using the CRISPR-Cas9 system. *Science*, *343*(6166), 80–84.

Wittrup, A., & Lieberman, J. (2015). Knocking down disease: A progress report on siRNA therapeutics. *Nature Reviews Genetics*, *16*(9), 543–552.

Yardley, D. A., Weaver, R., Melisko, M. E., Saleh, M. N., Arena, F. P., Forero, A., ... & Vahdat, L. T. (2015). EMERGE: A randomized phase II study of the antibody-drug conjugate glembatumumab vedotin in advanced glycoprotein NMB-expressing breast cancer. *Journal of Clinical Oncology: Official Journal of the American Society of Clinical Oncology*, *33*(14), 1609.

7 Phage Therapy as a Vanguard against Anti-Microbial Resistance
Prospects and Applications

Shikha Malik, Nishu Rawat, and Kiran Nehra

7.1 INTRODUCTION

This century has been witnessing a rapid acceleration in the rise and spread of antimicrobial resistance (AMR) in infectious disease-causing microorganisms throughout the globe, thus posing a substantial economic burden on the healthcare system. The battle against infectious diseases began with the discovery of the first antibiotic, penicillin, in 1928 and continued steadily until 2005 with the arrival of daptomycin, a cyclic lipopeptide. However, ever since, no new class of antibiotics has been developed by the scientific community (Malik et al., 2020). Several factors have collectively contributed in the failure of the existing antibiotics, mainly including an expensive process of antibiotic development, easy accessibility to the existing low-priced drugs with high efficacy and hence their overuse and misuse, and an avoidance of high-priced antibiotics by the healthcare system (Luepke et al., 2017). Hence, the discovery of novel and effective antimicrobial agents has become the need of the hour to minimize the reliance on antibiotics and control the subsequent emergence and spread of AMR.

In this direction, researchers have revived their interest in a century-old methodology named phage therapy. Phages are naturally occurring bacterial predators which require a susceptible host to complete their life cycle, and phage therapy is the practice of using lytic phages to selectively kill the clinically relevant bacterial pathogens (Lin et al., 2017). Phages offers numerous advantages over the antibiotic chemotherapy, such as self-amplification, low dosage, narrow host range, biofilm penetration, and minimal environmental impact (Malik et al., 2020). Moreover, phages have also shown efficacy against community-acquired, nosocomial MDR infections, collectively grouped as ESKAPE pathogens (*Enterococcus faecium*, *Staphylococcus aureus*, *Klebsiella pneumoniae*, *Acinetobacter baumannii*, *Pseudomonas aeruginosa*, and *Enterobacter* spp.) (Patil et al., 2021).

This chapter sheds light on the bacterial antibiotic conundrum, possible strategies, present status, and future perspectives of this century-old antibacterial agent, named bacteriophages.

DOI: 10.1201/9781032642567-9

7.2 BACTERIAL ANTIBIOTIC RESISTANCE: A GLOBAL CONCERN OF THE 21ST CENTURY

Antimicrobial resistance is a primitive and inherent characteristic of the microorganisms which provides them an edge during survival in similar ecological niches, and a majority of the antimicrobial compounds are natural products of microorganisms produced to obstruct the growth of other microbes (Clardy et al., 2009).

During the course of evolution spanning millions of years, microorganisms have procured highly convoluted drug-resistance mechanisms to bypass killing by antimicrobial agents. Substantially, bacteria can use an array of different biochemical pathways to achieve resistance against a particular antimicrobial class of agents. However, the key mechanisms underlying drug resistance in bacteria mainly include four major pathways: (i) alterations in the antibiotic molecule—one of the most effective approaches used by the bacteria to deal with antibiotics is to synthesize enzymes capable of adding certain chemical groups onto the antibiotic molecules or in some cases directly destroying the antimicrobial molecule, thus rendering the drug incompetent to bind to its target site; (ii) reduced drug penetration—another approach used by the bacteria (especially the gram-negatives) is to decrease the influx of drugs having intracellular or periplasmic target; (iii) increased drug efflux—in some cases the bacteria activate efflux pumps to eject the drug out of the cell; and (iv) alterations on the drug target site—bacteria have evolved themselves to circumvent the action of drugs by obstructing their target site (binding site of the drug), either by protecting the target site or by enzymatic modification of the target site, which subsequently reduces affinity for the drug molecule as shown in Figure 7.1 (Munita and Arias, 2016; Schaenzer and Wright, 2020).

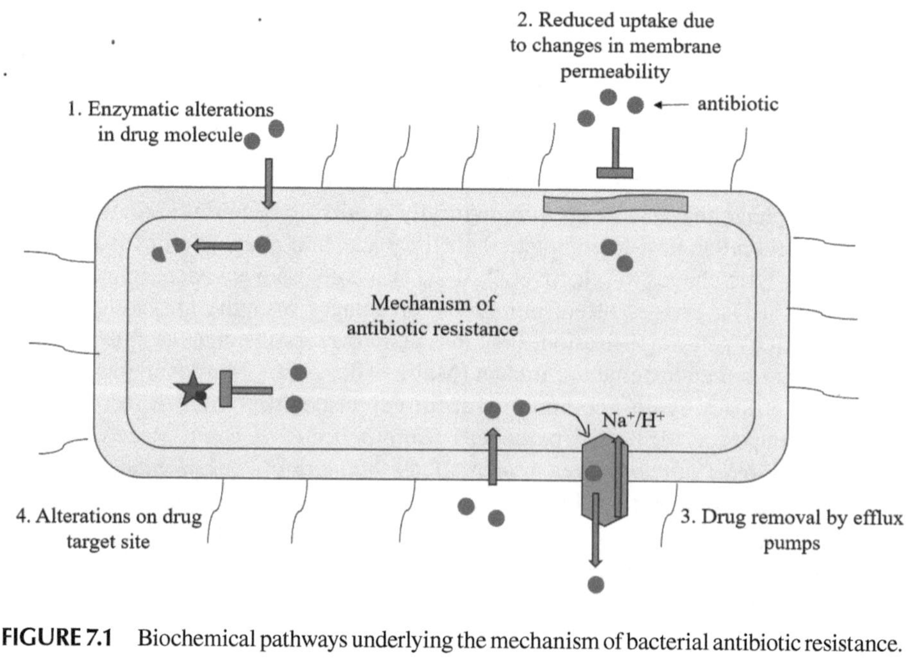

FIGURE 7.1 Biochemical pathways underlying the mechanism of bacterial antibiotic resistance.

TABLE 7.1
Potential Applications of Phage Therapy for Humans, Animals, and Environment

Phage Therapy Applications	References
In humans	
Phage as vectors for genes, drugs, and vaccines delivery	Hosseinidoust (2017), Ju et al. (2017)
Phage-encoded lysozymes as prophylactic agents	Adhya et al. (2014)
Phage-based detection methods for biodefense-related bacterial pathogens	Sozhamannan and Hofmann (2020)
Phages as grooming agents for gut microbiome	Strathdee et al. (2023)
In animals	
Phages as antimicrobials in veterinary and animal husbandry	Loponte et al. (2021)
Phages as alternative to antibiotics for treating livestock and companion animals	Ferriol-González and Domingo-Calap (2021)
In environment	
For wastewater treatment	Bolsan et al. (2022)
Phages as alternative to antibiotics in aquaculture and agriculture	Sieiro et al. (2020)
Phage for biocontrol and biopreservation of plant and animal-based foods	Garvey (2022)

The highly complex and sophisticated antibiotic-resistant mechanisms adopted by these living entities have complicated the existing treatment strategies available for the control of pathogenic microorganisms. Therefore, more effective and novel antimicrobial compounds, along with a complete knowledge of their underlying mechanisms, are required to encounter the drug-resistance conundrum. The past decade has witnessed a significant amount of research and clinical trials in the area of the utilization of phages as novel antimicrobial agents against various infectious diseases across different parts of the world (Xu et al., 2022).

Besides the therapeutic application of phages, research studies are also ongoing for using these bacterial viruses in humans, for veterinary and animal husbandry, and for controlling pathogens in the food and wastewater treatment plants (Table 7.1). Hence, it could be rightfully said that phages offer a diverse range of applications in the field of medical science and in modern biotechnology.

7.3 PHAGE THERAPY: STRATAGEM TO COMBAT AMR

Phages, in the form of non-antibiotic antimicrobial therapy, have been exploited in several ways to counter drug-resistant bacterial infections. These include the use of naturally occurring phages either alone or in a combination of two or more phages in the form of a cocktail, or using the phages in combination with antibiotics. More recently, various biotechnological advancements shown in Figure 7.2 have further disseminated the collection of possible phage therapeutics to comprise novel approaches using purified phage lytic enzymes (PLEs) and genetically modified phages.

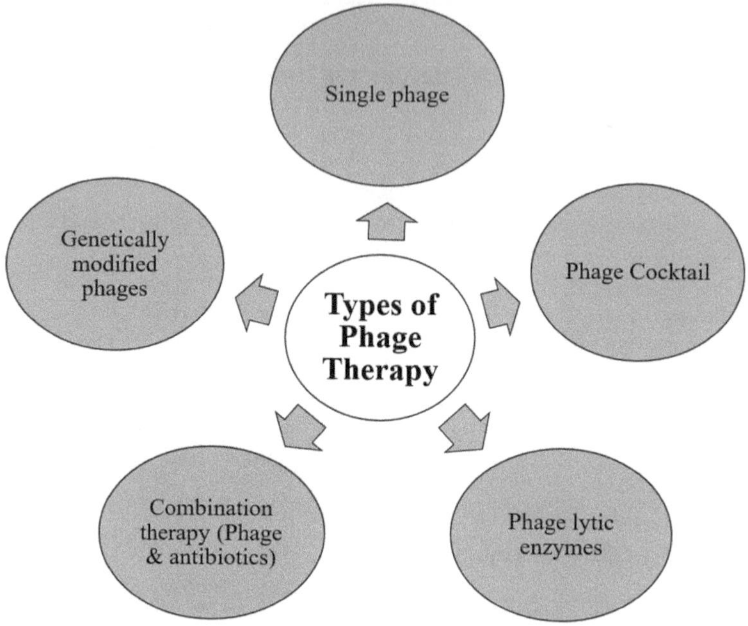

FIGURE 7.2 Strategies of using phage therapy to combat drug-resistant bacterial infections.

7.3.1 Cocktail Therapy

Conventionally, phage therapy involves the use of a single lytic phage (monophage therapy) against pathogenic bacterial infection, and several findings have reported positive results both in ex vivo and in vivo settings (Singh et al., 2022). However, at the beginning of the current millennium, the research concerning phage therapy shifted its attention from monophage to polyphage therapy owing to a plethora of advantages exhibited by phage cocktails (Chan et al., 2013). Phage cocktail broadens the host spectrum, slows down the rise of phage-resistant bacterial mutants, and aids in eradicating biofilm-related infections (Malik et al., 2020). Several studies supporting cocktail therapy against bacterial infections have been published (Salman and Abdulamir, 2014).

Apart from dealing with limitations like narrow host spectrum, phage-resistant mutants, and biofilms, cocktail therapy also presents some important safety apprehensions (Chan et al., 2013), mainly associated with the release of bacterial inflammatory proteins (named endotoxins) by the phages, which sometimes worsen the bacterial infections (Lin et al., 2022). Therefore, to mitigate the limitations associated with phage cocktails, other promising advances including the use of PLEs, phages in combination with antibiotics, and genetically engineered phages have been used to cure MDR infections.

7.3.2 Phage Lytic Enzymes

Owing to the advancement in the field of genomics, phage-encoded lytic proteins that are functionally identical to eukaryotic lysozyme are being considered as other prospective phage-based antimicrobial molecules to withstand challenges related to phage resistance and host specificity (Lin et al., 2017). The phage-encoded proteins include holin, endolysin (lysin), virion-associated peptidoglycan hydrolase (VAPGH) proteins, and EPS depolymerase. Phage protein holin is a transmembrane protein that allows the lysin to access the bacterial cell wall and lyse it; VAPGH protein disrupts the bacterial cell wall by binding to the antigens present on the cell surface, thereby allowing the phage to inject its DNA into the cell; and EPS depolymerase attacks the biofilm's EPS matrix and degrades it, enabling the phage to access the bacterial cells that are concealed therein (Drulis-Kawa et al., 2012). In a study conducted by Jun et al. (2013), the in vitro and in vivo antibacterial evaluation of a recombinant phage endolysin SAL-1 demonstrated rapid and efficient bactericidal action against planktonic and biofilm-forming S. aureus infections. Further, SAL200 also exhibited minimum adverse effects on human body when the first phase I randomized double blind clinical trial was conducted to estimate its tolerability, safety, pharmacodynamics, and pharmacokinetics in healthy male volunteers (Jun et al., 2017). Likewise, an array of studies have successfully demonstrated the in vitro antibacterial activity of phage lytic enzymes against MDR bacterial pathogens (Habibinava et al., 2021; Wang et al., 2022). Considering the above results, the highly evolved PLEs or endolysins can be considered a good drug candidate in the battle against drug-resistant bacterial pathogens.

PLEs have also been successfully combined with outer membrane (OM) permeabilizers (OMPs) and chelating substances like ethylene diamine tetraacetic acid disodium salt dihydrate (EDTA) to make them effective against gram-negatives having OM. Several reports of combining PLE with OMPs against gram-negative microbes such as *A. baumannii*, *E. coli*, and *K. pneumoniae* have been stated (Guo et al., 2021). However, more recently, several researchers have enhanced the muralytic activity of PLEs or endolysins against gram-negative pathogens by synergistically pairing them with various antibiotics (Blasco et al., 2020). These various combinatorial strategies, either in the form of endolysins and chelating agents or endolysins with antibiotics, have contributed significantly in reducing bacterial load both under in vitro and in vivo conditions.

7.3.3 Phages in Combination with Antibiotics

Another commonly employed approach to circumvent bacterial resistance and to reduce the dosage of antibiotics involves combining the antibiotics with phages to produce synergistic interactions. Phage-antibiotic synergy (PAS) is defined as the phenomenon wherein sublethal concentrations of antibiotics substantially boost the release of phage progeny from the bacterial host (Malik et al., 2021). The key mechanism of PAS phenomenon is based on an enhanced antibiotic susceptibility in the presence of phages, bacterial cell filamentation/elongation due to antibiotics, reduction in the MIC of antibiotics, and bacterial polysaccharide degradation by phage-encoded depolymerases that enables drug diffusion and cell penetration (Łusiak-Szelachowska et al., 2022). Initial studies on PAS were performed against uropathogenic *E. coli*, which were then followed by several studies on other biofilm-forming pathogens such as *K. penumoniae* and *P. aeruginosa*, wherein a significant decrease in the bacterial density was detected when phages and antibiotics were used synergistically (Malik et al., 2020).

Significant clinical improvements were reported in patients having orthopedic infections caused by drug-resistant *A. baumannii* and *K. pneumoniae*, cystic fibrosis patients with MDR *P. aeruginosa* pneumonia, lung-transplant recipients having MDR infections caused by *Burkholderia dolosa* and *P. aeruginosa*, patients having recurrent urinary tract infection (UTI) caused by *K. pneumoniae*, among several others (Łusiak-Szelachowska et al., 2022). However, more in vitro and in vivo studies on the combined action of both agents should be conducted in near future to determine the time of administration, dose, and type of each agent used.

7.3.4 Genetically Modified Phages

Owing to the broad array of opportunities offered by the recent advances in the field of genome engineering, these bacterial viruses have been genetically modified by various techniques of gene mutation, gene replacement, exogenous gene delivery, or phage capsid modification. These genetic modifications can minimize the effect of phages on the mammalian immune system, broaden their host range, modify antibiotic sensitivity, and boost bactericidal activity against MDR pathogens either by disturbing their drug-resistant pathways or by other means (Gibb et al., 2021; Guo et al., 2021). For example, Lu and Collins engineered *E. coli* phage T7 to express the dispersin B enzyme, which can degrade one of the essential components of bacterial biofilms within the bacterial host cell upon infection.

Although the field of phage engineering has taken a leap in boosting the antibacterial effects against drug-resistant bacteria, experiments on modified phages have been mostly done under in vitro conditions or on animal models. Very few studies related to clinical trials have been reported so far. Therefore, a series of large-scale clinical trials and safety assessment studies need to be conducted to use genetically modified phages in clinical settings (Guo et al., 2021).

7.4 PRESENT STATE AND FUTURE PERSPECTIVES OF PHAGE THERAPY

Phage therapy has been extensively utilized to treat a variety of bacterial infections and diseases such as skin and soft tissue infections, oral infections, gastrointestinal infections, respiratory infections, urinary tract infections, eye infections, ear infections, nasal infections, sepsis, inflammatory bowel diseases, diabetes mellitus, alcoholic hepatitis, breast cancer, colorectal cancer, schizophrenia, and Parkinson's disease (Xu et al., 2022).

Apart from these cases, several other reports listed in Table 7.2 successfully demonstrating phage therapy against various infectious diseases have also come up in the past few years.

Nevertheless, the future perspectives of phage therapy lie on the shoulders of the scientists and clinicians of the next generation. The global fight against constantly evolving superbugs would be tackled with undying determination and collaboration among researchers, increased funding, smooth regulatory clearances, and improvements in clinical trials.

TABLE 7.2
List of Representative In Vivo Phage Therapy Treatments for Various Infectious Diseases

S.no.	Disease	Target Organism	Bacteriophage	Object	Clinical Outcome	Year	Location	References
					Skin and soft tissue infections			
1.	Infected diabetic toe ulcers	*S. aureus*	Anti-staphylococcal bacteriophage Sb-1	Human	No tissue breakdown, adverse events, or recurrence of infection were reported	2018	United States	Fish et al. (2016)
2.	Diabetic wound infections	*S. aureus*	Phage cocktail AB-SA01	Balb/c mice	Phage-treated mice exhibited significant ($p < 0.001$) wound healing like vancomycin treatment. No mortality was recorded associated with infections, and no pathological lesions other than the skin wounds were reported in postmortem examinations.	2020	Australia	Kifelew et al. (2020)
					Gastrointestinal infections			
3.	Mild-to-moderate gastrointestinal distress	*E. coli*	Cocktail of Myoviridae (LH01, LL12, & T4D) and Siphoviridae (LL5) phages	Human	Significant increase in blood tCO_2 levels were observed which signified low bacterial load	2019	United States	Febvre et al. (2019)
					Oral infections			
4.	Dental caries	*Streptococcus mutans*	SMHBZ3 phage	BALB/c mice	Phage prohibited carious lesion formation in mice modal	2021	Israel	Wolfoviz-Zilberman et al. (2021)
5.	Endodontic infection	*Enterococcus faecalis*	ClyR, a chimeric lysin	Ex vivo dental model	Cly exhibited robust antimicrobial activity against both sessile and planktonic *E. faecalis*	2018	China	Li et al. (2018)

(Continued)

TABLE 7.2 (*Continued*)
List of Representative In Vivo Phage Therapy Treatments for Various Infectious Diseases

S.no.	Disease	Target Organism	Bacteriophage	Object	Clinical Outcome	Year	Location	References
				Respiratory infections				
6.	Chronic lung infection	*K. pneumoniae*	Phage PELP20	Balb/c mice	Phages effectively cleared bacteria from the lungs of 70% of mice against the established 6d lung infection and substantially reduced colony-forming unit counts in the remaining 30% compared with controls	2017	United Kingdom	Waters et al. (2017)
7.	Pneumonia and septic shock	*S. aureus*	Cocktail of myoviridae bacteriophages (AB-SA01)	Human	Investigational phage therapy with good manufacturing practice was considered safe, as no adverse events were reported post phage administration	2020	Australia	Fabijan et al. (2020)
				Urinary tract infections				
8.	Infection followed by transurethral resection of the prostate	*Enterococcus spp, E. coli, Proteus mirabilis, P. aeruginosa, Staphylococcus spp, and Streptococcus spp*	Pyophage cocktail	Human	Intravesical phage therapy was found effective and safe in treating UTI in patients	2021	Georgia	Leitner et al (2021)
9.	Recurrent UTI followed by combined liver/ kidney transplantation	ESBL-producing *K. pneumoniae*	Phage cocktail (Metamorpho, mineola, and pKp20)	Human	Intravenous phage therapy with no concomitant antibiotics effectively killed the bacteria within 4 weeks without recurrence for 1 year	2023	USA	Le et al. (2023)

(*Continued*)

TABLE 7.2 (*Continued*)
List of Representative In Vivo Phage Therapy Treatments for Various Infectious Diseases

S.no.	Disease	Target Organism	Bacteriophage	Object	Clinical Outcome	Year	Location	References
				Eye and ear infection				
10.	Keratoconjunctivitis	*P. aeruginosa*	Cystoviridaephage	Rabbits	Phage-carrying in situ gel-forming eye drops effectively treated keratoconjunctivitis	2021	Iran	Rahimzadeh et al. (2021)
11.	Recurrent ear infections and chronic rhinosinusitis	*P. aeruginosa* and methicillin-resistant S. *aureus* (MRSA)	Phage SeMN68phi1 and SaWIQ0456AØ1	Human	MRSA was successfully eliminated with the combination of systemic and intranasal phage therapy with a combination of antibiotics	2022	USA	Rodriguez et al. (2022)
				Bacteremia/sepsis				
12.	Neonatal septicemia	*E. coli, K. pneumoniae, Haemophilus influenzae, P. aeruginosa, Citrobacter freundii and Moraxella catarrhalis*	A total of 29 lytic phages specific against each bacterial strain	Mouse model	Monovalent and polyvalent phages effectively treated bacteremia post 45 min of bacterial infection and after intervals of 5 h, 8 h, 14 h, 18 h and 24 h followed by injection of bacterial pathogens	2019	Iraq	Kaabi et al. (2019)

7.5 CONCLUSION

Since antibiotic-resistant bacteria pose a life-threatening hazard to global health, phage therapy holds the capability to serve as an alternative therapy to alleviate the constantly rising burden of infectious diseases. Undoubtedly, phage therapy offers a series of advantages; however, the current knowledge seems inadequate to permit the use of phage therapy as an alternative to antibiotics. Well-controlled and properly planned clinical trials are required to determine the safety and role of phage therapy in clinical settings to treat infectious diseases. Although recent advancements in the field of phage genomics have contributed significantly in improving the reliability and efficacy of phage therapy, issues related to phage selection, suitable route and frequency of administration, phage resistance, dosage, and pharmacodynamic and pharmacokinetic properties of phage are needed to be addressed. Regardless of these challenges, phage therapy may play a critical role in protecting the overall health of humans in the future.

REFERENCES

Adhya, S., Merril, C. R., & Biswas, B. (2014). Therapeutic and prophylactic applications of bacteriophage components in modern medicine. *Cold Spring Harbor Perspectives in Medicine*, *4*(1), a012518. https://doi.org/10.1101/cshperspect.a012518.

Blasco, L., Ambroa, A., Trastoy, R., Bleriot, I., Moscoso, M., Fernández-Garcia, L., ... & Tomas, M. (2020). In vitro and in vivo efficacy of combinations of colistin and different endolysins against clinical strains of multi-drug resistant pathogens. *Scientific Reports*, *10*(1), 7163. https://doi.org/10.1038/s41598-020-64145-7.

Bolsan, A. C., Rodrigues, H. C., Abilhôa, H. C. Z., Hollas, C. E., Venturin, B., Gabiatti, N. C., ... & De Prá, M. C. (2022). Bacteriophages in wastewater treatment: Can they be an approach to optimize biological treatment processes?. *Environmental Science and Pollution Research*, *29*(60), 89889–89898. https://doi.org/10.1007/s11356-022-24000-w.

Chan, B. K., Abedon, S. T., & Loc-Carrillo, C. (2013). Phage cocktails and the future of phage therapy. *Future Microbiology*, *8*(6), 769–783. https://doi.org/10.2217/fmb.13.47.

Clardy, J., Fischbach, M. A., & Currie, C. R. (2009). The natural history of antibiotics. *Current Biology*, *19*(11), R437–R441. https://doi.org/10.1016/j.cub.2009.04.001.

Drulis-Kawa, Z., Majkowska-Skrobek, G., Maciejewska, B., Delattre, A. S., & Lavigne, R. (2012). Learning from bacteriophages-advantages and limitations of phage and phage-encoded protein applications. *Current Protein and Peptide Science*, *13*(8), 699–722.

Febvre, H. P., Rao, S., Gindin, M., Goodwin, N. D., Finer, E., Vivanco, J. S., ... & Weir, T. L. (2019). PHAGE study: Effects of supplemental bacteriophage intake on inflammation and gut microbiota in healthy adults. *Nutrients*, *11*(3), 666. https://doi.org/10.3390/nu11030666.

Ferriol-González, C., & Domingo-Calap, P. (2021). Phage therapy in livestock and companion animals. *Antibiotics*, *10*(5), 559. https://doi.org/10.3390/antibiotics10050559.

Fish, R., Kutter, E., Wheat, G., Blasdel, B., Kutateladze, M., & Kuhl, S. (2016). Bacteriophage treatment of intransigent diabetic toe ulcers: A case series. *Journal of Wound Care*, *25*(7), S27–S33. https://doi.org/10.12968/jowc.2016.25.Sup7.S27.

Garvey, M. (2022). Bacteriophages and food production: Biocontrol and bio-preservation options for food safety. *Antibiotics*, *11*(10), 1324. https://doi.org/10.3390/antibiotics11101324.

Gibb, B., Hyman, P., & Schneider, C. L. (2021). The many applications of engineered bacteriophages: An overview. *Pharmaceuticals*, *14*(7), 634. https://doi.org/10.3390/ph14070634.

Guo, D., Chen, J., Zhao, X., Luo, Y., Jin, M., Fan, F., … & Liu, Z. (2021). Genetic and chemical engineering of phages for controlling multidrug-resistant bacteria. *Antibiotics*, *10*(2), 202. https://doi.org/10.3390/antibiotics10020202.

Habibinava, F., Zolfaghari, M. R., Zargar, M., Shahrbabak, S. S., & Soleimani, M. (2021). vB-Ea-5: A lytic bacteriophage against multi-drug-resistant *Enterobacter aerogenes*. *Iranian Journal of Microbiology*, *13*(2), 225. https://doi.org/10.18502/ijm.v13i2.5984.

Hosseinidoust, Z. (2017). Phage-mediated gene therapy. *Current Gene Therapy*, *17*(2), 120–126. https://doi.org/10.2174/1566523217666170510151940.

Ju, Z., & Sun, W. (2017). Drug delivery vectors based on filamentous bacteriophages and phage-mimetic nanoparticles. *Drug Delivery*, *24*(1), 1898–1908.

Jun, S. Y., Jang, I. J., Yoon, S., Jang, K., Yu, K. S., Cho, J. Y., … & Kang, S. H. (2017). Pharmacokinetics and tolerance of the phage endolysin-based candidate drug SAL200 after a single intravenous administration among healthy volunteers. *Antimicrobial Agents and Chemotherapy*, *61*(6), 10–1128. https://doi.org/10.1128/AAC.02629-16.

Jun, S. Y., Jung, G. M., Yoon, S. J., Oh, M. D., Choi, Y. J., Lee, W. J., … & Kang, S. H. (2013). Antibacterial properties of a pre-formulated recombinant phage endolysin, SAL-1. *International Journal of Antimicrobial Agents*, *41*(2), 156–161.

Kaabi, S. A. G., & Musafer, H. K. (2019). An experimental mouse model for phage therapy of bacterial pathogens causing bacteremia. *Microbial Pathogenesis*, *137*, 103770. https://doi.org/10.1016/j.micpath.2019.103770.

Kifelew, L. G., Warner, M. S., Morales, S., Vaughan, L., Woodman, R., Fitridge, R., … & Speck, P. (2020). Efficacy of phage cocktail AB-SA01 therapy in diabetic mouse wound infections caused by multidrug-resistant *Staphylococcus aureus*. *BMC Microbiology*, *20*, 1–10. https://doi.org/10.1186/s12866-020-01891-8.

Le, T., Nang, S. C., Zhao, J., Yu, H. H., Li, J., Gill, J. J., … & Aslam, S. (2023). Therapeutic potential of intravenous phage as standalone therapy for recurrent drug-resistant urinary tract infections. *Antimicrobial Agents and Chemotherapy*, *67*(4), e00037–23.

Leitner, L., Ujmajuridze, A., Chanishvili, N., Goderdzishvili, M., Chkonia, I., Rigvava, S., … & Kessler, T. M. (2021). Intravesical bacteriophages for treating urinary tract infections in patients undergoing transurethral resection of the prostate: A randomised, placebo-controlled, double-blind clinical trial. *The Lancet Infectious Diseases*, *21*(3), 427–436. https://doi.org/10.1016/S1473-3099(20)30330-3.

Li, W., Yang, H., Gong, Y., Wang, S., Li, Y., & Wei, H. (2018). Effects of a chimeric lysin against planktonic and sessile *Enterococcus faecalis* hint at potential application in endodontic therapy. *Viruses*, *10*(6), 290. https://doi.org/10.3390/v10060290.

Lin, D. M., Koskella, B., & Lin, H. C. (2017). Phage therapy: An alternative to antibiotics in the age of multi-drug resistance. *World Journal of Gastrointestinal Pharmacology and Therapeutics*, *8*(3), 162. https://doi.org/10.4292/wjgpt.v8.i3.162.

Lin, J., Du, F., Long, M., & Li, P. (2022). Limitations of phage therapy and corresponding optimization strategies: A review. *Molecules*, *27*(6), 1857.

Loponte, R., Pagnini, U., Iovane, G., & Pisanelli, G. (2021). Phage therapy in veterinary medicine. *Antibiotics*, *10*(4), 421. https://doi.org/10.3390/antibiotics10040421.

Lu, T. K., & Collins, J. J. (2009). Engineered bacteriophage targeting gene networks as adjuvants for antibiotic therapy. *Proceedings of the National Academy of Sciences*, *106*(12), 4629–4634. https://doi.org/10.1073/pnas.0800442106.

Luepke, K. H., Suda, K. J., Boucher, H., Russo, R. L., Bonney, M. W., Hunt, T. D., & Mohr III, J. F. (2017). Past, present, and future of antibacterial economics: Increasing bacterial resistance, limited antibiotic pipeline, and societal implications. *Pharmacotherapy: The Journal of Human Pharmacology and Drug Therapy*, *37*(1), 71–84.

Łusiak-Szelachowska, M., Międzybrodzki, R., Drulis-Kawa, Z., Cater, K., Knežević, P., Winogradow, C., ... & Górski, A. (2022). Bacteriophages and antibiotic interactions in clinical practice: What we have learned so far. *Journal of Biomedical Science*, *29*(1), 23.

Malik, S., Nehra, K., & Rana, J. S. (2021). Bacteriophage cocktail and phage antibiotic synergism as promising alternatives to conventional antibiotics for the control of multi-drug-resistant uropathogenic *Escherichia coli*. *Virus Research*, *302*, 198496. https://doi.org/10.1016/j.virusres.2021.198496

Malik, S., Sidhu, P. K., Rana, J. S., & Nehra, K. (2020). Managing urinary tract infections through phage therapy: A novel approach. *Folia Microbiologica*, *65*(2), 217–231. https://doi.org/10.1007/s12223-019-00750-y.

Munita, J. M., & Arias, C. A. (2016). Mechanisms of antibiotic resistance. *Virulence Mechanisms of Bacterial Pathogens*, 481–511. https://doi.org/10.1128/microbiolspec.VMBF-0016-2015.

Patil, A., Banerji, R., Kanojiya, P., Koratkar, S., & Saroj, S. (2021). Bacteriophages for ESKAPE: Role in pathogenicity and measures of control. *Expert Review of Anti-Infective Therapy*, *19*(7), 845–865.https://doi.org/10.1080/14787210.2021.1858800

Rahimzadeh, G., Saeedi, M., Nokhodchi, A., Moosazadeh, M., Ghasemi, M., Rostamkalaei, S. S., ... & Hevelaee, E. N. (2021). Evaluation of in-situ gel-forming eye drop containing bacteriophage against *Pseudomonas aeruginosa* keratoconjunctivitis in vivo. *BioImpacts BI*, *11*(4), 281. https://doi.org/10.34172/bi.2021.10.

Rodriguez, J. M., Woodworth, B. A., Fackler, J., & Brownstein, M. J. (2022). Case Report: Successful use of phage therapy in refractory MRSA chronic rhinosinusitis. *International Journal of Infectious Diseases*, *121*, 14–16.

Salman, A. E., & Abdulamir, A. S. (2014). Assessment of bacteriophage cocktails used in treating multiple-drug resistant *Pseudomonas aeruginosa*. *International Journal of Current Microbiology and Applied Sciences*, *3*, 711–723.

Schaenzer, A. J., & Wright, G. D. (2020). Antibiotic resistance by enzymatic modification of antibiotic targets. *Trends in Molecular Medicine*, *26*(8), 768–782.

Sieiro, C., Areal-Hermida, L., Pichardo-Gallardo, Á., Almuiña-González, R., De Miguel, T., Sánchez, S., ... & Villa, T. G. (2020). A hundred years of bacteriophages: Can phages replace antibiotics in agriculture and aquaculture. *Antibiotics*, *9*(8), 493.

Singh, A., Padmesh, S., Dwivedi, M., & Kostova, I. (2022). How good are bacteriophages as an alternative therapy to mitigate biofilms of nosocomial infections. *Infection and Drug Resistance*, 503–532. https://doi.org/10.2147/IDR.S348700.

Sozhamannan, S., & Hofmann, E. R. (2020). The state of the art in biodefense related bacterial pathogen detection using bacteriophages: How it started and how it's going. *Viruses*, *12*(12), 1393. https://doi.org/10.3390/v12121393.

Strathdee, S. A., Hatfull, G. F., Mutalik, V. K., & Schooley, R. T. (2023). Phage therapy: From biological mechanisms to future directions. *Cell*, *186*(1), 17–31.

Wang, F., Xiao, Y., Lu, Y., Deng, Z. Y., Deng, X. Y., & Lin, L. B. (2022). Bacteriophage lytic enzyme P9ly as an alternative antibacterial agent against antibiotic-resistant *Shigella dysenteriae* and *Staphylococcus aureus*. *Frontiers in Microbiology*, *13*, 821989.

Waters, E. M., Neill, D. R., Kaman, B., Sahota, J. S., Clokie, M. R., Winstanley, C., & Kadioglu, A. (2017). Phage therapy is highly effective against chronic lung infections with Pseudomonas aeruginosa. *Thorax*, *72*(7), 666–667. https://doi.org/10.1136/thoraxjnl-2016–209265.

Wolfoviz-Zilberman, A., Kraitman, R., Hazan, R., Friedman, M., Houri-Haddad, Y., & Beyth, N. (2021). Phage targeting *Streptococcus mutans* in vitro and in vivo as a caries-preventive modality. *Antibiotics*, *10*(8), 1015. https://doi.org/10.3390/pharmaceutics14091936.

Xu, H. M., Xu, W. M., & Zhang, L. (2022). Current status of phage therapy against infectious diseases and potential application beyond infectious diseases. *International Journal of Clinical Practice*, 2022(1), 4913146. https://doi.org/10.1155/2022/4913146.

8 *Terminalia arjuna* for Treatment of Cardiovascular Diseases

From Traditional Usage to High-Throughput Drug Discovery

Shiwani Chahal, Inderjeet Singh, Heena Jain, Anita Rani, Anita Rani Gill, Akhilesh Mishra, and Priyanka Siwach

8.1 INTRODUCTION

Cardiovascular diseases (CVDs) are a wide spectrum of conditions related to abnormal functioning of the cardiovascular system shown in Figure 8.1. According to the World Health Organization (WHO), of ~17 million premature deaths from noncommunicable diseases, 38% are due to CVDs only, in 2019, and taking all diseases together, CVDs account for 17.9 million casualties estimating 32% of the total global deaths (WHO, 2021). While this trend continues, the American Heart Association (AHA) and the National Institutes of Health (NIH) have jointly published a recent statistical annual update for a significant rise in the wide range of conditions associated with CVDs causing economic and health burdens globally (Tsao et al., 2022). Recently, although several advanced treatment regimens and drugs are being practiced for different conditions, their success stories have mixed aspects (Virani et al., 2020). HF has been declared the largest unmet medical need of the 21st century, and developing new drugs for HF and all other types of CVDs is the need of the hour (Hong, 2022).

Nature-based compounds have been the major sources of new substances with great potential to treat human diseases for ages (Pourbagher-Shahri et al., 2021). However, natural product-based drug discovery and development is associated with many inherent challenges, physiological and environmental dependence on the metabolome, complex purification process, low bioavailability of the compounds, complex structures for chemical synthesis, complex metabolic pathways with endless

DOI: 10.1201/9781032642567-10

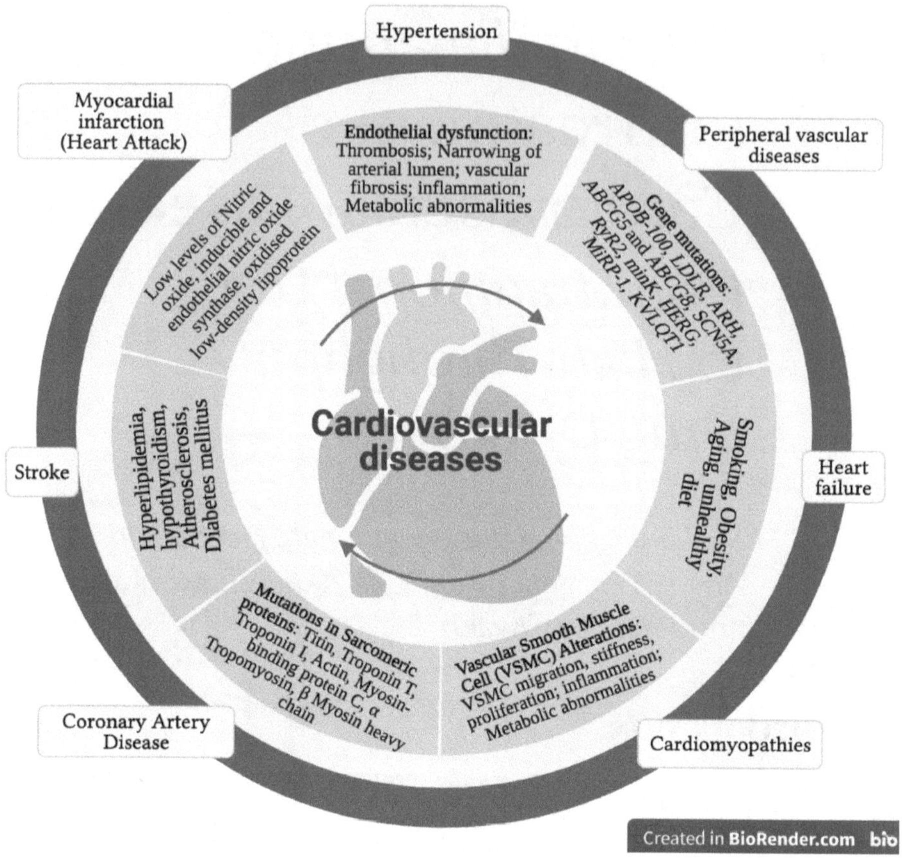

FIGURE 8.1 A general illustration of CVDs with types (rectangular boxes) and associated risks with pathological processes (gray quadrant) (created by BioRender, https://www.biorender.com).

regulatory points, biodiversity and conservation issues, and many more (Atanasov et al., 2015). Thus, in the late 1980s, when significant advancements in chemical synthesis began to take place, the pharmaceutical industries shifted their focus toward synthetic compound libraries and high-throughput screening (HTS) for the discovery of new drug leads. As per a recent review, of the 1881 new drugs approved by the FDA from January 1, 1981, to September 9, 2019, only 24.6% were purely synthetic while more than half were derived from nature; natural products biodiversity still holds the best option for finding new drug/lead molecule (Newman and Cragg, 2020).

Finding a suitable medicinal plant having drug molecules for any target disease can be a herculean task. On Earth, there exist around 250,000 flowering plant species holding a huge reservoir of drug molecules, and so the question is, what should be the approach? Ethnomedicinal information has always played a very significant role in this, and several examples exist which show a strong correlation between the current uses for the compounds isolated from any plant and the ethnomedicinal claims for that plant (Süntar, 2020). Among various plants documented in Indian Ayurvedic systems for treatment and cure of various types of cardiovascular disorders, *Terminalia arjuna* (Roxb.) Wight and Arn., often known as 'Arjuna,' hold a leading position. Several compounds have been isolated and purified from different parts of Arjuna and many have been investigated to show strong efficacy toward different conditions associated with CVDs (Kapoor et al., 2014), but none has been developed/used as a candidate drug molecule; this study is an effort to identify the gaps and develop the plan. Starting from exploring the ethnomedicinal claims, and going through the modern pharmacological and biochemical studies, this chapter investigates the status of its modern-day use for CVD drug development, discussing various steps involved in the process. It ends by proposing a design for the future path for using *T. arjuna* as a source of novel drug candidate molecule(s) for prevention and treatment of CVDs.

8.2 WORLDWIDE ETHNOMEDICINAL CLAIMS FOR GENUS *TERMINALIA*: A HISTORY OF USAGE FOR CVDS

The use of plants or plant extracts for healthcare is as old as mankind itself. Archaeological studies confirm the practice of herbal medicines as far back as 58,000 BC in Iraq (Leroi Gourhan, 1975). The first written evidence in Mesopotamia reporting 1,000 herbal medicines and in Egypt dates to 2600 BC (Atanasov et al., 2015). Documented records of traditional Chinese medicine have existed for thousands of years (Unschuld, 1986), and studies indicate the establishment of a well-developed Ayurveda-Indian medicinal system around the 1st millennium BC (Patwardhan, 2005). Pieces of evidence suggest the existence of the herbal medicine system in Greek and Roman culture too around the 1st century AD (Sneader, 2005). These systems contributed greatly to the development of modern drugs, the beginning can be credited to morphine from opium, back in 1817 (Sneader, 2005).

There are approximately 200–250 species belonging to the genus *Terminalia* (Family Combretaceae), broadly distributed all over the tropical and subtropical areas of Asia, Africa, and Australia (McGaw et al., 2001). The historical account of the usage of various species of Arjuna for different medicinal purposes is comprehensively presented in Figure 8.2.

Of the various species, *T. arjuna*, native to the Asian continent, has been known for its innumerable therapeutic properties with valuable evidence found in ancient texts of Ayurveda (*Vedas*, *Samhitas*, *Puranas* like *Agnipurana* and *Matasya purana*), old medicinal books and practices of tribal colonies (Mishra and Baarange, 2018). The use of Arjuna as medicine, as stated in *Atharvaveda* and *Kaushik sutra's* principal sutra volume (400–300 BC), may be the first written reference (Aghera et al., 2015). Since the rise of the Samhita period, Arjuna has been mentioned for the *hridroga* or heart-related diseases. Popular classical preparations using the Arjuna plant such as

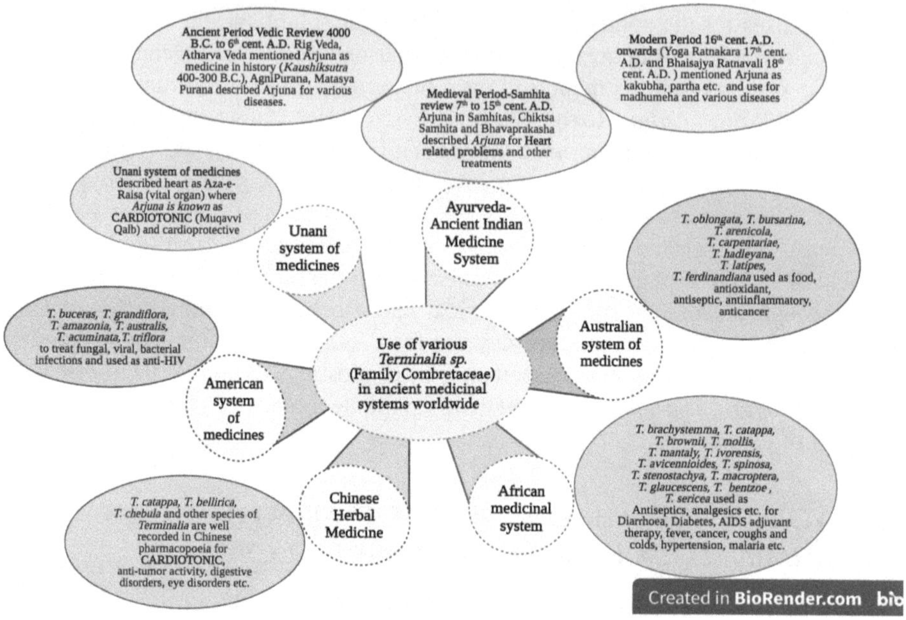

FIGURE 8.2 A comprehensive presentation of the usage of various species of the genus *Terminalia* in traditional medicine systems worldwide (Arman and Mahtab, 2013; Chaudhary et al., 2017; Cock, 2015; Mukherjee and Wahile, 2006; Zhang et al., 2019) (created by BioRender, https://www.biorender.com).

Kshirpaka and *Arjunarista* have been in practice for ages. Charaka Samhita given by Agnivesh (4th century AD) is an ancient medicinal treatise of India described as Arjuna by *kakubha* and *dhananjaya* (Chaudhary et al., 2017).

8.3 PHARMACOLOGICAL STUDIES OF VARIOUS EXTRACTS—SCIENTIFIC EVIDENCE FOR THE ETHNOMEDICINAL CLAIMS

The last two decades have witnessed a surge of pharmacological studies using different extracts prepared from various parts of the Arjuna tree. Some such studies reveal the significant potential of these extracts for various biological activities including cardioprotective, free radical scavenging, DNA damage protection, anti-inflammatory, antiatherosclerotic, and antioxidant, which are explained in the following sections.

Dedicated efforts, over the years, have revealed the efficacy of *T. arjuna* for the therapeutic benefits in heart failure, anti-ischemic effects, hypertension, lipid-lowering effects, and effects on endothelial dysfunction by directly modulating the consequences of CVDs and thereby suppress the progression of CVDs.

As this study concerns only the cardioprotective nature of *T. arjuna*, the pharmacological studies related to this aspect are discussed herewith in detail. Pharmacological studies reported that the alcoholic extract of the plant has strong antioxidant, cardio-protective, anti-inflammatory, and antihyperlipidemic properties and credited these to the presence of phenolic compounds, lactones, flavonoids, phytosterols, glycosides, and tannins present in the samples (Shah et al., 2019). The triterpenoids from the ethanolic extract of bark showed inhibitory action on angiotensin and thrombin enzymes tested on experimental rat models (Tahsin et al., 2021). Another study with the bark extract of Arjuna observed a reduction of oxidative stress in the case of myocardial ischemia induced by isoprenaline (Verma et al., 2021). A study showed the free radical scavenging and antioxidant effects from ethanolic and methanol extracts of *T. arjuna* (fruit and bark) under *in vitro* conditions (Halder et al., 2009). In addition, DNA nicking inhibiting property was also observed during the experiment. Due to hyperoxidation, affected basic cellular biomolecules result in an imbalance among reactive oxygen species (ROS), leading to strokes (Meena et al., 2021).

The hypolipidemic effect of arjunic acid and glycosides in different extracts of Arjuna was also reported in male Wistar rats and rabbits on a high-fat diet (Subramaniam et al., 2011). In a recent study, anti-atherosclerotic properties of the aqueous bark extract of *T. arjuna* were also reported (Bhat et al., 2018). Immunomodulatory, anti-inflammatory, and antioxidant activities were observed in *in vitro* culture conditions such as human monocytic (THP-1) and human aortic endothelial cells (HAECs) with *T. arjuna* stem bark (Kokkiripati et al., 2013). In recent studies, it has been reported that the intravenous administration of Arjuna aqueous extract in anesthetized dogs showed a decrease in blood pressure via modulating the autonomic responses (Bhatia et al., 1998). Bark extracts (alcoholic and aqueous) of Arjuna were found to increase the auricular and ventricular contractions, and coronary flow in different animal models (frogs, rats, and rabbits) (Radhakrishnan et al., 1993). One of the studies revealed the protective effects of bark extract of *T. arjuna* against doxorubicin-induced cardiotoxicity in male Wistar rats (Singh et al., 2008). Different dosages of bark powder of *T. arjuna* were found to be very effective in lowering the levels of total cholesterol, triglycerides, and LDL in broiler chicks (Pawar and Bhutani, 2005). Lipid-lowering effects and prevention of myocardial changes were reported in male Wistar rats following the use of various extracts of T. arjuna (Parveen et al., 2012).

8.4 CLINICAL STUDIES

A clinical study on 35 patients of chronic coronary artery diseases using bark extract of *T. arjuna* as an adjunct drug, that is, along with other conventional drugs such as aspirin, ACE inhibitor, beta-blockers, and statins showed improvement in anginal chest pain (Dwivedi et al., 2019). Twelve patients with refractory congestive heart failure were observed with marked improvement in heart failure signs and symptoms when treated with an aqueous extract of *T. arjuna* (Bharani et al., 1995). A significant reduction in the frequency of angina and ischemic mitral regurgitation with improvement in diastolic dysfunction was reported in 40 patients with coronary artery disease after administration of the aqueous extract of *T. arjuna* (Dwivedi et al., 2005). Thirty-six hypertensive patients assessed with the effect of bark powder of *T. arjuna* showed a greater decrease in both systolic blood pressure (SBP) and diastolic blood pressure (DBP) (Rao et al., 2001). Administration of bark powder of *T. arjuna* with statins showed a significant reduction in total cholesterol, low-density lipoprotein (LDL), lipoprotein, and nitrite levels associated with dyslipidemia, beta-thalassemia, and metabolic syndrome (Dwivedi and Kumar, 2007). A significant decrease in serum triglycerides and various inflammatory cytokines was also observed with the use of *T. arjuna* (Kapoor et al., 2014). Stem bark extract (aqueous) of *T. arjuna* has an advantageous preventive effect over other commonly used synthetic drugs such as Sildenafil (Meghwani et al., 2016).

8.5 BIOMOLECULES OF *T. ARJUNA*—THE POTENTIAL DRUG CANDIDATE MOLECULES

The therapeutic potential of any plant is because of its inherent wealth of secondary metabolites. Over the years, dedicated efforts to identify the nature of secondary metabolites in various parts of *T. arjuna* have led to the discovery of a plethora of phytoconstituents such as triterpenoids, flavonoids, tannins, saponins, sterols, polyphenols, and minerals, produced and distributed in its different parts like stem, roots, leaves, fruits, and seeds. Arjuna plant is reported to have strong cardioprotective activities (Dwivedi and Kumar, 2007). PubChem and Chembl ID of various compounds of arjuna, whose-so-ever available, have also been listed in Table 8.1.

The 2D structures of some of these compounds have been predicted and are available in PubChem, and the same has been retrieved and shown in Figure 8.3. These structures played an important role in identifying the target proteins which further leads to finding out their precise bioactivities. The docking studies using these compounds as ligands and target proteins from human proteome can reveal how strong the affinity is and have an analysis of other drug likelihood parameters.

8.6 CARDIOVASCULAR DRUG DISCOVERY AND DEVELOPMENT FROM *T. ARJUNA*

Cardiovascular drug discovery and development is facing several challenges including a decline of novel treatment procedures due to less-validated target molecules, hurdles, and expensive protocols with augmented failure rates of newer drugs at late-stage during clinical trials. And, hence, exploring the unexplored wealth of natural compounds of desired therapeutic value can be of great value. At present, no compound from *T. arjuna* is serving as a candidate drug lead molecule for CVDs, although several undefined supplements are available in the market as herbal products to augment different heart-related problems.

Modern drug discovery and development pipeline with in silico/high-throughput screening (HTS) based on docking studies at a molecular level is an efficient method to reduce the time and number of chemicals for pre-clinical trials in the process of

TABLE 8.1
Major Phytochemical Constituents Found in Different Parts of *T. arjuna*

Plant Part	Category of Phytochemicals	Phytoconstituents Present	References	PubChem ID	Chembl ID
Stem bark	Triterpenoids	Arjunin	Row et al. (1970)	102316370	–
		Arjunic acid	Honda et al. (1975)	15385516	515705
		Arjungenin	Singh et al. (2002a, b)	12444386	499598
		Terminic acid	Anjaneyulu and Prasad (1983)	132568257	–
		Termineltin	Singh et al. (1995)		–
		Arjunolic acid	Singh et al. (2002a, b)	73641	464466
	Ursane triterpenoids	2α,3β-Dihydroyurs-12,18-oic acid 28-O-β-D-glucopyranosyl ester	Wang et al. (2010)	–	–
		2α,3β,23-Trihydroxyurs-12,18-dien-28-oic acid 28-O-β-glucopyranosyl ester		–	–
		Qudranoside VIII		10675744	1077634
		Kajiichigoside F1		14019178	4128666
		2α,3β,23-Trihydroxyurs-23-trihydroxyurs-12,19-dien-28-oic acid 28-O-β-D-glucopyranosyl ester		–	–
	Glycosides	Arjunetin	Row et al. (1970)	–	–
		Arjunoside I, II	Singh et al. (2002 a, b)	21152828	3987111
		Arjunolone	Honda et al. (1976)	–	–
		Arjunolitin	Sharma et al. (1982)	–	–
		Arjunaphthanoloside	Tripathi et al. (1992)	–	–
		Arjunglucoside IV and V, Arjunasides A-E	Ali et al. (2003a, b)	–	–
		Olean-3b, 22-β-diol-12-en-28 b-D-glucopyranosie-oic acid	Wang et al. (2010a)	–	–

(*Continued*)

TABLE 8.1 (*Continued*)
Major Phytochemical Constituents Found in Different Parts of *T. arjuna*

Plant Part	Category of Phytochemicals	Phytoconstituents Present	References	PubChem ID	Chembl ID
		Terminarjunoside I and II	Patnaik et al. (2007)	102146986	–
		Terminoside A	Alam et al. (2008)	–	–
		Termionic acid	Ahmad et al. (1983)	–	–
	Flavonoids and phenolics	Arjunone	Sharma et al. (1982)	14034821	–
		Luteolin	Pettit et al. (1996)	5280445	151
		Baicalein	Anonymous (1999)	5281605	8260
		Ethyl gallate		13250	453196
		Gallic acid		370	288114
		Kempferol		5280863	150
		Oligomeric proanthocyanidins		122173182	–
		Pelargonidin		440832	–
		Quercetin		5280343	–
		(β)-catechin,	Saha and Pawar (2012)	9064	–
		(β)-gallocatechin and (−)-epigallocatechin		65084	
		Gallic acid, ellagic acid and its derivatives such		370	–
		as 3-O-methyl-ellagic acid 4-O-β-D-xylopyrano		5281855	–
		side,3-O-methyl ellagic acid 3-O-rhamnoside		–	–
				–	–

(*Continued*)

TABLE 8.1 (*Continued*)
Major Phytochemical Constituents Found in Different Parts of *T. arjuna*

Plant Part	Category of Phytochemicals	Phytoconstituents Present	References	PubChem ID	Chembl ID
		3-O-methyl ellagic acid 4-O-α-L-rhamnophranoside	Wang et al. (2010)	–	–
		(−)-epicatechin		72276	583912
	Tannin	Pyrocatechols	Takahashi et al. (1997)	289	280998
		Punicallin	Lin et al. (2000)	5388496	416615
		Castalagin	Kuo et al. (2005)	168165	607711
		Casuariin		9859098	509562
		Casuarinin		442673	4168974
		Punicalagin		16129869	–
		Terchebulin		16175789	–
		Terflavin C		101589227	–
	Minerals and trace elements	Calcium, magnesium, aluminum, zinc, copper, silica	Dwivedi and Udupa (1989)	–	–
	Other compounds	β-Sitosterol	Anjaneyulu and Prasad (1983)	222284	221542
Roots	Triterpenoids	Arjunoside I-IV	Anjaneyulu and Prasad (1982a, b)	–	–
		Arjunolic acid	Anjaneyulu and Prasad (1983)	–	–
		Oleanolic acid		–	–
		Terminic acid		–	–
		2a,19α-Dihydroxy-3Oxo-olean-12-en28-olic acid 28-O-β-D-glucopyranoside	Chouksey and Srivastava (2001)	–	–
		Arjunic acid	Singh et al. (2002a, b)	–	–

(*Continued*)

TABLE 8.1 (*Continued*)
Major Phytochemical Constituents Found in Different Parts of *T. arjuna*

Plant Part	Category of Phytochemicals	Phytoconstituents Present	References	PubChem ID	Chembl ID
	Glycosides	Arjunetosie (3-O-β-D-glucopyranosyl-2α,3β,19α-trihydroxyolean-12-en-28-oic acid 28-O-β-D-glucopyranoside)	Upadhyay et al. (2001)	–	–
Fruits	Triterpenoids and flavonoids	Arjunic acid, arjunone, arachidic stearate, cerasidin, ellagic acid, fridelin, gallic acid, hentriacontane, methyl oleaolate, myristyloleate, b-sitisterol	Rastogi and Mehrotra (1993)	–	–
Leaves and seeds	Flavonoids and glycosides	Luteolin, 14,16-dianhydrogitoxigenin 3-b-D-xylopyranosyl-(1>2)-O-β-D-galactopyranoside	Pettit et al. (1996) Yadava and Rathore (2000)	–	–

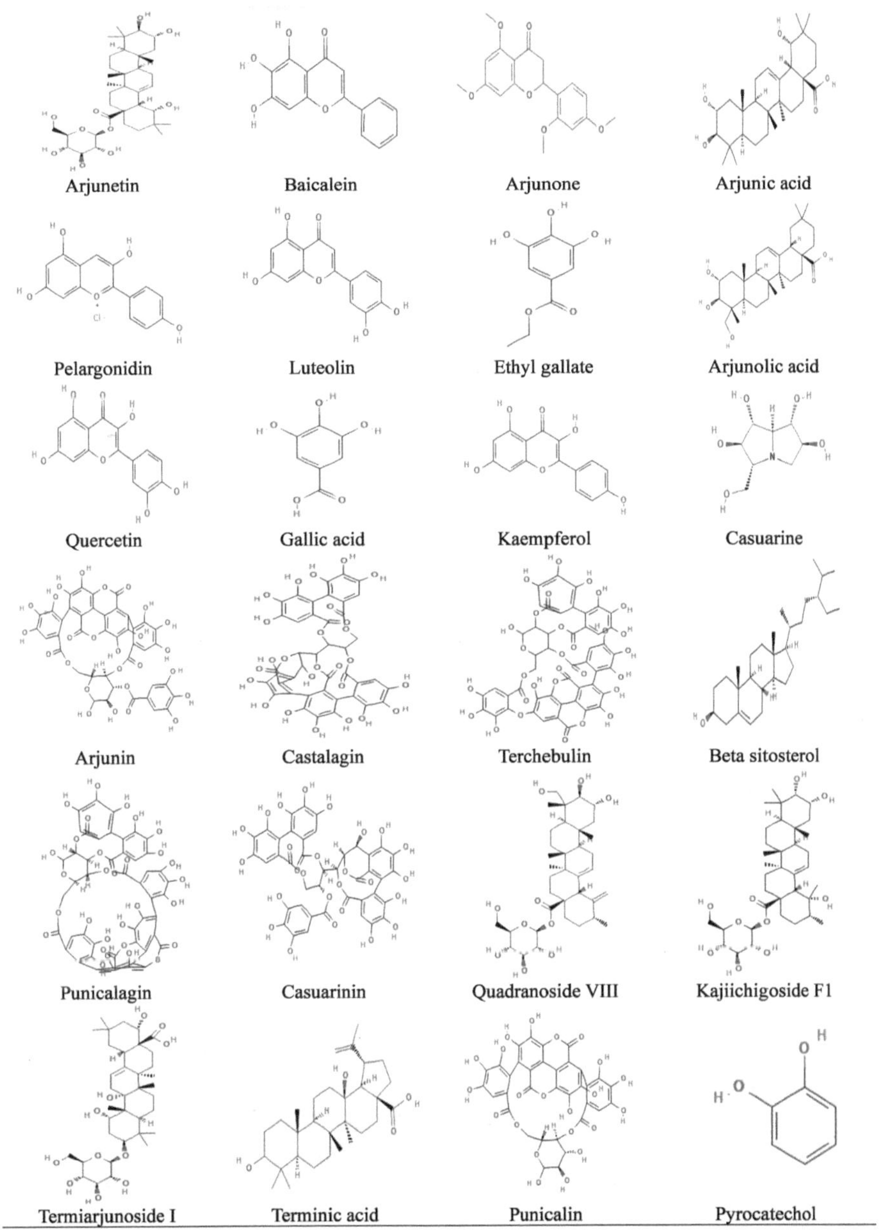

FIGURE 8.3 2-D structures of important phytochemical compounds of *T. arjuna*.

drug discovery as shown in Figure 8.4. Computational/*in silico* approaches can be a powerful method for the selection of a plant species/main chemical compound for the rationalization of bioactivity in the generation of new drug leads (Atanasov et al., 2015).

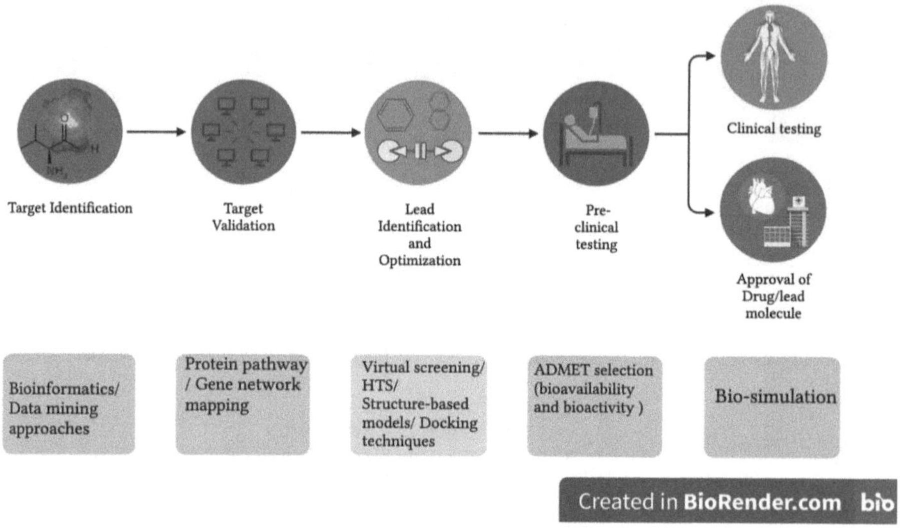

FIGURE 8.4 Modern drug discovery and development pathway (created by BioRender, https://www.biorender.com).

In recent years, several in silico studies have also been carried out to find targets for some phytochemicals of *T. arjuna*. The key targets identified by these studies are PDE5A, Na-K pump, ACE, human angiotensin receptor, P38 MAPK, HMGCR, β1AR, and β2AR, human C-reactive protein, COX-2, APOE4, viral proteases, SARS-CoV-2 3CL-protease, PL-protease and SARs-CoV-2 RdRP polymcrase, DPP-IV receptor, and Tyrosinase (as shown in Table 8.2). The renin–angiotensin system plays a pivotal role in regulating cardiovascular functions in which angiotensin-converting enzymes and angiotensin receptors act as important components of this system. Angiotensin-converting enzymes and angiotensin receptors can be targeted to treat hypertension and CVD progression. Additionally, the accumulation of another target protein, APOE4 (a plasma glycoprotein), increases cholesterol production and leads to atherosclerosis followed by coronary artery disease. Phosphodiesterase 5A (PDE5A) regulates calcium levels and vascular smooth muscle tone. Na-K pump acts as an important target protein to treat heart failure. HMGCR, one of the key enzymes of cholesterol biosynthesis, P38 MAPK, β-Adrenergic (a G protein-coupled receptor) with regulatory functions in the cardiovascular system play a crucial role via signaling (Ali et al., 2020). C-reactive protein, an inflammatory mediator, and COX-2 are known as potent risk indicators of CVDs. The inhibitory activity of DPP-IV-Dipeptidyl peptidase-IV enzyme can be targeted for cardioprotective effects (Mohanty et al., 2019).

Target identification is the first step in the drug discovery and development process, and several steps need to be followed to successfully develop the drug and get its approval. For all the compounds, mentioned in Table 8.2, having specific targets

TABLE 8.2
Few Important Examples of Lead Compound Discovery for Cardiovascular Targets from *T. arjuna* Using *in silico* Studies

Compound	Target Proteins	References
Casuarinin, castalagin, arjunin, quercetin, kaempferol, luteolin, punicalagin and pelargonidin	PDE5A, Na-K pump and βAR	Gaikwad and Jadhav (2021)
Arjunic acid, arjungenin and terminic acid	ACE, human angiotensin receptor, P38 MAPK, HMGCR, β1AR and β2AR, human C-reactive protein, COX-2	Kumar et al. (2020)
Arjunolic acid	APOE4	Hazarika et al. (2021)
Arjunetin	viral proteases	Arumugam et al. (2020)
Arjunetin	SARS-CoV-2 3CL-protease, PL-protease and SARs-CoV-2 RdRP polymerase	Arumugam et al. (2020)
Arjunetin, arjungenin, ellagic acid and arjunic acid	DPP-IV receptor	Mohanty et al. (2019)
Quadranoside VIII, gallic acid, ethyl gallate, pelargonidin	Tyrosinase	Gaikwad and Jadhav (2019)

PDE5A, phosphodiesterase 5A; ACE, angiotensin converting enzyme; DPP-IV, dipeptidyl peptidase-IV; APOE4, apolipoprotein-E4; MAPK P38, P38 mitogen activated protein kinase; HMGCR-3, hydroxy-3-methylglutaryl-coenzyme A reductase; β1AR and β2AR, β-adrenergic receptors.

related to CVDs, all the subsequent steps in drug discovery and development are yet to be done. Further, the other compounds of *T. arjuna*, as given in Figure 8.3, also need to be subjected to exhaustive target prediction in the human proteome.

9.7 FUTURE PROSPECTS

Considering the ethnomedicinal information and scientific evidence supporting the therapeutic value of *T. arjuna* for CVDs, the prospects of drug discovery and development seem very high. Recent advancements in the development of various in silico and molecular tools expand the spectrum of prospects. For example, statistically estimating computational methods based on predictive models are available for exploring the extent of binding affinity between compounds and their targets (Rifaioglu et al., 2019). For the development of new drugs for final product release, target prediction is not sufficient and drug formulation properties of the identified compound also need to be evaluated. Other important parameters of the compound, playing key roles in drug development, are chemical absorption, distribution, metabolism, excretion, and toxicity (ADMET). Several plant-based databases have been developed for imparting information related to the bioactivity of important metabolites, pharmacological utility, physiological data, ethnobotany, molecular information, active targets concerning a particular compound, genomics, and transcript-based data, and much more. In silico methods for drug formulations comprise various databases, and tools for ADMET predictions of phytomedicines, QSAR models, pharmacophore-based virtual screening (Rifaioglu et al., 2019). Applications of the above-mentioned tools to the secondary metabolome of *T. arjuna*, in a systematic and planned way, offer high stakes for the discovery of some novel drug(s) and/or strategies for controlling as well as treating various forms of CVDs.

9.8 CONCLUSION

The growing interest in green medicine worldwide has made the pharmaceutical industry pay attention to medicinal plants for desired utilization. At present, we are at a very advanced development stage in terms of molecular and technical research perspective, and therefore the speed and accuracy of finding and formulating drugs from natural resources can be amplified many times if planned properly. The traditional knowledge base from ancient times can be used in drug designing for modern pharmaceutics, but the major problem is an unstructured and scattered form of information related to important phytochemicals which are embedded in literature. Therefore, it becomes essential to adopt some data mining methodologies for the extraction, storage, and presentation of important formats. Further, the organization of all the available information generated out of the research efforts by different scientific groups worldwide is equally important to develop an overall picture and to identify the goals in the desired direction. The information organized above clearly identifies the gaps toward desired utilization of this plant and proposes various ways to fill the gaps. Exploration and scientific research on the therapeutic potential of phytochemicals present in Arjuna is very much required to understand biosynthetic pathways, molecular mechanisms, drug-target interactions and other toxicological studies.

REFERENCES

Aghera, H., Bundela, V., Shukla, V. J., Prajapati, P. K., & Naira, M. B. (2015). A notable review on *Terminalia arjuna* and its imperative ayurvedic formulations: An overview. *International Ayurvedic Medical Journal Research*, *3*, 1814–1821.

Ahmad, M. U., Mullah, K. B., Norin, T., & Ulla, J. K. (1983). Terminoic acid, A new trihydroxytriterpene carboxylic-acid from bark of Terminalia-arjuna. *Indian Journal of Chemistry Section B-organic Chemistry Including Medicinal Chemistry*, *22*(8), 738–740

Alam, M., Kaur, G., Ali, A., Hamid, H., Ali, M., & Athar, M. (2008). Two new bioactive oleanane triterpene glycosides from *Terminalia arjuna. Natural Product Research*, *22*(14), 1279–1288.

Ali, A., Kaur, G., Hamid, H., Abdullah, T., Ali, M., Niwa, M., & Alam, M. S. (2003). Terminoside A, a new triterpene glycoside from the bark of *Terminalia arjuna* inhibits nitric oxide production in murine macrophages. *Journal of Asian Natural Products Research*, *5*(2), 137–142.

Ali, A., Kaur, G., Hayat, K., Ali, M., & Ather, M. (2003). A novel naphthanol glycoside from *Terminalia arjuna* with antioxidant and nitric oxide inhibitory activities. *Die Pharmazie*, *58*(12), 932–934.

Ali, D. C., Naveed, M., Gordon, A., Majeed, F., Saeed, M., Ogbuke, M. I., … & Changxing, L. (2020). β-Adrenergic receptor, an essential target in cardiovascular diseases. *Heart Failure Reviews*, *25*, 343–354.

Anjaneyulu, A. S. R., & Prasad, A. R. (1982). Chemical examination of roots of *Terminalia arjuna* (Roxb) Wight and Arnot. Characterization of two new triterpenoid glycosides. *Indian Journal of Chemistry Section B-organic Chemistry Including Medicinal Chemistry*, *21*(6), 530–533.

Anjaneyulu, A. S. R., & Prasad, A. R. (1982). Chemical examination of the roots of *Terminalia arjuna*—the structures of arjunoside III and arjunoside IV, two new triterpenoid glycosides. *Phytochemistry*, *21*(8), 2057–2060.

Anjaneyulu, A. S. R., & Prasad, A. R. (1983). Structure of terminic acid, a dihydroxytriterpene carboxylic acid from *Terminalia arjuna*. *Phytochemistry*, *22*(4), 993–998.

Anonymous., 1999. *Terminalia arjuna*. *Alternative Medicine Review*, *4*, 436–437.

Arman, R. S., & Mahtab, A. S. (2013). Unani drugs in the management of cardiovascular diseases: A review on pharmacological studies on Arjun and Zafran. *International Journal of Research in Ayurveda and Pharmacy (IJRAP)*, *4*(3), 463–467, ref. 25.

Arumugam, S., Dash, S. S., Mitra, K., Doble, M., & Gummadi, S. N. (2020). Potential of Terminalia arjuna as a promising PhytoremedyAgainst COVID-19: DPPH scavenging, catalase inhibition and molecular docking studies. doi:10.26434/chemrxiv.12600587.v1.

Atanasov, A. G., Waltenberger, B., Pferschy-Wenzig, E. M., Linder, T., Wawrosch, C., Uhrin, P., … & Stuppner, H. (2015). Discovery and resupply of pharmacologically active plant-derived natural products: A review. *Biotechnology Advances*, *33*(8), 1582–1614.

Bharani, A., Ganguly, A., & Bhargava, K. D. (1995). Salutary effect of *Terminalia arjuna* in patients with severe refractory heart failure. *International Journal of Cardiology*, *49*(3), 191–199.

Bhat, O. M., Kumar, P. U., Rao, K. R., Ahmad, A., & Dhawan, V. (2018). *Terminalia arjuna* prevents Interleukin-18-induced atherosclerosis via modulation of NF-κB/PPAR-γ-mediated pathway in Apo E−/− mice. *Inflammopharmacology*, *26*, 583–598.

Bhatia, J., Bhattacharya, S. K., Mahajan, P., Dwivedi, S., 1998. Effect of *Terminalia arjuna* on coronary flow: An experimental study (Abstract). *Indian Journal of Pharmacology*, *30*, 118–120.

Chaudhary, S. P., Kirtika, S. A., & Ram, B. (2017). An Ayurvedic review of Arjuna from Samhita. *International Journal of Research in Ayurveda and Pharmacy*, *8*, 1–9.

Chou, C. Y., Jen, W. P., Hsieh, Y. H., Shiao, M. S., & Chang, G. G. (2006). Structural and functional variations in human apolipoprotein E3 and E4. *Journal of Biological Chemistry*, *281*(19), 13333–13344.

Chouksey, B. K., & Srivastava, S. K. (2001). Antifungal agent from *Terminalia arjuna*. *Indian Journal Chemistry B*, *40*, 354–6.

Cock, I. E. (2015). The medicinal properties and phytochemistry of plants of the genus Terminalia (Combretaceae). *Inflammopharmacology*, *23*, 203–229.

Dwivedi, S., & Kumar, V. (2007). Beta-thalassemia, hyperlipoproteinemia (a), and metabolic syndrome: Its low-cost holistic therapy. *The Journal of Alternative and Complementary Medicine*, *13*(2), 287–290.

Dwivedi, S., & Udupa, N. (1989). *Terminalia arjuna*: Pharmacognosy, phytochemistry, pharmacology and clinical use: A review. *Fitoterapia*, *60*(5), 413–420.

Dwivedi, S., Singh, S., Dwivedi, M., & Aggarwal, A. (2005, November). Effect of *Terminalia arjuna* on lipoprotein (a) in patients of coronary artery disease. In: *Proceedings of the 15th Annual Conference of the Indian Society of Hypertension Abstract Book*, UCMS-GTB Hospital, Delhi, India (pp. 12–13).

Gaikwad, D. T., & Jadhav, N. R. (2021). Discovery of potential inhibitors for phosphodiesterase 5A, sodium-potassium pump and beta-adrenergic receptor from *Terminalia arjuna*: In silico approach. *Journal of Biomolecular Structure and Dynamics*, *39*(5), 1754–1765.

Gaikwad, D., & Jadhav, N. (2019). Development of stable emulsified formulations of *Terminalia arjuna* for topical application: Evaluation of antioxidant activity of final product and molecular docking study. *Drug Development and Industrial Pharmacy*, *45*(11), 1740–1750.

Halder, S., Bharal, N., Mediratta, P. K., Kaur, I., & Sharma, K. K. (2009). Anti-inflammatory, immunomodulatory and antinociceptive activity of *Terminalia arjuna* Roxb bark powder in mice and rats. *Indian Journal of Experimental Biology*, *47*, 577–583.

Hazarika, L., Sen, S., Doshi, J., 2021. Molecular docking analysis of arjunolic acid from *Terminalia arjuna* with a coronary artery disease target APOE4. *Bioinformation*, *17*(11), 949–958.

Honda, T., Murae, T., Tsuyuki, T., & Takahashi, T. (1976). The structure of arjungenin. A new sapogenin from *Terminalia arjuna*. *Chemical and Pharmaceutical Bulletin*, *24*(1), 178–180.

Hong, C. C. (2022). The grand challenge of discovering new cardiovascular drugs. *Frontiers in Drug Discovery*, *2*, 1027401.

Kapoor, D., Vijayvergiya, R., & Dhawan, V. (2014). *Terminalia arjuna* in coronary artery disease: Ethnopharmacology, pre-clinical, clinical & safety evaluation. *Journal of Ethnopharmacology*, *155*(2), 1029–1045.

Kokkiripati, P. K., Kamsala, R. V., Bashyam, L., Manthapuram, N., Bitla, P., Peddada, V., … & Tetali, S. D. (2013). Stem-bark of *Terminalia arjuna* attenuates human monocytic (THP-1) and aortic endothelial cell activation. *Journal of Ethnopharmacology*, *146*(2), 456–464.

Kumar, V., Sharma, N., Sourirajan, A., Khosla, P. K., & Dev, K. (2020). Docking studies in targeting proteins involved in cardiovascular disorders using phytocompounds from *Terminalia arjuna*. *BioRxiv*. https://doi.org/10.1101/2020.06.22.164129.

Kuo, P. L., Hsu, Y. L., Lin, T. C., Lin, L. T., Chang, J. K., & Lin, C. C. (2005). Casuarinin from the bark of *Terminalia arjuna* induces apoptosis and cell cycle arrest in human breast adenocarcinoma MCF-7 cells. *Planta Medica*, *71*(03), 237–243.

Leroi-Gourhan, A. (1975). The flowers found with Shanidar IV, a Neanderthal burial in Iraq. *Science*, *190*(4214), 562–564.

Lin, T. C., Chien, S. C., Chen, H. F., & Hsu, F. L. (2000). Tannins and related compounds from Combretaceae plants. *Chinese Pharmaceutical Journal*, *52*(1), 1–26.

McGaw, L. J., Rabe, T., Sparg, S. G., Jäger, A. K., Eloff, J. N., & Van Staden, J. (2001). An investigation on the biological activity of Combretum species. *Journal of Ethnopharmacology*, *75*(1), 45–50.

Meena, D. K., Sahoo, A. K., Srivastava, P. P., Sahu, N. P., Jadhav, M., Gandhi, M., … & Das, B. K. (2021). On valorization of solvent extracts of *Terminalia arjuna* (arjuna) upon DNA scission and free radical scavenging improves coupling responses and cognitive functions under in vitro conditions. *Scientific Reports*, *11*(1), 10656.

Meghwani, H., Prabhakar, P., Mohammed, S. A., Seth, S., Hote, M. P., Banerjee, S. K., … & Maulik, S. K. (2017). Beneficial effects of aqueous extract of stem bark of *Terminalia arjuna* (Roxb.), An ayurvedic drug in experimental pulmonary hypertension. *Journal of Ethnopharmacology*, *197*, 184–194.

Mishra, K., & Baarange, V. (2018). An ayurvedic compilation of arjuna from vedas to nighantus. *International Ayurvedic Medical Journal* *6*(8), 1699–1705.

Mohanty, I. R., Borde, M., & Maheshwari, U. (2019). Dipeptidyl peptidase IV Inhibitory activity of *Terminalia arjuna* attributes to its cardioprotective effects in experimental diabetes: In silico, in vitro and in vivo analyses. *Phytomedicine*, *57*, 158–165.

Mukherjee, P. K., & Wahile, A. (2006). Integrated approaches towards drug development from Ayurveda and other Indian system of medicines. *Journal of Ethnopharmacology*, *103*(1), 25–35.

Newman, D. J., & Cragg, G. M. (2020). Natural products as sources of new drugs over the nearly four decades from 01/1981 to 09/2019. *Journal of Natural Products*, *83*(3), 770–803.

Parveen, A., Babbar, R., Agarwal, S., Kotwani, A., & Fahim, M. (2012). *Terminalia arjuna* enhances baroreflex sensitivity and myocardial function in isoproterenol-induced chronic heart failure rats. *Journal of Cardiovascular Pharmacology and Therapeutics*, *17*(2), 199–207.

Patnaik, T., Dey, R. K., & Gouda, P. (2007). Isolation of triterpenoid glycoside from bark of *Terminalia arjuna* using chromatographic technique and investigation of pharmacological behavior upon muscle tissues. *Journal of Chemistry*, *4*, 474–479.

Patwardhan, B. (2005). Ethnopharmacology and drug discovery. *Journal of Ethnopharmacology*, *100(1–2)*, 50–52.

Pawar, R. S., & Bhutani, K. K. (2005). Effect of oleanane triterpenoids from *Terminalia arjuna*—a cardioprotective drug on the process of respiratory oxyburst. *Phytomedicine*, *12*(5), 391–393.

Pettit, G. R., Hoard, M. S., Doubek, D. L., Schmidt, J. M., Pettit, R. K., Tackett, L. P., & Chapuis, J. C. (1996). Antineoplastic agents 338. The cancer cell growth inhibitory. Constituents of *Terminalia arjuna* (Combretaceae). *Journal of Ethnopharmacology*, *53*(2), 57–63.

Pourbagher-Shahri, A. M., Farkhondeh, T., Ashrafizadeh, M., Talebi, M., & Samargahndian, S. (2021). Curcumin and cardiovascular diseases: Focus on cellular targets and cascades. *Biomedicine & Pharmacotherapy*, *136*, 111214.

Radhakrishnan, R., Wadsworth, R. M., & Gray, A. I. (1993). *Terminalia arjuna*, an Ayurvedic cardiotonic, increases contractile force of rat isolated atria. *Phytotherapy Research*, *7*(3), 266–268.

Rao, B. C. S., Singh, R. H., & Tripathi, K. (2001). Effect of *Terminalia arjuna* (W&A) on regression of LVH in hypertensives: A clinical study. *Journal of Research in Ayurveda and Siddha*, *22(3–4)*, 216–227.

Rastogi, R. P., Mehrotra, B. N., 1993. *Compendium of Indian Medicinal Plants*, vol. *3*. CSIR, New Delhi.

Rifaioglu, A. S., Atas, H., Martin, M. J., Cetin-Atalay, R., Atalay, V., & Doğan, T. (2019). Recent applications of deep learning and machine intelligence on in silico drug discovery: Methods, tools and databases. *Briefings in Bioinformatics*, *20*(5), 1878–1912.

Row, L. R., Murty, P. S., Rao, G. S., Sastry, C. P., & Rao, K. V. J. (1970). Chemical examination of Terminalia species. XII. Isolation & structure determination of arjunic acid, a new trihydroxytriterpene carboxylic acid from *Terminalia arjuna* bark. *Indian Journal of Chemistry*, *8*(9), 772–775.

Saha, A., Pawar, V. M., & Jayaraman, S. (2012). Characterisation of polyphenols in *Terminalia arjuna* bark extract. *Indian Journal of Pharmaceutical Sciences*, *74*(4), 339.

Shah, S. M. A., Akram, M., Riaz, M., Munir, N., & Rasool, G. (2019). Cardioprotective potential of plant-derived molecules: A scientific and medicinal approach. *Dose-Response*, *17*(2), 1559325819852243.

Sharma, P. N., Shoeb, A. B. O. O., Kapil, R. S., & Popli, S. P. (1982). Arjunolone–a new flavone from stem bark of *Terminalia arjuna*. *Indian Journal of Chemistry B*, *21*, 263.

Singh, B., Singh, V. P., Pandey, V. B., & Rücker, G. (1995). A new triterpene glycoside from *Terminalia arjuna*. *Planta Medica*, *61*(06), 576–577.

Singh, D. V., Verma, R. K., Gupta, M. M., & Kumar, S. (2002a). Quantitative determination of oleane derivatives in *Terminalia arjuna* by high-performance thin layer chromatography. *Phytochemical Analysis: An International Journal of Plant Chemical and Biochemical Techniques*, *13*(4), 207–210.

Singh, D. V., Verma, R. K., Singh, S. C., & Gupta, M. M. (2002b). RP-LC determination of oleane derivatives in Terminalia arjuna. *Journal of Pharmaceutical and Biomedical Analysis, 28(3–4)*, 447–452.

Singh, G., Singh, A. T., Abraham, A., Bhat, B., Mukherjee, A., Verma, R., … & Burman, A. C. (2008). Protective effects of *Terminalia arjuna* against Doxorubicin-induced cardiotoxicity. *Journal of Ethnopharmacology, 117*(1), 123–129.

Sneader, W. (2005). *Drug Discovery: A History*. John Wiley & Sons, New York.

Subramaniam, S., Ramachandran, S., Uthrapathi, S., Gnamanickam, V. R., & Dubey, G. P. (2011). Anti-hyperlipidemic and antioxidant potential of different fractions of *Terminalia arjuna* Roxb. bark against PX-407 induced hyperlipidemia. *Indian Journal of Experimental Biology, 49*, 282–288.

Süntar, I. (2020). Importance of ethnopharmacological studies in drug discovery: Role of medicinal plants. *Phytochemistry Reviews, 19*(5), 1199–1209.

Tahsin, M. R., Sultana, A., Khan, M. S. M., Jahan, I., Mim, S. R., Tithi, T. I., … & Aktar, F. (2021). An evaluation of pharmacological healing potentialities of *Terminalia arjuna* against several ailments on experimental rat models with an in-silico approach. *Heliyon, 7*(11), e08225.

Takahashi, S., Tanaka, H., Hano, Y., Ito, K., Nomura, T., & Shigenobu, K. (1997). Hypotensive effect in rats of hydrophilic extract from *Terminalia arjuna* containing tannin-related compounds. *Phytotherapy Research: An International Journal Devoted to Medical and Scientific Research on Plants and Plant Products, 11*(6), 424–427.

Tripathi, V. K., Pandey, V. B., Udupa, K. N., & Ru, G. (1992). Arjunolitin, a triterpene glycoside from *Terminalia arjuna. Phytochemistry, 31*(1), 349–351.

Tsao, C. W., Aday, A. W., Almarzooq, Z. I., Alonso, A., Beaton, A. Z., Bittencourt, M. S., … & American Heart Association Council on Epidemiology and Prevention Statistics Committee and Stroke Statistics Subcommittee. (2022). Heart disease and stroke statistics—2022 update: A report from the American Heart Association. *Circulation, 145*(8), e153–e639.

Unschuld, P. U. (1986). *Medecine in China: A History of Pharmaceutics*. University of California Press.

Upadhyay, R. K., Pandey, M. B., Jha, R. N., Singh, V. P., & Pandey, V. B. (2001). Triterpene glycoside from *Terminalia arjuna. Journal of Asian Natural Products Research, 3*(3), 207–212.

Verma, T., Sinha, M., Bansal, N., Yadav, S. R., Shah, K., & Chauhan, N. S. (2021). Plants used as antihypertensive. *Natural Products and Bioprospecting, 11*, 155–184.

Virani, S. S., Alonso, A., Benjamin, E. J., Bittencourt, M. S., Callaway, C. W., Carson, A. P., … & American Heart Association Council on Epidemiology and Prevention Statistics Committee and Stroke Statistics Subcommittee. (2020). Heart disease and stroke statistics—2020 update: A report from the American Heart Association. *Circulation, 141*(9), e139–e596.

Wang, W., Ali, Z., Li, X. C., Shen, Y., & Khan, I. A. (2010). Triterpenoids from two Terminalia species. *Planta Medica, 76*(15), 1751–1754.

Wang, W., Ali, Z., Shen, Y., Li, X. C., & Khan, I. A. (2010a). Ursane triterpenoids from the bark of *Terminalia arjuna. Fitoterapia, 81*(6), 480–484.

WHO Fact sheet on Cardiovascular Diseases (CVDs), Updated June 2021, https://www.who.int/news-room/fact-sheets/detail/cardiovascular-diseases-(cvds) (assessed on 15 December, 2023)

Yadava, R. N., & Rathore, K. (2000). A new cardenolide from the seeds of *Terminalia arjuna* (W&A). *Journal of Asian Natural Products Research, 2*(2), 97–101.

Zhang, X. R., Kaunda, J. S., Zhu, H. T., Wang, D., Yang, C. R., & Zhang, Y. J. (2019). The genus Terminalia (Combretaceae): An ethnopharmacological, phytochemical and pharmacological review. *Natural Products and Bioprospecting, 9*, 357–392.

9 Nutrigenomics and Personalized Diet

Recent Advancements and Applications

Sanju Bala Dhull, Jyoti Rani, Pawan Kumar Rose, Mohd. Kashif Kidwai, Anita Singh Kirrolia, Gulden Goksen, and Natoavina Faliarizao

9.1 INTRODUCTION

Food holds a multifaceted role beyond being a fundamental energy source; it embodies cultural significance and serves as a focal point for social interactions. Its influence, along with the environment, on human health or illness, cannot be underestimated (Ronteltap et al., 2008). The World Health Organization (WHO) emphasizes the crucial role of maintaining an active lifestyle and embracing a balanced diet in preventing noncommunicable chronic diseases, such as diabetes, cardiovascular diseases, and cancers, which account for over 70% of global mortality (WHO, 2008a). A well-structured diet can significantly reduce disease risk, making it a primary preventive measure. Numerous dietetics studies have enhanced our understanding of maintaining health across diverse populations (Cozzolino and Cominetti, 2013). Success hinges on recognizing genetic variability in food metabolism, necessitating tailored dietary approaches aligned with individual health conditions and metabolism (Figure 9.1).

The Human Genome Project (HGP) has illuminated the profound influence of dietary nutrition on human health, revealing the complex interplay among genetic composition, metabolic processes, and environment. Nutrigenomics, coined in light of this endeavor, deepens our understanding of these dynamics (Sanhueza and Valenzuela, 2012). Nutrigenomics studies primarily delve into the ways nutrients present in food influence the metabolome, genome, and proteome. This approach explores gene-nutrient interactions at cellular levels, revealing diverse responses and supporting personalized nutrition tailored by sex, age, and pregnancy (Lewis and Dwyer, 2020). Cutting-edge technology enhances nutrigenetics, facilitating personalized approaches that consider individual genetic variances (Betts and Gonzalez, 2016). Integrating genetics into personalized nutrition holds promise but faces

DOI: 10.1201/9781032642567-11

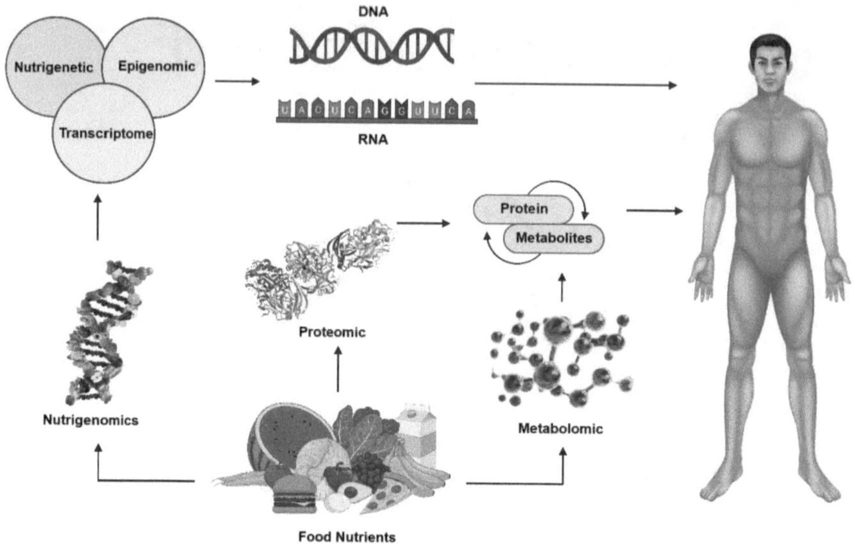

FIGURE 9.1 Relationship between nutrition, health, and disease.

challenges in translating scientific progress into practical strategies by addressing diet-related health issues and regulating dietary patterns, as well as supporting preventive measures against chronic diseases.

9.2 THE CONVERGENCE OF NUTRITION AND GENOMICS

Genetic variations shape how our bodies absorb and process nutrients, influencing our nutritional status and overall health. This interaction is termed nutritional genomics. Hippocrates, around 400 BC, proposed that maintaining optimal body temperature is crucial for food metabolism. In the 18th century, Lavoisier contributed to understanding food metabolism, while Liebig in the 19th century linked high body temperature maintenance to macronutrients (Gangotia et al., 2021). The period between 18th and 20th centuries marked a significant advancement in chemical and analytical nutrition, ushering in a new era of understanding in this field.

Antoine Lavoisier's research linked food metabolism to breathing and energy. The 19th century shifted focus to nutrition's role in preventing chronic diseases like cardiovascular conditions, cancer, and neurodegenerative disorders (Leclair, 2023). In the post-genomic era, there's a focus on understanding the intricate relationship between nutrients and the genome. Nutritional genomics encompasses nutrigenetics and nutrigenomics. Nutrigenomics investigates how dietary nutrients or bioactive compounds influence protein levels, metabolite quantities, metabolism, gene expression, and subsequently affect disease risk and overall health. An exemplification of nutrigenomics can be found in the presence of resveratrol in contemporary health foods like red wine. Resveratrol in red wine activates genes protecting tissues from free radical damage. Folate in green vegetables and fruits is vital for diet, genetics, and DNA synthesis. Nutrients like folate directly control gene expression as nuclear transcription factors and indirectly through DNA methylation and miRNA activity (Mierziak et al., 2021).

9.3 SIGNIFICANCE OF PERSONALIZED NUTRITION

Dietitians worldwide are increasingly integrating genetic testing, specifically nutrigenomics, to customize dietary recommendations based on individual genetic makeup. Personalized nutrition employs methodologies like epigenetics, blood biomarkers, proteins, genetics, metabolites, and gut microbiome analysis to enhance health by tailoring dietary plans to unique individual needs (Mozaffarian, 2016). Nutritional guidance is contingent upon factors such as age, physiological status, and the presence of chronic diseases, necessitating specialized diets in specific cases (Roberts and Barnard, 2005). Studies show that customizing nutrition based on genetic profiles significantly impacts the prevention and management of cardiovascular diseases, diabetes, obesity, and certain cancers. Personalized nutrition strategies optimize nutrient intake, targeting specific risk factors and considering individual nutrient absorption, metabolism, and utilization for better health outcomes.

9.4 FOUNDATION OF NUTRIGENOMICS

Nutrigenomics, merging nutrition, genetics, and genomics, explores genetic makeup's interaction with dietary response, unveiling how genetic variations influence nutritional needs, metabolism, and overall health. This knowledge forms the basis for personalized dietary recommendations, enabling targeted strategies to optimize health and manage diseases. With the potential to revolutionize nutrition and healthcare, nutrigenomics tailors dietary guidance to genetic predispositions, advocating for a precise, personalized well-being approach (Caradonna et al., 2020). Nutrigenomics, akin to pharmacogenomics but with natural dietary bioactive agents, tailors dietary advice to genetic predispositions, fostering a proactive health approach and promising advancements in chronic disease prevention by unveiling genetic-dietary interplay.

9.4.1 NUTRITION AND GENETIC VARIATION

Genes and genetic variation are central to nutrigenomics, exploring the intricate interplay between an individual's genetic makeup and their response to dietary components. Genetic variation influences how the body metabolizes nutrients and responds to diet, pivotal in understanding nutritional needs, metabolism, and disease susceptibility. This understanding is crucial in unraveling dietary nutrition's impact on metabolic pathways in nutrigenomics (Lal et al., 2022). Research within this field is dedicated to identifying how genes contribute to the development of diet-related diseases, elucidating the underlying mechanisms that constitute these genetic predispositions. Advancements in "omic" technologies, rooted in a deep comprehension of nutrition, genetics, and biochemistry, are propelling research and insights in this fascinating field.

9.4.2 Genomic Variation and Diet Response

Genomic variation encompasses genetic variances affecting how individuals react to diet and nutrition, impacting nutrient metabolism and overall health. In nutrigenomics, tailoring dietary recommendations to individual genetic composition is crucial for optimizing health outcomes, requiring a nuanced understanding of genomic diversity. Genomic investigations have revealed correlations between genetic variations and food responses, such as lactase gene variants influencing dairy consumption. Similarly, Goni et al. (2015) studied the genetic changes in the FTO gene can affect and the extent to which dietary interventions work to manage weight. Additionally, a person's susceptibility to several diseases linked to nutrition might be affected by chromosomal differences. The significance of individualized dietary approaches in disease prevention was highlighted by Kawaguchi et al. (2018) who demonstrated how chromosomal variants in the PNPLA3 gene are connected with the course of fatty liver disease in response to dietary variables.

Understanding genetic diversity enables the creation of personalized nutritional advice. Nutrigenomics considers genetic variation in detecting gene-nutrient interactions, and tailoring dietary programs for optimal health outcomes. Genetic variation significantly influences how individuals respond to diet and nutrition, and personalized dietary advice offered by nutrigenomics enables more effective methods for enhancing health and preventing diet-related disorders. A number of aspects as mentioned in Table 9.1 are to be considered for personalized nutrition as a part of precise nutrition strategy.

TABLE 9.1
Aspects to be Considered for Personalized Nutrition: Precision Nutrition Strategy

Sr. No.	Aspect	Parameters
1.	Genetic makeup	Understanding genetic variations' impact on nutrient metabolism and dietary needs.
2.	Nutrient requirements	Identifying nutrient needs based on age, sex, activity level, and health.
3.	Dietary preferences and restrictions	Considering dietary preferences, allergies, and cultural or ethical restrictions.
4.	Metabolic rate	Determining individual metabolic rate and energy expenditure to tailor calorie and macronutrient intake.
5.	Health conditions	Taking into account existing health conditions, medical history, and their nutritional implications
6.	Behavioral factors	Considering individual behavior, eating habits, and social environment in nutrition planning.
7.	Lifestyle	Assessing daily activity, sedentary behavior, and lifestyle for personalized planning.

9.5 TOOLS AND TECHNOLOGIES

Recent advancements in nutrigenomics and personalized nutrition use high-throughput DNA sequencing to analyze genetic makeup, guiding personalized food recommendations via bioinformatics algorithms. Technology tracks diet and lifestyle for tailor-made plans, empowering healthcare providers to offer precise advice, and optimizing health outcomes by considering genetic composition and lifestyle choices.

9.5.1 DNA Sequencing and Genotyping

In nutrigenomics and personalized nutrition, DNA sequencing and genotyping offer vital insights into individual genetic composition and dietary responses. DNA sequencing examines the genome for variants affecting nutritional metabolism, while genotyping identifies genetic markers linked to food responses and health outcomes, essential for understanding genetic influences on dietary needs and susceptibility to diet-related disorders. The discovery of genetic variances related to nutritional metabolism is paving the way for the realization of personalized nutrition. Moreover, research by Rowland et al. (2018), highlights how genetic data obtained through DNA sequencing and genotyping can guide tailored dietary recommendations. Nutrigenomics, integrating DNA sequencing and genotyping, utilizes an individual's genetic profile to tailor their diet, enhancing overall health and well-being. These technologies offer remarkable potential for personalized dietary approaches, shaping the future of nutrition toward optimized individual health through precise interventions.

9.5.2 Bioinformatics and Data Analysis

Nutrigenomics and personalized nutrition leverage bioinformatics and data analysis to translate large amounts of genomic and nutritional data into actionable insights. High-throughput genetic data undergoes thorough processing and analysis using computational methods, revealing genetic variations impacting individual nutrition responses (Livingstone et al., 2022). Bioinformatics streamlines the analysis of complex genetic data, linking specific genes to dietary requirements and health outcomes. It investigates genetic alterations related to dietary preferences, nutrient metabolism, and disease susceptibility, crucial for making sense of extensive genetic data generated by high-throughput technologies. Using complex algorithms, bioinformatics identifies genetic markers for crafting personalized dietary recommendations aligned with an individual's genetic profile and dietary responses.

Moreover, data analysis technologies facilitate the integration of multi-omics data, encompassing genomics, proteomics, and metabolomics. This integration offers a comprehensive understanding of an individual's biological processes related to nutrition (Mengelkoch et al., 2023). The fusion of bioinformatics and data analysis drives advancements in nutrigenomics and personalized nutrition, facilitating precision dietary approaches tailored to individual genetic and biological makeup, as underscored in numerous published studies. Musa et al. (2018) highlighted the application of advanced computational techniques to analyze genetic and nutritional data, aiding in the discovery of gene-diet correlations crucial for personalized nutrition. Bashiardes et al. (2018) highlighted the importance of bioinformatics in integrating multi-omics data to personalize dietary recommendations based on individual genetic and metabolic profiles. This synergy facilitates understanding the interplay between genetics and nutrition, enhancing nutrigenomics by providing tailored dietary recommendations for optimal health outcomes.

9.5.3 Metabolomics and Proteomics in Nutrigenomics

Metabolomics and proteomics have advanced nutrigenomics by revealing how diet impacts metabolic processes and protein expression. Metabolomics studies metabolites in biological samples, providing insights into an organism's metabolic state (Mishra et al., 2022). This approach aids in identifying metabolic fingerprints linked to dietary changes and genetic variations, deepening our comprehension of how specific foods affect metabolism and health. By identifying metabolic biomarkers predicting food responses, metabolomics guides personalized nutrition based on individual metabolic profiles. Proteomics systematically examines proteins, shedding light on their expression and modifications influenced by diet and genetics, clarifying how genetic composition impacts reactions to food components (Trujillo et al., 2006). The integration of metabolomics and proteomics substantially enriches our comprehension of nutrigenomics, paving the way for tailored dietary recommendations aligned with an individual's unique metabolic and protein expression profiles.

Integrating metabolomics and proteomics is vital for comprehending the interplay among genetics, metabolism, protein expression, and dietary reactions, forming the foundation for individualized dietary recommendations. Combining these data with other omics data and adopting a systems biology approach enables a deeper understanding of the complex relationships among genes, proteins, metabolites, and diet (De Roos and Brennan, 2017). Metabolomics aids in individualized dietary advice by uncovering metabolite profiles linked to food patterns. Integrating proteomics and metabolomics enhances understanding of how dietary interventions influence protein expression and metabolism. In nutrigenomics, merging these disciplines provides insight into how nutrition interacts with genetic makeup, crucial for designing precise dietary recommendations promoting optimal health and disease prevention.

9.6 GENETIC VARIANTS AND DIET RESPONSE

Nutrigenomics investigates how genetic variances influence an individual's response to nutrition, with metabolomics aiding in personalized dietary advice by revealing metabolite profiles linked to food patterns. The integration of proteomics and metabolomics is crucial for understanding how dietary interventions impact protein expression and metabolism. Merging metabolomics and proteomics in nutrigenomics offers insight into how nutrition interacts with genetic makeup, essential for designing precise dietary recommendations promoting optimal health and disease prevention. The Human Genome Project has propelled nutritional genomics by unveiling the intricate connections between genes, nutrition, and diverse health conditions (Chen et al., 2018). Genome sequencing unveiled vast genetic diversity, including millions of single nucleotide polymorphisms (SNPs) associated with dietary effects. These SNPs impact individual responses to nutrition, medication metabolism, environmental factors, and dietary processing (Arballo et al., 2021). The interplay between genes and nutrition, vital for bodily equilibrium via metabolism and gene expression, is complex. Nutritional disorders stem from multiple gene interactions with various nutrients, not solely from a single gene, with distinct nutritional responses influenced by genetic makeup at the core of nutrigenetics.

9.6.1 Nutritional Genes and SNPs

In nutritional genomics, genes and SNPs play crucial roles. SNPs greatly impact genes related to nutrient metabolism, influencing the body's processing and utilization of dietary nutrients. These genes produce enzymes vital for nutrient breakdown, absorption, and utilization, with SNP variations altering enzyme activity (Carey et al., 2022). SNPs possess the potential to enhance or impede enzyme function, thereby directly affecting nutritional metabolism. Research focused on SNPs and nutritional genes has significantly broadened our understanding of how genetic variations can shape an individual's response to diet and susceptibility to nutrition-related ailments. For instance, the APOE gene, which is crucial in lipid metabolism and dietary fat processing, illustrates this impact. Specific SNPs within this gene, such as the APOε4 variant, are associated with variations in cholesterol levels and lipid metabolism. The APOε4 variant is linked to an elevated risk of hypercholesterolemia and cardiovascular disease, especially with high dietary fat intake (Munoz et al., 2006). NPs can affect the efficiency of nutrient transporters and receptors, impacting nutrient absorption and utilization, thereby shaping individual nutritional needs. Variations in taste receptor genes like TAS2R38, TAS1R2, and TAS1R3 can alter taste perception and dietary preferences, influencing food choices and nutritional intake (Frayling et al., 2007).

SNPs can also influence metabolic pathways related to specific nutrients. For example, SNPs in genes involved in carbohydrate metabolism, such as TCF7L2, can affect blood sugar levels and dietary recommendations. Variants in TCF7L2 have been linked to an increased risk of type 2 diabetes and varying reactions to dietary carbohydrates, impacting susceptibility to diabetes (Xie et al., 2009).

9.6.2 Impact of Genetic Variants on Diet and Health

Genetic variants significantly impact an individual's dietary response and overall health, central to nutrigenomics. These variations influence nutrient metabolism and utilization, potentially leading to sensitivities or intolerances like gluten or lactose intolerance. Understanding these predispositions is crucial for modifying diets to prevent adverse reactions maintain digestive health and lower the risk of diet-related disorders like obesity, diabetes, heart disease, and certain cancers. Studies link specific FTO gene variations to an increased obesity risk, especially with high-calorie diets (Kohlmeier et al., 2016). Different reactions to the ingestion of omega-3 fatty acids have been linked to variations in the PPARG gene. According to several researches, some PPARG genotypes may benefit from omega-3 fatty acid intake for the heart more than others (Ginsburg and Willard, 2009). The CYP1A2 gene is involved in caffeine metabolism, with variations influencing individuals' rates of caffeine metabolism and their response to caffeine intake, including stimulant effects and potential health risks (Becerra-Tomás et al., 2018). These findings underscore the necessity of personalized dietary recommendations based on genetic profiles, demonstrating how genetic variants interact with diet to impact health outcomes. Understanding these connections is critical for tailoring dietary strategies to promote health and mitigate the risk of nutrition-related disorders according to an individual's genetic makeup.

9.7 PERSONALIZED DIET RECOMMENDATIONS

Personalized diet recommendations are based on an individual's unique genetic makeup, health status, lifestyle, preferences, and goals. The intricate relationship among genetic predispositions, biological intricacies, cultural nuances, and environmental factors influencing personalized nutrition is shown in Figure 9.2. Tailoring dietary advice to specific genetic variants can help optimize nutrition, improve health outcomes, and reduce the risk of diet-related diseases.

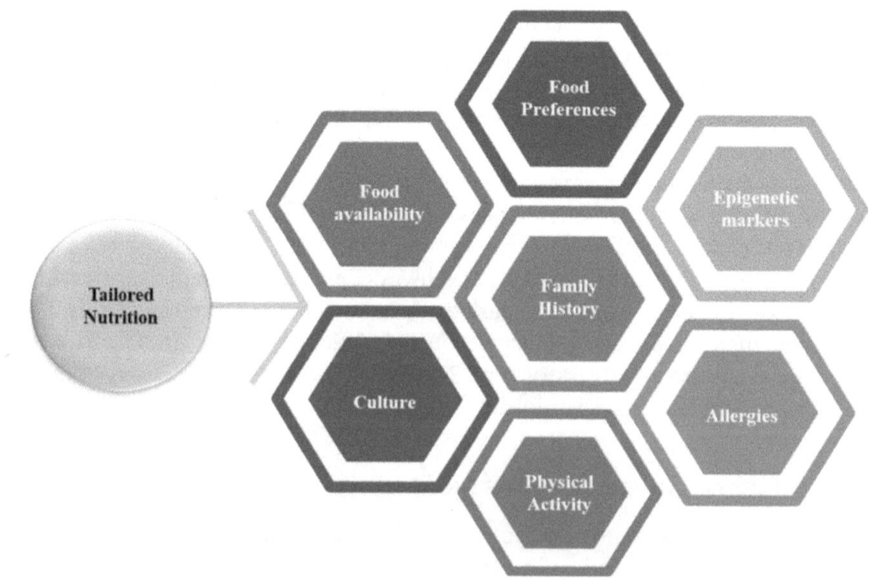

FIGURE 9.2 The complex interplay of genetic predispositions, biology, culture, and environment shapes personalized nutrition.

9.7.1 Development of Personalized Nutrition Plans

Personalized nutrition plans tailor dietary strategies to individual needs, considering genetics, lifestyle, medical history, preferences, and goals. Genetic testing provides insights into disease risk, food sensitivities, and metabolism, guiding customized plans based on genetic variants and comprehensive health assessments (Bischoff et al., 2017). Tailoring a nutrition plan to an individual's energy requirements and overall well-being involves considering lifestyle, daily activities, exercise, and stress levels. Clear objectives, like weight management or disease prevention, are established based on this information. Notable research, like the PREDIMED trial focusing on personalized nutrition plans, such as the Mediterranean diet, showcases the development and application of such approaches to prevent cardiovascular disease. It showed how a Mediterranean diet considerably decreased the risk of serious cardiovascular events in people with high cardiovascular risk (Manore, 2015).

9.7.2 Tailoring Macronutrient Intake

Tailoring macronutrient intake optimizes physiological function and energy production based on individual needs, lifestyle, and activity level, considering diverse objectives like weight management, muscle gain, sports performance, and medical conditions (Napoleao et al., 2021). Customizing macronutrient ratios enables individuals to optimize dietary choices for specific goals like weight loss or athletic performance. For example, increasing protein intake can enhance satiety and muscle preservation during calorie reduction, while prioritizing carbohydrates sustains energy for athletes. Consulting healthcare professionals or dietitians is essential for tailoring macronutrient intake effectively, considering lifestyle, metabolism, and individual needs for a personalized dietary approach (Ahmadieh and Arabi, 2011). These studies collectively underscore the necessity of personalized macronutrient adjustments, considering individual needs, activity levels, and health objectives, to craft an effective and science-based dietary approach tailored to the unique requirements of each person.

9.7.3 Micronutrient Optimization

Micronutrient optimization involves adjusting one's diet to ensure sufficient intake of essential vitamins, minerals, and trace elements. Research highlights the crucial role of micronutrients in immune function, metabolism, cognitive function, and disease prevention. The potential effect of some micronutrient deficiency on DNA damage is presented in Table 9.2. Micronutrient deficiencies can lead to various health problems, emphasizing the need for a balanced and varied diet. Studies link adequate micronutrient intake to reduced chronic disease risk. For instance, sufficient vitamin D supports bone health and immunity, reducing osteoporosis risk. Antioxidants like vitamins C and E, and minerals like selenium, help lower the risk of chronic illnesses such as cardiovascular disease and certain cancers by scavenging free radicals (Raymond et al., 2022). Optimal health necessitates adjusting micronutrient intake based on age, gender, health, and preferences, achieved through a diverse diet

TABLE 9.2
Micronutrient Deficiency May Impact DNA Damage

Sr. No.	Micronutrient	Effects on DNA Damage When Deficient
1	Vitamin C	Increased risk of oxidative DNA damage
2	Vitamin E	Higher susceptibility to DNA damage by free radicals
3	Vitamin D	Impaired DNA repair mechanisms
4	Folate	Increased risk of DNA strand breaks
5	Vitamin B12	Altered DNA synthesis and repair
6	Iron	Reduced DNA repair capacity
7	Zinc	Impaired DNA repair and stability
8	Selenium	Increased susceptibility to DNA damage
9	Copper	Altered DNA structure and function
10	Magnesium	Disruption of DNA stability and repair

including fruits, vegetables, whole grains, lean proteins, nuts, seeds, and, if needed, supplementation. A range of nutrient-rich foods and, if necessary, micronutrient supplements can be incorporated into one's diet to improve health outcomes and lower the chance of developing several chronic diseases.

9.8 DISEASE PREVENTION AND MANAGEMENT

Nutrigenomics, a developing field, integrates personalized nutrition recommendations based on genetic makeup and individual nutrient responses for disease prevention and management. It explores how genetic differences influence dietary response, nutrient metabolism, and disease development, aiming to create tailored dietary programs for improved outcomes.

9.8.1 Nutrigenomics in Chronic Disease Prevention

Numerous studies have shown the potential of nutrigenomics in chronic disease prevention. For instance, research has revealed that specific genetic variations can influence an individual's response to dietary fat intake and subsequently impact the risk of developing cardiovascular diseases (Kim et al., 2023). Another study demonstrated that individuals with certain genetic polymorphisms may have an increased risk of developing type 2 diabetes when consuming a diet high in sugary beverages (Kelly et al., 2013). These findings highlight the significance of personalized dietary interventions, guided by genetic predispositions, in reducing disease risk. Nutrigenomics can also inform dietary recommendations for managing inflammation, influenced by genetic variants. Understanding genetic variations can help tailor food plans to manage inflammation and mitigate chronic inflammatory diseases. Nutrigenomics offers personalized dietary recommendations based on genetic profiles, aiding in chronic disease prevention and enhancing long-term health outcomes through tailored nutritional guidance.

9.8.2 Applications in Cardiovascular Health

Nutrigenomic analysis pinpoints genetic markers associated with cardiovascular risk factors, allowing personalized nutrition to mitigate risks. Genetic variations influence fat processing, impacting heart health; specific mutations benefit from low-saturated fat diets, reducing LDL cholesterol. Tailored dietary interventions based on genetic predispositions are vital for managing cholesterol levels and minimizing cardiovascular risk, extending to nutritional guidance for hypertension by addressing genetic differences in salt sensitivity affecting blood pressure responses to sodium intake (de Toro-Martín et al., 2017). Customized diets, tailored to individual traits, are essential for promoting cardiovascular health, considering unique health markers, genetics, lifestyle, and dietary preferences. Recent studies demonstrate the effectiveness of personalized dietary strategies in reducing cardiovascular risk factors. For example, modifying dietary recommendations based on individual responses to meals high in saturated fats can improve blood lipid profiles and mitigate the risk of cardiovascular disease (Weissberg-Benchell and Shapiro, 2017). Tailoring dietary recommendations based on genetic markers for lipid metabolism and adjusting salt intake according to individual sodium tolerance reduce cardiovascular risk. Carter et al. (2014) underline the significance of customized diets in cardiovascular disease management and prevention. Understanding genetic influences enables tailored salt dietary advice, crucial for blood pressure control and reducing cardiovascular event risk. Nutrigenomics offers personalized dietary guidance based on genetic profiles, significantly impacting cardiovascular health. Healthcare providers can optimize cardiovascular outcomes by aligning dietary recommendations with genetic predispositions.

9.8.3 Role of Personalized Nutrition in Diabetes Management

Personalized nutrition is vital for effective diabetes management, aligning dietary recommendations with individual factors like genetics, metabolism, and lifestyle. Tailored approaches improve glucose control and overall health, as supported by research (Hall and Chung, 2018). According to a study conducted by Carter et al. (2014), individuals with diabetes who received personalized nutritional guidance experienced superior blood sugar management and significantly lower HbA1c levels compared to those following generalized dietary advice. Another study by Douglas et al. (2018) explored the benefits of a customized low-carbohydrate diet for controlling type 2 diabetes, and the findings revealed that tailoring carbohydrate intake to an individual's metabolic response substantially improved glycemic control and facilitated weight loss.

Personalized nutrition shows promise in enhancing diabetes management by adapting dietary advice to individual needs. Ongoing advancements in technology and science may lead to a future where personalized diets play a crucial role in improving glycemic control and overall well-being for individuals with diabetes. Realizing this potential requires continued research, education, and healthcare technology advancements.

9.9 CHALLENGES AND ETHICAL CONSIDERATIONS

Personalized nutrition holds promise for improving health outcomes but faces challenges and ethical dilemmas. These include data access and privacy concerns, equitable distribution of benefits, and addressing socioeconomic barriers. Ethical considerations, such as informed consent and transparency, are vital, as highlighted by Aruoma et al. (2019). Successfully integrating personalized nutrition into healthcare requires balancing benefits and ethical issues while overcoming challenges. It is crucial to safeguard individuals' autonomy, privacy, and access to personalized nutrition while maximizing its advantages. This evolving field must prioritize ethical principles to ensure equitable and effective implementation of individualized dietary approaches.

9.10 FUTURE DIRECTIONS AND EMERGING TRENDS

The future of healthcare and nutrition looks promising with rapid advancements in nutrigenomics and personalized dietary approaches. Nutrigenomics studies how genetic variations influence responses to nutrients, enabling highly customized diets based on an individual's genetic profile. Advancements in genetic testing afford deeper insights into genes' impact on metabolism and health. Precision nutrition considers genetics, lifestyle, medical history, and gut microbiota. Collaboration among geneticists, nutritionists, dietitians, and data scientists is increasing to develop algorithms for personalized dietary advice, accessible through web platforms or mobile apps. Ongoing research in this field is exploring the potential of nutrigenomics to optimize dietary recommendations. Some studies highlighted the prospect of dietary therapies tailored to an individual's unique genetic profile. We envision a future where individuals have the opportunity to adopt highly personalized dietary plans that consider their genetic makeup and other distinctive characteristics, ultimately promoting better health and well-being. The integration of nutrigenomics into personalized diets is an evolving and promising field.

9.11 CONCLUSION

Nutrition has undergone a profound transformation with the rise of nutrigenomics and personalized dietary approaches, replacing traditional one-size-fits-all guidelines with tailored strategies based on individuals' genetic makeup. Advanced sequencing technology and bioinformatics enable detailed exploration of genetic profiles, identifying variations linked to dietary needs and responses. Personalized diets optimize nutrient intake, metabolic functions, and food preferences, considering individuals' ability to metabolize nutrients, susceptibility to diet-related diseases, and reactions to foods. They hold the potential to enhance overall well-being, manage chronic illnesses, and improve health outcomes by aligning dietary recommendations with genetic predispositions.

Personalized diets are valuable in disease management and prevention, as nutrigenomics identifies genetic markers for health conditions, facilitating specialized dietary plans. Tailored regimens benefit athletes and fitness enthusiasts by optimizing performance, aiding recovery, and reducing injury risks. Ongoing research and technological advancements promise deeper insights into genetic influences on dietary responses. Integrating machine learning into nutrigenomic analysis enhances the precision and applicability of personalized recommendations, envisioning a future where personalized nutrition becomes standard practice, revolutionizing diet and nutrition approaches for a healthier society.

REFERENCES

Ahmadieh, H., & Arabi, A. (2011). Vitamins and bone health: Beyond calcium and vitamin D. *Nutrition Reviews*, *69*(10), 584–598.

Arballo, J., Amengual, J., & Erdman Jr, J. W. (2021). Lycopene: A critical review of digestion, absorption, metabolism, and excretion. *Antioxidants*, *10*(3), 342.

Aruoma, O. I., Hausman-Cohen, S., Pizano, J., Schmidt, M. A., Minich, D. M., Joffe, Y., … & Brady, D. M. (2019). Personalized nutrition: Translating the science of nutrigenomics into practice: Proceedings from the 2018 American College of Nutrition Meeting. *Journal of the American College of Nutrition*, *38*(4), 287–301.

Bashiardes, S., Godneva, A., Elinav, E., & Segal, E. (2018). Towards utilization of the human genome and microbiome for personalized nutrition. *Current Opinion in Biotechnology* *51*, 57–63.

Becerra-Tomás, N., Díaz-López, A., Rosique-Esteban, N., Ros, E., Buil-Cosiales, P., Corella, D., … & Alba, M. B. (2018). Legume consumption is inversely associated with type 2 diabetes incidence in adults: A prospective assessment from the PREDIMED study. *Clinical Nutrition*, *37*(3), 906–913.

Betts, J. A., & Gonzalez, J. T. (2016). Personalised nutrition: What makes you so special. *Nutrition Bulletin* *41*(4),353–359.

Bischoff, S. C., Boirie, Y., Cederholm, T., Chourdakis, M., Cuerda, C., Delzenne, N. M., … & Barazzoni, R. (2017). Towards a multidisciplinary approach to understand and manage obesity and related diseases. *Clinical Nutrition*, *36*(4), 917–938.

Caradonna, F., Consiglio, O., Luparello, C., & Gentile, C. (2020). Science and healthy meals in the world: Nutritional epigenomics and nutrigenetics of the mediterranean diet. *Nutrients*, *12*(6), 1748.

Carey, R. M., Kim, T., Cohen, N. A., Lee, R. J., & Nead, K. T. (2022). Impact of sweet, umami, and bitter taste receptor (TAS1R and TAS2R) genomic and expression alterations in solid tumors on survival. *Scientific Reports*, *12*(1), 8937.

Carter, P., Achana, F., Troughton, J., Gray, L. J., Khunti, K., & Davies, M. J. (2014). AM editerranean diet improves H b A 1c but not fasting blood glucose compared to alternative dietary strategies: A network meta-analysis. *Journal of Human Nutrition and Dietetics*, *27*(3), 280–297.

Chen, Y., Michalak, M., & Agellon, L. B. (2018). Focus: Nutrition and food science: Importance of nutrients and nutrient metabolism on human health. *The Yale Journal of Biology and Medicine*, *91*(2), 95.

Cozzolino, S. M. F., & Cominetti, C. (2013). *Biochemical and Physiological Bases of Nutrition in Different Stages of Life in Health and Disease*. Monole, Sao Paulo, Brazil,.

De Roos, B., & Brennan, L. (2017). Personalised interventions: A precision approach for the next generation of dietary intervention studies. *Nutrients*, *9*(8), 847.

Douglas, P. S., Rzeszut, A. K., Merz, C. N. B., Duvernoy, C. S., Lewis, S. J., Walsh, M. N., & Gillam, L. (2018). Career preferences and perceptions of cardiology among US internal medicine trainees: Factors influencing cardiology career choice. *JAMA Cardiology*, *3*(8), 682–691.

Frayling, T. M., Timpson, N. J., Weedon, M. N., Zeggini, E., Freathy, R. M., Lindgren, C. M., ... & McCarthy, M. I. (2007). A common variant in the FTO gene is associated with body mass index and predisposes to childhood and adult obesity. *Science*, *316*(5826), 889–894.

Gangotia, D., Gupta, A., & Mani, I. (2021). Role of bioinformatics in biological sciences. In V. Singh, & A. Kumar (Eds.), *Advances in Bioinformatics*. Springer Singapore, pp. 37–57.

Ginsburg, G. S., & Willard, H. F. (2009). Genomic and personalized medicine: Foundations and applications. *Translational Research*, *154*(6), 277–287.

Goni, L., Cuervo, M., Milagro, F. I., & Martínez, J. A. (2015). Future perspectives of personalized weight loss interventions based on nutrigenetic, epigenetic, and metagenomic data. *The Journal of Nutrition*, *146*(4), 905S–912S.

Hall, K. D., & Chung, S. T. (2018). Low-carbohydrate diets for the treatment of obesity and type 2 diabetes. *Current Opinion in Clinical Nutrition & Metabolic Care*, *21*(4), 308–312.

Hill, L. E., & El-Sohemy, A. (2012). Genetic Variation and Nutrient Metabolism. *Present Knowledge in Nutrition*, *27*.

Kawaguchi, T., Shima, T., Mizuno, M., Mitsumoto, Y., Umemura, A., Kanbara, Y., ... & Okanoue, T. (2018). Risk estimation model for nonalcoholic fatty liver disease in the Japanese using multiple genetic markers. *PLoS One*, *13*(1), e0185490.

Kelly, T. N., Takeuchi, F., Tabara, Y., Edwards, T. L., Kim, Y. J., Chen, P., ... & He, J. (2013). Genome-wide association study meta-analysis reveals transethnic replication of mean arterial and pulse pressure loci. *Hypertension*, *62*(5), 853–859.

Kim, S., Jeon, H. K., Lee, G., Kim, Y., & Yoo, H. Y. (2023). Associations between the genetic heritability of dyslipidemia and dietary patterns in korean adults based on sex differences. *Nutrients*, *15*(20), 4385.

Kohlmeier, M., De Caterina, R., Ferguson, L. R., Görman, U., Allayee, H., Prasad, C., ... & Martinez, J. A. (2016). Guide and position of the international society of nutrigenetics/ nutrigenomics on personalized nutrition: Part 2-ethics, challenges and endeavors of precision nutrition. *Lifestyle Genomics*, *9*(1), 28–46.

Lal, M. K., Sharma, E., Tiwari, R. K., Devi, R., Mishra, U. N., Thakur, R., ... & Sahu, S. K. (2022). Nutrient-mediated perception and signalling in human metabolism: A perspective of nutrigenomics. *International Journal of Molecular Sciences*, *23*(19), 11305.

Leclair, P.(2023). *Landmark Experiments in Protein Science*. CRC Press.

Lewis, J. L., & Dwyer, J. T. (2020). Establishing nutrient intake values. In: B. P. Marriott, D. F. Birt, V. A. Stallings, & A. A. Yates (Eds.), *Present Knowledge in Nutrition* (pp. 267–288). Academic Press.

Livingstone, K. M., Ramos-Lopez, O., Perusse, L., Kato, H., Ordovas, J. M., & Martínez, J. A. (2022). Reprint of: Precision nutrition: A review of current approaches and future endeavors. *Trends in Food Science & Technology*, *130*, 51–62.

Manore, M. M. (2015). Weight management for athletes and active individuals: A brief review. *Sports Medicine*, *45*(Suppl 1), 83–92.

Mengelkoch, S., Rose, S. M. S. F., Lautman, Z., Alley, J. C., Roos, L. G., Ehlert, B., ... & Slavich, G. M. (2023). Multi-omics approaches in psychoneuroimmunology and health research: Conceptual considerations and methodological recommendations. *Brain, Behavior, and Immunity*, *114*, 475–487.

Mierziak, J., Kostyn, K., Boba, A., Czemplik, M., Kulma, A., & Wojtasik, W. (2021). Influence of the bioactive diet components on the gene expression regulation. *Nutrients*, *13*(11), 3673.

Mishra, U. N., Jena, D., Sahu, C., Devi, R., Kumar, R., Jena, R., ... & Kumar, A. (2022). Nutrigenomics: An inimitable interaction amid genomics, nutrition and health. *Innovative Food Science & Emerging Technologies*, *82*, 103196.

Mozaffarian, D. (2016). Dietary and policy priorities for cardiovascular disease, diabetes, and obesity: A comprehensive review. *Circulation*, *133*(2), 187–225.

Munoz, J., Lok, K. H., Gower, B. A., Fernandez, J. R., Hunter, G. R., Lara-Castro, C., ... & Garvey, W. T. (2006). Polymorphism in the transcription factor 7-like 2 (TCF7L2) gene is associated with reduced insulin secretion in nondiabetic women. *Diabetes*, *55*(12), 3630–3634.

Musa, A., Ghoraie, L. S., Zhang, S. D., Glazko, G., Yli-Harja, O., Dehmer, M., ... & Emmert-Streib, F. (2018). A review of connectivity map and computational approaches in pharmacogenomics. *Briefings in Bioinformatics*, *19*(3), 506–523.

Napoleao, A., Fernandes, L., Miranda, C., & Marum, A. P. (2021). Effects of calorie restriction on health span and insulin resistance: Classic calorie restriction diet vs. ketosis-inducing diet. *Nutrients*, *13*(4), 1302.

Ordovas, J. M. (2006). Genetic interactions with diet influence the risk of cardiovascular disease. *The American Journal of Clinical Nutrition*, *83*(2), 443S–446S.

Raymond, J. L., & Morrow, K. (2022). *Krause and Mahan's Food and the Nutrition Care Process, 16e, E-Book: Krause and Mahan's Food and the Nutrition Care Process, 16e, E-Book*. Elsevier Health Sciences.

Toro-Martín, D., Arsenault, B. J., Després, J. P., & Vohl, M. C. (2017). Precision nutrition: A review of personalized nutritional approaches for the prevention and management of metabolic syndrome. *Nutrients*, *9*(8), 913.

Roberts, C. K., & Barnard, R. J. (2005). Effects of exercise and diet on chronic disease. *Journal of Applied Physiology*, *98*(1), 3–30.

Rochfort, S. (2005). Metabolomics reviewed: A new "omics" platform technology for systems biology and implications for natural products research. *Journal of Natural Products*, *68*(12), 1813–1820.

Ronteltap, A., Van Trijp, J. C. M., & Renes, R. J. (2008). Consumer acceptance of nutrigenomics-based personalised nutrition. *British Journal of Nutrition*, *101*(1), 132–144.

Rowland, I., Gibson, G., Heinken, A., Scott, K., Swann, J., Thiele, I., & Tuohy, K. (2018). Gut microbiota functions: Metabolism of nutrients and other food components. *European Journal of Nutrition*, *57*, 1–24.

Sanhueza, C., & Valenzuela, B. (2012). Nutrigenomics: Revealing molecular aspects of a personalized nutrition. *Revista Chilena de Nutrición*, *39*(1), 71–85.

Trujillo, E., Davis, C., & Milner, J. (2006). Nutrigenomics, proteomics, metabolomics, and the practice of dietetics. *Journal of the American Dietetic Association*, *106*(3), 403–413.

Weissberg-Benchell, J., & Shapiro, J. B. (2017). A review of interventions aimed at facilitating successful transition planning and transfer to adult care among youth with chronic illness. *Pediatric Annals*, *46*(5), e182–e187.

WHO. (2018a). Noncommunicable diseases. https://www.who.int/publications/i/item/9789241514620 (accessed 1 October 2023).

Xie, S., Lee, Y. F., Kim, E., Chen, L. M., Ni, J., Fang, L. Y., ... & Chang, C. (2009). TR4 nuclear receptor functions as a fatty acid sensor to modulate CD36 expression and foam cell formation. *Proceedings of the National Academy of Sciences*, *106*(32), 13353–13358.

10 Biosensor-Based Diagnosis of Infectious Diseases

Khushboo Sethi, Ruma Rani, Manju Bernela, and Anita Rani Gill

10.1 INTRODUCTION

Infectious diseases continue to pose significant challenges to global health, causing substantial morbidity and mortality worldwide (Bloom and Cadarette, 2019). Pathogenic microorganisms, which include bacteria, viruses, fungi, and parasites, are the source of infectious diseases. A few of the diseases listed by the World Health Organization (WHO) comprise tuberculosis, meningitis, malaria, AIDS, pneumonia, polio, hepatitis, SARS-CoV-2, dengue, Chagas disease, leprosy, toxoplasmosis, and leishmaniasis. The spread of these diseases is influenced by urbanization, deficient urban planning, inadequate sanitation, population density, and disruption of ecosystems (Neiderud et al., 2015). Despite vaccination efforts, the rise in certain disease incidences underscores challenges related to intentional under-vaccination, emphasizing the need for public health education programs (Van Panhuis et al., 2013). Moreover, new and recurrent diseases caused by emerging or existing pathogens have been documented (Clements and Casani, 2016). So, there is a high demand for novel immunization strategies and alternatives to combat pathogen resistance to antibiotics (Dye, 2014). Early and accurate diagnosis plays a crucial role in effective disease management, enabling timely treatment initiation, implementation of infection control measures, and prevention of further transmission. In recent years, biosensor-based diagnosis has emerged as a promising approach for the detection and monitoring of infectious diseases (Pohanka, 2018).

Biosensors act as analytical tools that use biological components (such as enzymes, antibodies, or whole cells) in conjunction with a transducer to detect and convert biological responses into measurable signals. This provides novel solutions for rapid and efficient infectious disease diagnosis due to their simplicity, real-time analysis capabilities, and potential for miniaturization (Yoo and Lee, 2016). The field of biosensor technology has witnessed remarkable progress, with various biosensor platforms being developed for the diagnosis of infectious diseases. Some emerging systems have shown promise by enabling pathogen detection with minimal sample manipulation, which is especially crucial for samples infected with certain pathogens. This review addresses various biosensor principles, their types, and how they can be used to identify infectious disorders brought on by bacteria and viruses, along with future perspectives.

DOI: 10.1201/9781032642567-12

10.2 PRINCIPLES AND TYPES OF BIOSENSING TECHNOLOGY

A biosensor works by identifying a target biomarker that is peculiar to a given infection. The employment of a bioreceptor, an immobilized sensing component, enables this recognition. A bioreceptor may be a monoclonal antibody, DNA, RNA, enzyme, tissue, glycan, lectin, or even an entire cell. The bioreceptor element is coupled with a transducer, a crucial part of the biosensor system, which converts the biochemical interaction between the biomarker and the bioreceptor into a quantifiable signal at the end. The signal generated by the transducer is typically read out and processed using instrumentation, such as spectrophotometers, potentiostats, or microcontrollers. The basic principle of biosensors is depicted in Figure 10.1.

The biochemical characteristics of the bioreceptor are crucial because they guarantee excellent sensitivity and selectivity in detecting the biomarker while avoiding interference from other chemicals or microorganisms present. To effectively identify

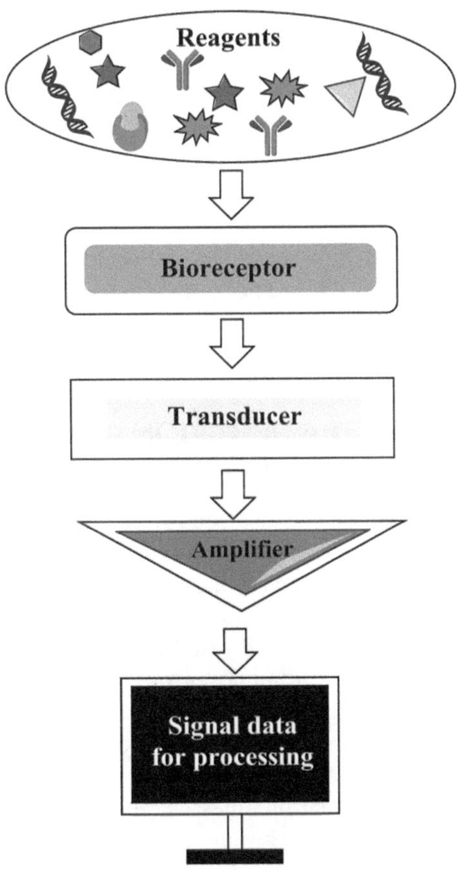

FIGURE 10.1 Schematic representation of the basic principle of the biosensor by detection of the reagents by the receptor sites followed by the transduction of the signal and processing of signal data.

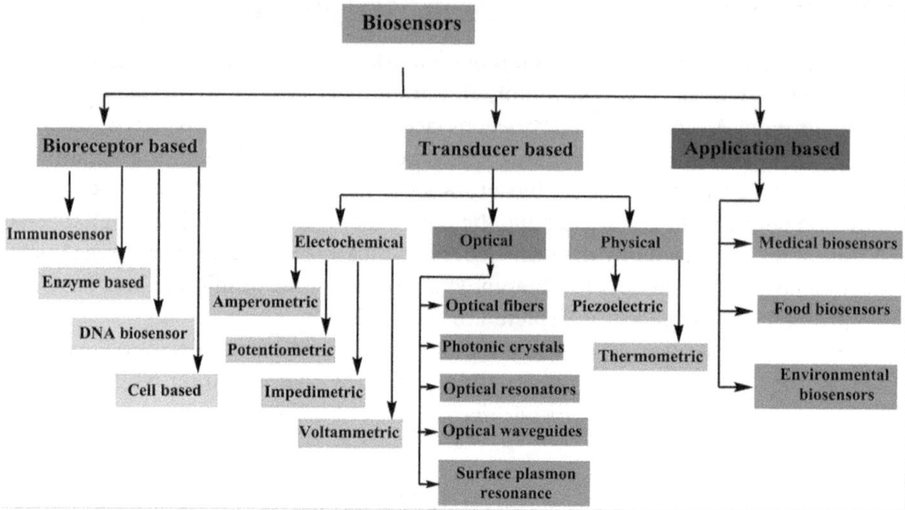

FIGURE 10.2 Different types of biosensors based on bioreceptor, transducer, and application.

and quantify pathogens, biosensors must record and display signals. Moreover, in diagnosing infectious diseases, high sensitivity is crucial because pathogens can rapidly spread even before clinical signs appear in animals or humans.

Biosensors can be classified into different types based on the type of bioreceptor, transducer, and intended application, as shown in Figure 10.2.

10.3 BIOSENSORS FOR THE DETECTION OF INFECTIOUS DISEASES

In the field of medical diagnostics, biosensors for the detection of infectious diseases have grown in significance. These biosensors are intended to quickly and accurately identify particular pathogens or biomarkers linked to infectious diseases. They are essential tools for managing and controlling infectious disease outbreaks, particularly in environments with limited resources. In this situation, a wide range of biosensors can be used, such as immunosensors that target particular antibodies or antigens linked to the infectious agent, aptamer-based biosensors that use specific RNA or DNA sequences for pathogen detection, and nucleic acid-based biosensors for detecting viral or bacterial DNA/RNA. These biosensors are essential for early diagnosis, tracking the development of illnesses, and stopping the spread of infectious diseases like COVID-19, human immunodeficiency virus (HIV), tuberculosis, and among others.

This review will primarily focus on using biosensors in detecting infectious diseases, including the detection of pathogenic viruses and bacteria, whose details are given in the following sections.

10.3.1 Detection of Pathogenic Virus

Viruses, which are infectious agents, including human papillomavirus (HPV), hepatitis virus, dengue virus, influenza virus, ebolavirus, HIV, Hantavirus, and SARS-CoV-2 (Severe Acute Respiratory Syndrome Coronavirus 2) can give rise to numerous diseases. HPV, with its 100 genotypes, is linked to cancer, particularly affecting the cervix and anus. However, due to its low specificity, HPV diagnosis is difficult. Numerous experiments have been conducted to develop biosensors that are HPV-resistant. A very sensitive electrochemical biosensor for the detection of the human papillomavirus (HPV) was created by Huang et al. (2015). Functionalization of a glassy carbon electrode with graphene, gold nanorods, and a polymeric layer was used to build the biosensor, which used DNA probes. Electrochemical impedance spectroscopy (EIS) and the redox marker 1,10-phenanthroline ruthenium dichloride ($Ru(phen)_3^{2+}$) were used to increase the electrochemical signal. This biosensor impressively displayed effective viral DNA detection, successfully locating the target in samples of human serum with a remarkably low detection limit of 4.03×10^{-14} M. In another study, a portable innovative microsystem biosensor was designed to simultaneously conduct 98 tests for HPV detection by small sample volume. The 4-aminophenol monolayer and a gold nanolayer served as the biosensor's structural support system on a poly (methyl methacrylate) surface using EIS (Urrego et al., 2015). Chaibun et al. (2022) developed a genosensor that employs a sandwich hybridization approach, where the HPV target binds to both the silica-redox dye reporter probe and the capture probe, followed by electrochemical detection. This sensor exhibits exceptional specificity and sensitivity, capable of detecting HPV-16 at a limit of 22 femtomolar (fM) and HPV-18 at 20 fM. Rasouli et al. (2023) reported an electrochemical DNA-nano biosensor for the detection of cervical cancer-causing HPV-16. It entails applying green Fe_3O_4-Au core–shell nanoparticles (NPs) on screen-printed carbon electrodes (SPCEs). Then, single-strand DNA (ssDNA) probes for HPV DNA sequences are functionalized onto these modified electrodes. $[Fe(CN)6]^{3-/4-}$ was used as the redox indicator, while for tracking the hybridization events with the immobilized probe DNA, two methods *viz.* cyclic voltammetry (CV) and differential pulse voltammetry (DPV) were used. The findings depicted that the modified electrodes can distinguish between $[Fe(CN)6]^{3-/4-}$ redox current signals in the presence or absence of immobilized probe DNA. Additionally, they can quantitatively estimate the concentration of the probe DNA. The best performance was achieved with a probe DNA concentration of 5 μM, resulting in a detection limit of 0.1 nM and a sensitivity of 2.4 μA/nM for the HPV biosensor.

Another infectious disease, Hepatitis B, is caused by a DNA virus that targets liver hepatocytes. This global health issue has impacted approximately 2 billion people, which is roughly one-third of the world's population, with positive hepatitis B serology. Hepatitis B virus infection poses a significant threat to the liver, increasing the risk of conditions like liver cirrhosis, cancer, and even resulting in death (Yang et al., 2014). Given its potential seriousness, diligent monitoring during the chronic phase is crucial to prevent the progression of conditions such as cirrhosis, liver failure, and hepatocellular carcinoma. As a novel alternative to traditional methodologies, Castro et al. (2014) introduced an electrochemical biosensor designed to identify a particular DNA sequence associated with the hepatitis B virus using a DPV technique for detection purposes. The outcomes of their study showed the efficacy of this device in diagnosing individuals infected with the virus, primarily relying on serum samples, and had a detection limit of 2.61 nM. In another study, Shourian and colleagues (2015) developed a colorimetric immunosensor designed for the identification of the surface antigen associated with the hepatitis B virus based on a sandwich immunoassay technique that included luminol and biotin-functionalized gold NPs. The immunosensor demonstrated a linear range of detection spanning from 1.7 to 1,920 pg/mL, having a detection limit of 0.358 pg/mL.

Tin-doped WO_3/In_2O_3 nanowires were used as a photoelectrode by Shariati and Sadegh (2020) to develop an ultrasensitive DNA biosensor for hepatitis B virus detection. The biosensor used electrochemical impedance measurements to identify DNA hybridization without labels at extremely low concentrations. With a linear response from 0.1 pM to 10 μM and a spectacular 1 fM detection limit, laser amplification increased sensitivity. Cheng et al. (2023) used a novel technique involving surface etching of individual helical gold nanorods (HGNRs) under a dark-field microscope (DFM) to create a novel and highly sensitive biosensor for detecting HBV-DNA. The optical probe in this work was an HGNR, recognized for its high surface activity. They were etched, becoming smoother gold nanorods or even gold nanospheres, which changed the color and light scattering of the individual particles noticeably. The sensitivity of this method was further improved by combining the DFM technique with hybridization chain reaction amplification. It obtained a limit of detection (LOD) of 30.15 fM (3 times the standard deviation over the slope) and enabled the sensitive detection of HBV-DNA within a concentration range spanning from 0.05 to 10 pM.

Another serious global public health concern is presented by dengue fever since it could affect roughly 2.5 billion people worldwide. It is prevalent in 98 tropical and subtropical countries and is largely spread by infected female mosquitoes, particularly *Aedes aegypti* and *Aedes albopictus.* Reverse Transcription Polymerase Chain Reaction (RT-PCR) and immunoassay approaches like ELISA are being used to identify viral RNA in dengue diagnostic procedures. These techniques rely on the identification of antibodies created by virus-infected people. However, employing these methods to make an early diagnosis can be costly, time-consuming, and may not have the required sensitivity, which could lead to false-positive results (Nguyen et al., 2012; Zhang et al., 2015). Silva and their research team (2014) used modified graphite ink that was loaded with thiophene and a coating of gold NPs to create a screen-printed electrode. On this electrode, they mounted anti-NS1 antibodies, and they looked for amperometric reactions to the dengue virus's NS1 protein. CV was used to detect the presence of ferrocyanide and ferricyanide. The biosensor displayed a detection limit of 0.015 μg/mL, and the resulting linear detection range ranged from 0.04 to 0.6 μg/mL. Additionally, Cecchetto et al. (2015) developed an immunosensor that targets the Non-Structural Protein of Dengue (NS1) utilizing gold electrodes, which were modified with a self-assembled monolayer of mercaptoundecanoic acid. Covalent bonds were used to immobilize anti-NS1 antibodies on the surface. EIS and the calibration curve detection methods were used, resulting in a linear range of 0.01–2.00 μg/mL in phosphate-buffered saline (PBS) and 0.01–1.00 μg/mL in serum. The biosensor detection limits in PBS and serum were 3 and 30 ng/mL, respectively.

In a different investigation, Kim et al. (2019) used chemically modified affinity peptides to create an electrochemical biosensor for detecting dengue virus NS1. They created several peptides, produced them, and covalently immobilized them on a gold sensor surface. Square wave voltammetry (SWV) and EIS were used to evaluate the sensor's performance. Using the 3-sigma rule, the peptide DGV BP1 demonstrated promise as a recognition peptide, with a detection limit for NS1 of 1.49 μg/mL. A different research team created a colorimetric genosensor for the quick, precise, sensitive, and affordable detection of the dengue virus in clinical samples. This cutting-edge sensing system depends on the precise interaction of DNA-conjugated gold nanoparticles (AuNPs) to the dengue virus genomic RNA. The ruby red color of the gold colloid changes to blue because of the creation of a DNA–RNA heteroduplex structure as a result of nano-aggregation in an acidic environment. With the naked eye or using absorbance, it is simple to see this hue change. The genomic RNA of all four dengue serotypes has a conserved area, which is the target of the DNA probe used in this genosensor. The designed genosensor demonstrates exceptional specificity and selectivity, with an impressive detection limit of ~1 pg/μL (Duyen et al., 2023).

Influenza virus is a common and dangerous pathogen known for its easy transmission and constant mutations, making early diagnosis difficult (Nidzworski et al., 2014; Tepeli et al., 2018). Jiang et al. (2015) created a polydiacetylene-based biosensor that uses H5N1 (avian influenza) antibody detection. Upon virus detection, the polydiacetylene vesicles displayed a remarkable color change, going from blue to red. For the extremely specific and sensitive detection of *Haemophilus influenza* in human biofluids, Hassanpour et al. (2018) created a unique optical biosensor made of bioconjugated citrate-capped silver nanoparticles (AgNPs) connected to pDNA. An alternative strategy was used by Lee et al. (2019), who created a label-free localized surface plasmon resonance (SPR) biosensor for H5N1 detection with a low LOD of 1 pM (10^{-12} M). The system used hollow, spike-like gold nanoparticles mounted on a multifunctional DNA 3-way junction.

Ebolavirus, HIV, and Hantavirus are other hazardous viruses that have an impact on people all over the world (Rojas et al., 2020; Kharsany et al., 2020; Mittler et al., 2019). The first one is an Ebola-causing negative strand-RNA virus that is a member of the Filoviridae family. Hemorrhagic fever is believed to be the most lethal of the symptoms that this agent causes in its victims (Nicastri et al., 2019; Mérens et al., 2018; Murphy, 2019). No vaccination or specific treatment is available to date (Walldorf et al., 2019). The creation of biosensors for detecting this disease has, however, been highlighted in several research (Kaushik et al., 2016). Enzyme-amplified detection was used by Ilkhani and Farhad (2018) to create a unique electrochemical-based DNA biosensor, which increased the sensitivity and selectivity of the device for the pathogen.

HIV is a retrovirus that targets the immune system of a patient, impairing their ability to fight various illnesses and, if untreated, leading to death. Although early detection can also save many lives and slow the spread of HIV, clinical therapies are still the most effective way to reduce mortality from the disease (Tavakoli et al., 2017, Nandi et al., 2020). Additionally, Gong et al. (2019) developed an electrical DNA biosensor using a nanocomposite of polyaniline and graphene (PAN/GN) made *via* reverse-phase polymerization with excellent selectivity and sensitivity for the detection of the HIV-1 gene fragment.

A cluster of viruses known as the Hantavirus belongs to the *Bunyaviridae* family. Contact with substances infected with mouse excrement, such as drinks, food, or particles, triggers the spread of the virus. Within 2–7 days of infection, it results in hemorrhagic fever, respiratory failure, and heart failure. Gogola et al. (2019) created an electrochemical immunosensor using a virus antigen or protein to chemically modify the gold surface. In a subsequent investigation, biochar (BC) was used to build a fast electrochemical biosensor because of its highly functionalized surface for covalent interaction with biomolecules (Martins et al., 2019). Both studies have developed devices that are used as promising and suitable tools for Hantavirus clinical detection.

Emerging viruses with pandemic potential offer serious risks because of their quick spread, difficult diagnosis, and lack of vaccinations and therapies. SARS-CoV-2-caused COVID-19, which has a high contagiousness and asymptomatic cases, is an example of this. Although RT-PCR is the primary diagnostic technique, it has drawbacks such as time requirements and accessibility issues in remote locations. Rapid point-of-care technologies that can effectively identify the virus are urgently needed to solve this (Panghal et al., 2020; Al-Rohaimi et al., 2020; Ji et al., 2020). Lab-on-a-chip-based biosensors for detecting SARS-CoV-2 have been developed by several research teams (Choi, 2020; Yin et al., 2021). This method eliminates the requirement for specialized employees, boosts manufacturing effectiveness, and lowers assay costs by integrating microfluidic components into the biosensor (Cady et al., 2017). ID NOW®, Filmarray®, GeneXpert®, and RTisochip® are a few commercialized point-of-care equipment based on this microfluidic technology that has been essential in managing the pandemic (Zhuang et al., 2020). The diagnosis of COVID-19 has also benefited significantly from cell-based biosensors. Using membrane-engineered mammalian cells armed with human chimeric spike S1 antibodies, Mavrikou et al. (2020) created a biosensor. The SARS-CoV-2 S1 spike protein can be selectively detected by this technology. The protein changes the cellular bioelectric characteristics, which are assessed by the Bioelectric Recognition Assay when it binds to the membrane-bound antibodies. This biosensor delivers data in roughly 3 minutes and achieves an outstanding LOD of 1 fg/mL. The biosensor also includes a mobile read-out device that may be used with a smartphone. Another research team created a dual-functional plasmonic biosensor that makes use of both the localized surface plasmon resonance (LSPR) and plasmonic photothermal (PPT) effect. A two-dimensional gold nanoisland functionalized with complementary DNA receptors enables the selective detection of particular SARS-CoV-2 sequences using nucleic acid hybridization. Furthermore, the PPT effect can raise the in situ hybridization temperature, which helps two closely related gene sequences to vary from one another. The lower LOD for this biosensor was 0.22 pM, demonstrating its exceptional sensitivity (Qiu et al., 2020).

10.3.2 Detection of Pathogenic Bacteria

Pathogenic bacteria are important detection targets in many industries, including medicine and food safety. *Mycobacterium tuberculosis*, *Mycobacterium leprae*, *Streptococcus*, and other pathogenic bacteria are responsible for several serious diseases that affect people all over the world, including tuberculosis, leprosy, and meningitis, respectively (Sanvicens et al., 2009; Wang et al., 2012; Singh et al., 2014;

Ibrahim et al., 2015). *E. coli, Salmonella typhi, and Clostridium perfringens* are examples of common pathogenic bacteria that can cause a variety of illnesses in people, animals, and plants (Wang et al., 2012).

The pathogenic bacteria *Mycobacterium tuberculosis*, which causes tuberculosis, is the leading infectious cause of mortality and a major global public health concern (Garcia-Monco, 2014). In 2013, there were around 5.7 million TB cases reported globally, according to the WHO. Various biosensor platforms have been developed recently for the detection of tuberculosis, using a variety of biological recognition components and transducers. Liu et al. (2014), for example, created an electrochemical genosensor for *M. tuberculosis*. This novel method required immobilizing a specific IS6110 gene sequence on an electrode modified with reduced graphene oxide (GO) gold nanoparticles, which served as a sensing platform. For the objective of amplifying the tracer, polyaniline-gold nanoparticles were utilized. The sensor linear response ranges from 1×10^{-15} to 1×10^{-9} M. To immobilize a DNA probe, Costa and colleagues introduced an electrochemical genosensor in 2014 that used self-assembled monolayers of mercaptobenzoic acid and magnetite nanoparticles (Fe_3O_4NPs) on an unaltered gold electrode. The detection threshold of the genosensor was as low as 6 ng/μL. Another method was reported by Zhang et al. (2015), who used CdSe quantum dots as labels in conjunction with MspI endonuclease and gold nanoparticles to improve selectivity and magnify the signal. This sensor demonstrated a linear response range from 1×10^{-14} to 1×10^{-9} M, efficiently and with exceptional selectivity distinguishing mismatched DNA for *M. tuberculosis*. Various other detection systems have been documented, including SPR has been extensively utilized in the development of genosensors for *M. tuberculosis* (Hsu et al., 2013; Xiang et al., 2015), as well as immunoassays (Hsieh et al., 2012; Mukundan et al., 2012; Hiraiwa et al., 2015). A highly sensitive electrochemical DNA biosensor for *Mycobacterium tuberculosis* (MTB) detection was created by a different research team in 2020. By combining MPA-Fe_3O_4 nanoparticles with NCC/CTAB-modified nanocellulose, they were able to form a composite material that was then dropped onto a SPCE. They next used EDC/NHS coupling to immobilize an MTB DNA probe on the modified SPCE. They employed the hybridization labels $Ru(bpy)_3^{2+}$, which were characterized by DPV, to improve signal amplification and electrochemical responsiveness. With a wide detection range from 1.0×10^{-6} to 1.0×10^{-12} M, this biosensor detected target DNA at concentrations as low as 7.96×10^{-13} M (Zaid et al., 2020). An improved silicon nanowire field-effect transistor (SiNW-FET) biosensor was created by Ma et al. (2021) for the quick and accurate detection of *Mycobacterium tuberculosis*. They improved the biosensor's surface functional characteristics, which led to outstanding sensitivity and a detection limit for proteins of 0.01 fg/mL. Within the range of 1 fg/mL to 100 μg/mL, the current change showed a linear relationship with protein concentration. A label-free DNA biosensor was recently created by Taufiq et al. (2023) using commercial screen-printed electrodes (SPEs) from DropSens and Zensors. Through DNA hybridization with the IS6110 gene marker, this biosensor successfully detected tuberculosis (TB) in both raw sputum samples and purified DNA. Achieving a high R^2 value of 0.993 throughout a DNA concentration range of 0.5–10 nM, the biosensor demonstrated great selectivity for *M. tuberculosis* over *S. typhimurium*, and *Escherichia coli*. It also had a low LOD of 1.90 nM.

Leprosy, a chronic disease caused by *Mycobacterium leprae*, is another disease that requires attention. A total of 215,000 leprosy cases were recorded worldwide in 2013, according to the WHO, underscoring the significance of early detection to stop transmission and prevent serious patient suffering. In this regard, Afonso et al. (2010) developed an electrochemical genosensor that was based on the immobilization of a single-stranded DNA oligonucleotide onto a graphite electrode that has been treated with poly (4-aminophenol). This technique revealed a linear detection range for *M. leprae* that ranged from 0.35 to 35.0 ng/µL. Furthermore, Paula et al. (2013) created a quick and accurate leprosy test. This technique involves immobilizing two different antigens on nitrocellulose membranes to find leprosy-related IgM and IgG antibodies. Another research group created an electrochemical DNA biosensor for *Mycobacterium leprae* by detecting PCR products. Methylene blue was utilized as a hybridization indicator. Using poly(4-aminophenol)-modified graphite electrodes, the capture probe was immobilized. Using atomic force microscopy, they described the electrode surface morphologically. They used linear voltammetry to measure the amount of methylene blue on the DNA biosensor, and they were able to do so with a LOD of 1×10^{-10} mol/L (Afonso et al., 2021). Huang et al. (2023) created the ML-MCDA-lateral flow biosensor (LFB) rapid ML (*M. leprae*) detection method in 2023. This approach combines multiple cross-displacement amplification (MCDA) with a nanoparticle-based LFB to visually amplify and identify ML-specific genetic material. The assay was improved by the use of several nucleic acids, and 150 fg of ML genomic DNA could be found in each vessel. The entire process, including isothermal amplification and outcome validation, was finished in just 40 minutes. This method makes early clinical diagnosis and field leprosy screening possible by quickly, sensitively, and visibly detecting ML.

Numerous infections, including bacteria, fungi, viruses, and parasites, can cause meningitis, which is characterized by meningeal inflammation. *Streptococcus pneumoniae*, *Neisseria meningitidis*, *Haemophilus influenzae*, *Listeria monocytogenes*, and *Streptococcus* species are the bacteria that contribute most frequently among the many types (Abio et al., 2013; Bahr et al., 2014). *Neisseria meningitidis* has the potential to cause major epidemics, so should be given importance. The virulence gene, Omp85, in *N. meningitides* codes for a conserved outer membrane protein, and has been studied with a focus on detection targets. Reddy et al. (2012) created an immunosensor using antibodies that target the *N. meningitidis* cell surface outer membrane protein 85 as the biological recognition element and a quartz crystal microbalance as a transducer. Furthermore, electrochemical genosensors have advanced thanks to the use of certain oligonucleotides that target virulence factors. One of these methods involves immobilizing thiol-labeled DNA probes on gold electrode surfaces to promote hybridization with the ctrA gene's complementary sequences. The sensor exhibited sensitivities of 115.8 µA/ng with a 0.917 regression coefficient (Patel et al., 2009), 0.0115 µA/ng with a 0.999 regression coefficient (Patel et al., 2010), and 168.64 µA/ng with a 0.98 regression coefficient (Tak et al., 2014). Another genosensor developed by Dash et al. (2013) has a sensitivity of 9.508 µA/ng and uses particular oligonucleotides complementary to the rmpM (reduction-modifiable protein M) virulence gene. Tak et al. (2014) created a sensitive DNA biosensor by building a nanocomposite matrix of zinc oxide (ZnO) and multiwalled carbon nanotubes (CNTs) on an indium tin oxide (ITO)-coated glass platform. They used DPV to immobilize a 23-base single-stranded DNA probe onto the electrode, which allowed for the rapid and accurate identification of *N. meningitidis* within 45 seconds. The biosensor showed potential for early-stage meningitis diagnosis and had a linear response over a wide concentration range (5–180 ng/μL), raising the possibility of a portable diagnostic instrument. In 2020, Gupta et al. used conductive ink to develop a paper-based genochip with built-in electrodes. They mounted a particular DNA sequence on the WS2 QDs/PAD electrode using tungsten disulfide quantum dots (WS2 QDs) that had been studied using a variety of methods. The resulting biosensor, which is characterized by CV, identified the target DNA by hybridizing it with an indicator, methylene blue. It has a low LOD of 1 nM and a large linear response range from 1 nM to 100 μM. Tavakoli and colleagues in 2022 unveiled a novel microfluidic paper-based analytical device (FPAD) that combines single-stranded DNA-functionalized GO nano-biosensors with loop-mediated isothermal amplification (LAMP). This novel method provides a straightforward, speedy, economical, and quantitative method of identifying *N. meningitidis*, a major cause of meningitis. The technology has a low LOD of just six DNA copies per detecting zone, and the findings can be read in less than an hour.

Another prominent cause of foodborne infections and a serious public health problem is *Salmonella*. It can contaminate food at many points during the steps of production, processing, distribution, and marketing. Given its frequent occurrence, it is urgently necessary to develop quick and precise techniques for early identification, detection, and monitoring of Salmonella to protect public health. Starodub et al. (2013) developed highly specialized biosensors for *Salmonella* spp. detection using Total Internal Reflection Ellipsometry (TIRE) and SPR in this direction. These biosensors utilized surface binding layers as the reactive element and Ag–Ab reactions. The TIRE biosensor demonstrated even better sensitivity, capable of detecting multiple cells in 10 mL, while the SPR biosensor showed sensitivity in the range of 10^1–10^6 cells/mL. For the detection of *Salmonella* bacteria in food, Vaisocherová et al. (2016) used an SPR biosensor with an ultra-low fouling poly (carboxibetaine acrylamide) surface. They reported an average reaction time of 80 minutes and a LOD of 7.4×10^3 CFU/mL. This biosensor also worked well for concurrently identifying *E. coli*. In 2023, Konstantinou et al. created a reliable and accurate biosensor system that uses cell-based biosensor technology (BERA) and the BioElectric Diagnostics portable equipment by EMBIO Diagnostics to detect *Salmonella* spp. in meat products in less than 3 minutes. With a high accuracy rate of 86.1%, this technique showed that it could identify between samples containing and not containing *Salmonella* within 24 hours of a 3-minute analysis. In all food substrates, the LOD was as low as 10 CFU/g. Using LAMP, clustered regularly interspaced short palindromic repeats (CRISPR), and CRISPR-associated protein 12a (Cas12a), Lee et al. (2023) proposed a low-cost LFB to prevent false positives, they designed a CRISPR ribonucleic acid (crRNA) with a 21-base pair protospacer that did not overlap with the LAMP primers. They used a single-stranded DNA probe that was biotin and 6-carboxyfluorescein (6-FAM) tagged to recognize the LFB. The entire process, which consisted of DNA extraction (20 minutes), LAMP (60 minutes), Cas12a cleavage

(5 minutes), and LFB (5 minutes), could be used to detect a limit of 1.22×10^{0} CFU/mL in pure culture.

Another important bacterium, *Vibrio cholerae*, is the cause of the deadly Cholera disease, which spreads quickly through tainted food and beverages and has a 50%–60% death rate. Early diagnosis is therefore essential for disease management (Cecchini et al., 2016). Narmani et al. used gold nanoparticles (AuNPs) and magnetic nanoparticles (NPs) to produce an ultrasensitive and selective fluorescent DNA biosensor to detect the O1 OmpW gene of the bacteria (Narmani et al., 2018). A very sensitive electrochemical DNA biosensor for *Vibrio cholerae* detection was created by Ali et al. (2021). It made use of a DNA carrier matrix, a 3-aminopropyltriethoxysilane (APTES) modified glassy carbon electrode, and a gold nanocube. The biosensor demonstrated an outstanding low LOD of 7.41×10^{-30} mol/L, a broad linear response range for target DNA from 10^{-8} to 10^{-27} mol/L, and selective detection among bacterial cultures. An ultrasensitive electrochemical DNA biosensor for rapid *V. cholerae* DNA detection in environmental samples was developed by Futra et al. (2023). The biosensor achieved a low LOD of 1.25×10^{-18} M using a sandwich DNA hybridization method. Additionally, it showed long-term stability by continuing to function for up to 55 days.

10.4 CONCLUSION AND FUTURE PROSPECTS

Over the years, there have been major advancements in biosensor-based infectious disease diagnosis, providing quick, accurate, and targeted detection techniques that are critical for prompt disease management. These developments include the creation of biosensors that are more compact, portable, and affordable as well as the incorporation of cutting-edge technologies like nanomaterials and microfluidics. These innovations have the potential to revolutionize disease diagnosis, especially in resource-limited settings. However, several challenges must be addressed to fully realize the potential of biosensors in infectious disease diagnosis. These challenges include the need for improved sensitivity and specificity, the standardization of testing procedures, and the incorporation of biosensors into existing healthcare infrastructure. Additionally, the emergence of new infectious diseases and the development of drug-resistant strains require constant adaptation of biosensor technologies.

The future of biosensor-based infectious disease diagnosis is bright. Molecular biology, nanotechnology, and data analytics advances are expected to improve biosensor performance and expand the range of detectable pathogens. Furthermore, the incorporation of biosensors into telemedicine platforms can facilitate remote monitoring and reporting, thereby increasing access to healthcare. Biosensors may play an important role in global health security in the long run, allowing for rapid responses to disease outbreaks and efficient resource allocation. Continuous research and collaboration among scientists, clinicians, and industry partners will be critical in overcoming current obstacles and propelling the field forward. Biosensor-based diagnosis, with these advancements, can significantly contribute to early disease detection, personalized treatment, and the prevention of infectious disease transmission.

REFERENCES

Abio, A., Neal, K. R., & Beck, C. R. (2013). An epidemiological review of changes in meningococcal biology during the last 100 years. *Pathogens and Global Health*, *107*(7), 373–380.

Afonso, A. S., Goulart, L. R., Goulart, I. M., Machado, A. E., Madurro, J. M., & Brito-Madurro, A. G. (2010). A promising bioelectrode based on gene of Mycobacterium leprae immobilized onto poly (4-aminophenol). *Journal of Applied Polymer Science*, *118*(5), 2921–2928.

Afonso, A. S., Madurro, J. M., & Brito-Madurro, A. G. (2021). Electrochemical DNA Biosensor for *Mycobacterium leprae* Identification. *Brazilian Archives of Biology and Technology*, *64*, e21210030.

Ali, M. R., Bacchu, M. S., Setu, M. A. A., Akter, S., Hasan, M. N., Chowdhury, F. T., ... & Khan, M. Z. H. (2021). Development of an advanced DNA biosensor for pathogenic *Vibrio cholerae* detection in real sample. *Biosensors and Bioelectronics*, *188*, 113338.

Al-Rohaimi, A. H., & Al Otaibi, F. (2020). Novel SARS-CoV-2 outbreak and COVID19 disease: A systemic review on the global pandemic. *Genes & Diseases*, *7*(4), 491–501.

Bahr, N. C., & Boulware, D. R. (2014). Methods of rapid diagnosis for the etiology of meningitis in adults. *Biomarkers in Medicine*, *8*(9), 1085–1103.

Bloom, D. E., & Cadarette, D. (2019). Infectious disease threats in the twenty-first century: Strengthening the global response. *Frontiers in Immunology*, *10*, 445106.

Cady, N. C., Fusco, V., Maruccio, G., Primiceri, E., & Batt, C. A. (2017). Micro-and nanotechnology-based approaches to detect pathogenic agents in food. In: A. M. Grumezescu (Ed.), *Nanobiosensors* (pp. 475–510). Academic Press.

Castro, A. C. H., Franca, E. G., de Paula, L. F., Soares, M. M., Goulart, L. R., Madurro, J. M., & Brito-Madurro, A. G. (2014). Preparation of genosensor for detection of specific DNA sequence of the hepatitis B virus. *Applied Surface Science*, *314*, 273–279.

Cecchetto, J., Carvalho, F. C., Santos, A., Fernandes, F. C., & Bueno, P. R. (2015). An impedimetric biosensor to test neat serum for dengue diagnosis. *Sensors and Actuators B: Chemical*, *213*, 150–154.

Cecchini, F., Fajs, L., Cosnier, S., & Marks, R. S. (2016). *Vibrio cholerae* detection: Traditional assays, novel diagnostic techniques and biosensors. *Trac Trends in Analytical Chemistry*, *79*, 199–209.

Chaibun, T., Thanasapburachot, P., Chatchawal, P., Lee, S. Y., Jiaranuchart, S., Jearanaikoon, P., & Lertanantawong, B. (2022). Electrochemical DNA biosensor for the multianalyte detection of HPV-16 and HPV-18 in oral and cervical lesions. https://doi.org/10.21203/rs.3.rs-1179126/v1

Cheng, R., Li, L. T., Huang, M., Zhu, F., Li, Q., Liu, H., & Wang, J. (2023). Highly sensitive plasmonic biosensor for hepatitis B virus DNA based on the surface etching of the active helical gold nanorods. *Chemical Engineering Journal*, *468*, 143627.

Choi, J. R. (2020). Development of point-of-care biosensors for COVID-19. *Frontiers in Chemistry*, *8*, 517.

Clements, B. W., & Casani, J. A. P. (2016). Emerging and reemerging infectious disease threats. In: Disasters and Public Health. Butterworth-Heinemann, 245.

Costa, M. P., Andrade, C. A., Montenegro, R. A., Melo, F. L., & Oliveira, M. D. (2014). Self-assembled monolayers of mercaptobenzoic acid and magnetite nanoparticles as an efficient support for development of tuberculosis genosensor. *Journal of Colloid and Interface Science*, *433*, 141–148.

Dash, S. K., Sharma, M., Khare, S., & Kumar, A. (2013). rmpM genosensor for detection of human brain bacterial meningitis in cerebrospinal fluid. *Applied Biochemistry and Biotechnology*, *171*, 198–208.

Duyen, V. T. C., Van Toi, V., Van Hoi, T., & Truong, P. L. (2023). A novel colorimetric biosensor for rapid detection of dengue virus upon acid-induced aggregation of colloidal gold. *Analytical Methods*, *15*(32), 3991–3999.

Dye, C. (2014). After 2015: Infectious diseases in a new era of health and development. *Philosophical Transactions of the Royal Society B: Biological Sciences*, *369*(1645), 20130426.

Futra, D., Tan, L. L., Lee, S. Y., Lertanantawong, B., & Heng, L. Y. (2023). An ultrasensitive voltammetric genosensor for the detection of bacteria *vibrio cholerae* in vegetable and environmental water samples. *Biosensors*, *13*(6), 616.

Garcia-Monco, J. C. (2014). Tuberculosis. *Handbook of Clinical Neurology*, *121*, 1485–1499.

Gogola, J. L., Martins, G., Caetano, F. R., Ricciardi-Jorge, T., dos Santos, C. N. D., Marcolino-Junior, L. H., & Bergamini, M. F. (2019). Label-free electrochemical immunosensor for quick detection of anti-hantavirus antibody. *Journal of Electroanalytical Chemistry*, *842*, 140–145.

Gong, Q., Han, H., Yang, H., Zhang, M., Sun, X., Liang, Y., ... & Qiao, J. (2019). Sensitive electrochemical DNA sensor for the detection of HIV based on a polyaniline/graphene nanocomposite. *Journal of Materiomics*, *5*(2), 313–319.

Gupta, R., Valappil, M. O., Sakthivel, A., Mathur, A., Pundir, C. S., Murugavel, K., ... & Alwarappan, S. (2020). Tungsten disulfide quantum dots based disposable paper based lab on genochip for specific meningitis DNA detection. *Journal of the Electrochemical Society*, *167*(10), 107501.

Hassanpour, S., Baradaran, B., Hejazi, M., Hasanzadeh, M., Mokhtarzadeh, A., & de la Guardia, M. (2018). Recent trends in rapid detection of influenza infections by bio and nanobiosensor. *TrAC Trends in Analytical Chemistry*, *98*, 201–215.

Hiraiwa, M., Kim, J. H., Lee, H. B., Inoue, S., Becker, A. L., Weigel, K. M., ... & Chung, J. H. (2015). Amperometric immunosensor for rapid detection of *Mycobacterium tuberculosis*. *Journal of Micromechanics and Microengineering*, *25*(5), 055013.

Hsieh, S. C., Chang, C. C., Lu, C. C., Wei, C. F., Lin, C. S., Lai, H. C., & Lin, C. W. (2012). Rapid identification of *Mycobacterium tuberculosis* infection by a new array format-based surface plasmon resonance method. *Nanoscale Research Letters*, *7*, 1–6.

Hsu, S. H., Lin, Y. Y., Lu, S. H., Tsai, I. F., Lu, Y. T., & Ho, H. T. (2013). *Mycobacterium tuberculosis* DNA detection using surface plasmon resonance modulated by telecommunication wavelength. *Sensors*, *14*(1), 458–467.

Huang, H., Bai, W., Dong, C., Guo, R., & Liu, Z. (2015). An ultrasensitive electrochemical DNA biosensor based on graphene/Au nanorod/polythionine for human papillomavirus DNA detection. *Biosensors and Bioelectronics*, *68*, 442–446.

Huang, J., Tong, Y., Chen, Y., Yang, X., Wei, X., Chen, X., ... & Li, S. (2023). Highly sensitive and rapid determination of *Mycobacterium leprae* based on real-time multiple cross displacement amplification. *BMC Microbiology*, *23*(1), 272.

Ibrahim, F., Thio, T. H. G., Faisal, T., & Neuman, M. (2015). The application of biomedical engineering techniques to the diagnosis and management of tropical diseases: A review. *Sensors*, *15*(3), 6947–6995.

Ilkhani, H., & Farhad, S. (2018). A novel electrochemical DNA biosensor for Ebola virus detection. *Analytical Biochemistry*, *557*, 151–155.

Ji, T., Liu, Z., Wang, G., Guo, X., Lai, C., Chen, H., ... & Zhou, Q. (2020). Detection of COVID-19: A review of the current literature and future perspectives. *Biosensors and Bioelectronics*, *166*, 112455.

Jiang, L., Luo, J., Dong, W., Wang, C., Jin, W., Xia, Y., ... & He, H. (2015). Development and evaluation of a polydiacetylene based biosensor for the detection of H5 influenza virus. *Journal of Virological Methods*, *219*, 38–45.

Kaushik, A., Tiwari, S., Jayant, R. D., Marty, A., & Nair, M. (2016). Towards detection and diagnosis of Ebola virus disease at point-of-care. *Biosensors and Bioelectronics*, *75*, 254–272.

Kharsany, A. B., McKinnon, L. R., Lewis, L., Cawood, C., Khanyile, D., Maseko, D. V., ... & Toledo, C. (2020). Population prevalence of sexually transmitted infections in a high HIV burden district in KwaZulu-Natal, South Africa: Implications for HIV epidemic control. *International Journal of Infectious Diseases*, *98*, 130–137.

Kim, J. H., Cho, C. H., Ryu, M. Y., Kim, J. G., Lee, S. J., Park, T. J., & Park, J. P. (2019). Development of peptide biosensor for the detection of dengue fever biomarker, nonstructural 1. *PLoS One*, *14*(9), e0222144.

Konstantinou, L., Varda, E., Pempetsiou, S., Apostolou, T., Loizou, K., Dougiakis, L., Inglezakis, A. and Hadjilouka, A. (2023). A cell-based bioelectric biosensor for *Salmonella* spp. detection in food. *Engineering Proceedings*, *35*(1), 4.

Lee, S. Y., & Oh, S. W. (2023). Lateral flow biosensor based on LAMP-CRISPR/Cas12a for sensitive and visualized detection of *Salmonella* spp. *Food Control*, *145*, 109494.

Lee, T., Kim, G. H., Kim, S. M., Hong, K., Kim, Y., Park, C., … & Min, J. (2019). Label-free localized surface plasmon resonance biosensor composed of multi-functional DNA 3 way junction on hollow Au spike-like nanoparticles (HAuSN) for avian influenza virus detection. *Colloids and Surfaces B: Biointerfaces*, *182*, 110341.

Liu, C., Jiang, D., Xiang, G., Liu, L., Liu, F., & Pu, X. (2014). An electrochemical DNA biosensor for the detection of *Mycobacterium tuberculosis*, based on signal amplification of graphene and a gold nanoparticle–polyaniline nanocomposite. *Analyst*, *139*(21), 5460–5465.

Ma, J., Du, M., Wang, C., Xie, X., Wang, H., Li, T., … & Wu, M. (2021). Rapid and sensitive detection of *Mycobacterium tuberculosis* by an enhanced nanobiosensor. *ACS Sensors*, *6*(9), 3367–3376.

Martins, G., Gogola, J. L., Caetano, F. R., Kalinke, C., Jorge, T. R., Santos, C. N., … &Marcolino-Junior, L. H. (2019). Quick electrochemical immunoassay for Hantavirus detection based on biochar platform. *Talanta*, *204*, 163–171.

Mavrikou, S., Moschopoulou, G., Tsekouras, V., &Kintzios, S. (2020). Development of a portable, ultra-rapid and ultra-sensitive cell-based biosensor for the direct detection of the SARS-CoV-2 S1 spike protein antigen. *Sensors*, *20*(11), 3121.

Mérens, A., Bigaillon, C., & Delaune, D. (2018). Ebola virus disease: Biological and diagnostic evolution from 2014 to 2017. *Médecine et Maladies Infectieuses*, *48*(2), 83–94.

Mittler, E., Dieterle, M. E., Kleinfelter, L. M., Slough, M. M., Chandran, K., & Jangra, R. K. (2019). Hantavirus entry: Perspectives and recent advances. *Advances in Virus Research*, *104*, 185–224.

Mukundan, H., Kumar, S., Price, D. N., Ray, S. M., Lee, Y. J., Min, S., … & Swanson, B. I. (2012). Rapid detection of *Mycobacterium tuberculosis* biomarkers in a sandwich immunoassay format using a waveguide-based optical biosensor. *Tuberculosis*, *92*(5), 407–416.

Murphy, C. N. (2019). Recent advances in the diagnosis and management of Ebola virus disease. *Clinical Microbiology Newsletter*, *41*(21), 185–189.

Nandi, S., Mondal, A., Roberts, A., & Gandhi, S. (2020). Biosensor platforms for rapid HIV detection. *Advances in Clinical Chemistry*, *98*, 1–34.

Narmani, A., Kamali, M., Amini, B., Kooshki, H., Amini, A., & Hasani, L. (2018). Highly sensitive and accurate detection of *Vibrio cholera* O1 OmpW gene by fluorescence DNA biosensor based on gold and magnetic nanoparticles. *Process Biochemistry*, *65*, 46–54.

Neiderud, C. J. (2015). How urbanization affects the epidemiology of emerging infectious diseases. *Infection Ecology & Epidemiology*, *5*(1), 27060.

Nguyen, B. T. T., Peh, A. E. K., Chee, C. Y. L., Fink, K., Chow, V. T., Ng, M. M., & Toh, C. S. (2012). Electrochemical impedance spectroscopy characterization of nanoporous alumina dengue virus biosensor. *Bioelectrochemistry*, *88*, 15–21.

Nicastri, E., Kobinger, G., Vairo, F., Montaldo, C., Mboera, L. E., Ansunama, R., & Ippolito, G. (2019). Ebola virus disease: Epidemiology, clinical features, management, and prevention. *Infectious Disease Clinics*, *33*(4), 953–976.

Nidzworski, D., Pranszke, P., Grudniewska, M., Król, E., &Gromadzka, B. (2014). Universal biosensor for detection of influenza virus. *Biosensors and Bioelectronics*, *59*, 239–242.

Panghal, A., & Flora, S. J. S. (2020). Viral agents including threat from emerging viral infections. In: S.J.S. Flora & Vidhu Pachauri (Eds.), *Handbook on Biological Warfare Preparedness* (pp. 65–81). Academic Press.

Patel, M. K., Solanki, P. R., Kumar, A., Khare, S., Gupta, S., & Malhotra, B. D. (2010). Electrochemical DNA sensor for Neisseria meningitidis detection. *Biosensors and Bioelectronics*, *25*(12), 2586–2591.

Patel, M. K., Solanki, P. R., Seth, S., Gupta, S., Khare, S., Kumar, A., & Malhotra, B. D. (2009). CtrA gene based electrochemical DNA sensor for detection of meningitis. *Electrochemistry Communications*, *11*(5), 969–973.

Paula Vaz Cardoso, L., Dias, R. F., Freitas, A. A., Hungria, E. M., Oliveira, R. M., Collovati, M., & Martins Araújo Stefani, M. (2013). Development of a quantitative rapid diagnostic test for multibacillary leprosy using smart phone technology. *BMC Infectious Diseases*, *13*, 1–10.

Pohanka, M. (2018). Overview of piezoelectric biosensors, immunosensors and DNA sensors and their applications. *Materials*, *11*(3), 448.

Qiu, G., Gai, Z., Tao, Y., Schmitt, J., Kullak-Ublick, G. A., & Wang, J. (2020). Dual-functional plasmonic photothermal biosensors for highly accurate severe acute respiratory syndrome coronavirus 2 detection. *ACS Nano*, *14*(5), 5268–5277.

Rasouli, E., Basirun, W. J., Johan, M. R., Rezayi, M., Mahmoudian, M. R., &Poenar, D. P. (2023). Electrochemical DNA-nano biosensor for the detection of cervical cancer-causing HPV-16 using ultrasmall Fe_3O_4-Au core-shell nanoparticles. *Sensing and Bio-Sensing Research*, *40*, 100562.

Reddy, S. B., Mainwaring, D. E., Al Kobaisi, M., Zeephongsekul, P., &Fecondo, J. V. (2012). Acoustic wave immunosensing of a meningococcal antigen using gold nanoparticle-enhanced mass sensitivity. *Biosensors and Bioelectronics*, *31*(1), 382–387.

Rojas, M., Monsalve, D. M., Pacheco, Y., Acosta-Ampudia, Y., Ramírez-Santana, C., Ansari, A. A., & Anaya, J. M. (2020). Ebola virus disease: An emerging and re-emerging viral threat. *Journal of Autoimmunity*, *106*, 102375.

Sanvicens, N., Pastells, C., Pascual, N., & Marco, M. P. (2009). Nanoparticle-based biosensors for detection of pathogenic bacteria. *TrAC Trends in Analytical Chemistry*, *28*(11), 1243–1252.

Shariati, M., & Sadeghi, M. (2020). Ultrasensitive DNA biosensor for hepatitis B virus detection based on tin-doped WO 3/In 2 O 3 heterojunction nanowire photoelectrode under laser amplification. *Analytical and Bioanalytical Chemistry*, *412*, 5367–5377.

Shourian, M., Ghourchian, H., &Boutorabi, M. (2015). Ultra-sensitive immunosensor for detection of hepatitis B surface antigen using multi-functionalized gold nanoparticles. *Analytica Chimica Acta*, *895*, 1–11.

Silva, M. M. S., Dias, A. C. M. S., Cordeiro, M. T., Marques Jr, E., Goulart, M. O. F., & Dutra, R. F. (2014). A thiophene-modified screen printed electrode for detection of dengue virus NS1 protein. *Talanta*, *128*, 505–510.

Singh, R., Mukherjee, M. D., Sumana, G., Gupta, R. K., Sood, S., & Malhotra, B. D. (2014). Biosensors for pathogen detection: A smart approach towards clinical diagnosis. *Sensors and Actuators B: Chemical*, *197*, 385–404.

Starodub, N. F., Ogorodniichuk, J., Lebedeva, T., & Shpylovyy, P. (2013). Immune biosensors based on the SPR and TIRE: Efficiency of their application for bacteria determination. *Biophotonics—Riga*, *9032*, 211–214.

Tak, M., Gupta, V., & Tomar, M. (2014). Flower-like ZnO nanostructure based electrochemical DNA biosensor for bacterial meningitis detection. *Biosensors and Bioelectronics*, *59*, 200–207.

Taufiq, S., Waqar, M., Sharif, M. N., & Abbas, S. R. (2023). Towards portable rapid TB biosensor: Detecting *Mycobacterium tuberculosis* in raw sputum samples using functionalized screen printed electrodes. *Bioelectrochemistry*, *150*, 108353.

Tavakoli, A., Karbalaie Niya, M. H., Keshavarz, M., Ghaffari, H., Asoodeh, A., Monavari, S. H., & Keyvani, H. (2017). Current diagnostic methods for HIV. *Future Virology*, *12*(3), 141–155.

Tavakoli, H., Hirth, E., Luo, M., Timilsina, S. S., Dou, M., Dominguez, D. C., & Li, X. (2022). A microfluidic fully paper-based analytical device integrated with loop-mediated isothermal amplification and nano-biosensors for rapid, sensitive, and specific quantitative detection of infectious diseases. *Lab on a Chip*, *22*(23), 4693–4704.

Tepeli, Y., & Ülkü, A. (2018). Electrochemical biosensors for influenza virus a detection: The potential of adaptation of these devices to POC systems. *Sensors and Actuators B: Chemical*, *254*, 377–384.

Urrego, L. F., Lopez, D. I., Ramirez, K. A., Ramirez, C., & Osma, J. F. (2015). Biomicrosystem design and fabrication for the human papilloma virus 16 detection. *Sensors and Actuators B: Chemical*, *207*, 97–104.

Vaisocherová-Lísalová, H., Víšová, I., Ermini, M. L., Špringer, T., Song, X. C., Mrázek, J., … & Homola, J. (2016). Low-fouling surface plasmon resonance biosensor for multi-step detection of foodborne bacterial pathogens in complex food samples. *Biosensors and Bioelectronics*, *80*, 84–90.

Van Panhuis, W. G., Grefenstette, J., Jung, S. Y., Chok, N. S., Cross, A., Eng, H., … & Burke, D. S. (2013). Contagious diseases in the United States from 1888 to the present. *The New England Journal of Medicine*, *369*(22), 2152.

Walldorf, J. A., Cloessner, E. A., Hyde, T. B., MacNeil, A., Bennett, S. D., Carter, R. J., … & Marston, B. J. (2019). Considerations for use of Ebola vaccine during an emergency response. *Vaccine*, *37*(48), 7190–7200.

Wang, Y., Ye, Z., & Ying, Y. (2012). New trends in impedimetric biosensors for the detection of foodborne pathogenic bacteria. *Sensors*, *12*(3), 3449–3471.

Xiang, Y., Zhu, X., Huang, Q., Zheng, J., & Fu, W. (2015). Real-time monitoring of mycobacterium genomic DNA with target-primed rolling circle amplification by a Au nanoparticle-embedded SPR biosensor. *Biosensors and Bioelectronics*, *(66)*, 512–519.

Yang, D., Zuo, C., Wang, X., Meng, X., Xue, B., Liu, N., … & Zhu, H. (2014). Complete replication of hepatitis B virus and hepatitis C virus in a newly developed hepatoma cell line. *Proceedings of the National Academy of Sciences*, *111*(13), E1264–E1273.

Yin, L., Man, S., Ye, S., Liu, G., & Ma, L. (2021). CRISPR-Cas based virus detection: Recent advances and perspectives. *Biosensors and Bioelectronics*, *193*, 113541.

Yoo, S. M., & Lee, S. Y. (2016). Optical biosensors for the detection of pathogenic microorganisms. *Trends in Biotechnology*, *34*(1), 7–25.

Zaid, M. H. M., Abdullah, J., Yusof, N. A., Wasoh, H., Sulaiman, Y., Noh, M. F. M., & Issa, R. (2020). Reduced graphene oxide/tempo-nanocellulose nanohybrid-based electrochemical biosensor for the determination of mycobacterium tuberculosis. *Journal of Sensors*, 4051474, 11 p.

Zhang, B., Salieb-Beugelaar, G. B., Nigo, M. M., Weidmann, M., & Hunziker, P. (2015). Diagnosing dengue virus infection: Rapid tests and the role of micro/nanotechnologies. *Nanomedicine: Nanotechnology, Biology and Medicine*, *11*(7), 1745–1761.

Zhuang, J., Yin, J., Lv, S., Wang, B., & Mu, Y. (2020). Advanced "lab-on-a-chip" to detect viruses: Current challenges and future perspectives. *Biosensors and Bioelectronics*, *163*, 112291.

11 Targeting Hallmarks of Cancer Using a Combo of Traditional Medicine and Modern Approaches

Basanti Brar, Kiran Bala, Kaushlesh Ranjan, Seema Joshi, Sumnil Marwaha, Anju Manuja, and Minakshi Prasad

11.1 INTRODUCTION

Cancer cells exhibit flaws in the regulatory processes that dictate their rate of division, as well as in the feedback loops that oversee these regulatory mechanisms, resulting in disruptions in homeostasis.

In contrast, normal cells display a well-controlled growth pattern. They only proliferate when prompted by growth factors and possess a molecular brake system that halts their division when they sustain damage. If the damage is irreparable, they undergo programmed cell death, known as apoptosis. Normal cells also have a limited number of divisions they can undergo and maintain their designated location within a tissue structure. Moreover, they rely on blood supply for growth.

To develop into a cancerous state, a cell must surmount all these regulatory mechanisms. In each of these processes, several proteins play a crucial role, and a malfunction in a key protein is required for cancer development. These proteins lose their functionality or exhibit altered functionality due to damage to the DNA sequences in their genes, brought about by acquired or somatic mutations (Hanahan and Weinberg, 2000).

Cancer poses an enduring challenge for researchers worldwide. There have been numerous attempts to understand the unique behavior and properties of tumor cells, and at least grasp the fundamental mechanisms driving their formation and development. Hanahan and Weinberg presented a significant editorial highlighting that malignant cells in every form of cancer share six distinct traits. These traits allocate healthy cells to transition into tumors (Hanahan and Weinberg, 2000). They involve maintaining proliferative signaling, eluding growth suppression, escaping immune system detection, attaining replicative immortality, fostering inflammation related to tumors, and triggering invasion and metastasis. This influential study opened new pathways toward comprehending the puzzling aspects of this aggressive disease.

Nonetheless, in 2011, the same authors advocated for eight fundamental features, which involve the developed abilities to maintain continuous signaling for cell

DOI: 10.1201/9781032642567-13

proliferation, elude mechanisms that inhibit growth, withstand cell death, achieve replicative immortality, initiate or access blood vessel formation, activate metastasis and invasion, reconfigure cellular metabolism, and escape immune detection. In their most recent expansion of this concept, the regulation of cellular metabolism and evasion of immune recognition were identified as newly recognized hallmarks (Hanahan and Weinberg, 2011).

11.2 CANCER HALLMARKS

The progress of targeted therapy has developed alongside the unraveling of cancer's fundamental mechanisms. Presently, with a grasp of the 14 core hallmarks of cancer and the crucial changes disrupting regular cell growth pathways, the implementation of precise treatments has become viable in a considerable array of cases as shown in Figure 11.1. This review aims to explore the notion of cancer hallmarks, which has provided the emergence of besieged therapies, and examine how this strategy has led to significant enhancements in patient survival rates.

11.2.1 Sustaining Proliferative Signaling

In normal cells, growth and division are contingent upon external cues. However, in the context of cancer cell growth, this regulatory mechanism undergoes modifications and deregulation (Hanahan and Weinberg, 2011). A central aspect of these hallmark features revolves around the mitogen-activated protein kinases (MAPK)/extracellular signal-regulated kinases (ERK) mechanisms, commonly called as the RAS-RAF-MEK-ERK mechanism (Hanahan and Weinberg, 2011). This method comprises a sequence of signaling events governing various cellular functions like growth, differentiation, and delicate equilibrium between cellular survival and apoptosis (Zhang and Liu, 2002) (Figure 11.1).

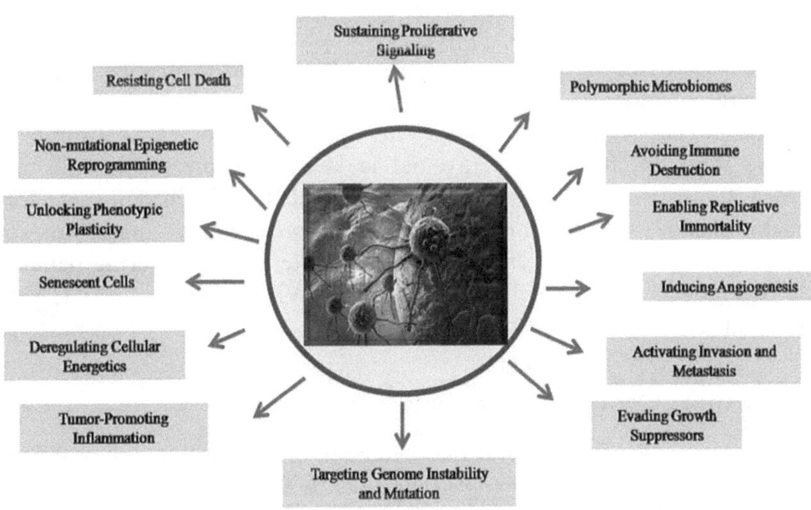

FIGURE 11.1 Core hallmarks of cancer and the crucial changes disrupting regular cell growth pathways.

Irregularities or malfunctions within MAPK pathway or its constituents can trigger tumor genesis, enabling cancer cells to maintain proliferative signals while evading fostering angiogenesis, apoptosis and facilitating metastasis. Given the paramount significance of this trail, extensive research has been devoted to inhibiting its activity. An essential constituent of this pathway is BRAF (Davies et al., 2002). BRAF mutations are present in about 8% of all cancers, rising to nearly 50% in higher melanoma cases, around 40% in papillary thyroid cancer, approximately 10% in colorectal cancers, and lower percentages in lung cancer. Additionally, BRAF mutations are evident in various hematological malignancies, including plasma cell myeloma and hairy cell leukemia (Tol et al., 2009; Holderfield et al., 2014).

Two prominent inhibitors, Dabrafenib and Vemurafenib, specifically target mutant BRAF and are primary drugs in managing melanoma. Vemurafenib, for instance, has significantly enhanced overall survival and progression-free survival (PFS) in advanced melanoma patients (Sullivan and Flaherty, 2013). Dabrafenib, another BRAF mutant inhibitor, has shown a PFS advantage with a lower incidence of adverse effects (Hauschild et al., 2012). BRAF mutations are linked to an unfavorable prognosis in cancer, associated with established poor prognostic factors in colorectal tumors, including elderly patients, female gender, and a limited response to epidermal growth factor receptor (EGFR) inhibitor and chemotherapy medications (Clarke and Kopetz, 2015).

In the context of papillary thyroid cancer with the BRAF V600E mutation, Vemurafenib has demonstrated a significant objective response rate of up to 38.5% in patients with cases refractory to radioactive iodine (Brose et al., 2016). Considerate downstream signaling events and their connection to "sustaining proliferative signaling" hallmark has paved the way for these targeted therapeutic advancements. Scientific progress has illuminated the potential role of downstream pathways in the development of drug resistance. Trametinib, an MEK inhibitor, has been utilized alongside BRAF inhibitors, allowing the concurrent targeting of two distinct mutations (Marani et al., 2023). This combined treatment approach has unveiled several advantages, including prolonged overall survival when comparison to BRAF inhibition alone, an increased response rate, extended PFS, all accompanied by a notable reduction in dermatological side effects (Flaherty et al., 2012).

A pivotal moment in medical history was the introduction of imatinib, a tyrosine kinase inhibitor, for chronic myeloid leukemia treatment. Imatinib targets the BCR-ABL fusion gene responsible for the Philadelphia chromosome formation. Imatinib stands as a testament to success in cancer management, fundamentally altering the prognosis of chronic myeloid leukemia and offering patients an extended overall survival through targeted therapy (Villanueva et al., 2011; Mughal and Goldman, 2006).

11.2.2 Evading Growth Suppressors

In the normal physiological context, the regulation of cell growth and proliferation is meticulously orchestrated, encompassing a delicate equilibrium between signals that promote growth and those that inhibit it. Tumor cells must disrupt this balance to facilitate their growth and survival (Hanahan and Weinberg, 2011). Two crucial proteins, namely retinoblastoma (pRb) and P53, govern these antigrowth signals. pRb remains active when unphosphorylated, a state achieved through its association with the E2F transcription factor to form an RB/E2F complex.

The cellular processes involved in growth suppression are intricate, with the retinoblastoma (pRb) protein playing a pivotal role. This protein complex hinders the transition of cells into the next S stage, which corresponds to replication of DNA in the normal cell cycle (Zhang and Zhu, 2023). Additionally, it exerts inhibitory control over cyclin-dependent kinases (CDKs) during the G1 phase. However, in response to outer stimuli, accumulated CDKs can phosphorylate pRb, thereby releasing it from its association with E2F. This phosphorylation event promotes cell division, rendering cells unresponsive to antigrowth signals (Dick and Rubin, 2013; Nevins, 2001).

The P53 gene holds significant importance in cancer research due to its involvement in various cancer types, including tumors of head and neck, colorectal cancer, and esophageal cancer where mutations or loss of this gene have been identified. Exposure to risk factors like radiation and chemotherapy can lead to mutations or loss of these genes, causing uncontrolled cell division (Sherr and McCormick, 2002). Another key regulator in this practice is the transforming growth factor-beta (TGF-beta), which functions as a negative controller of cell growth. It regulates pRb signaling by preventing RB phosphorylation, thereby inhibiting cell proliferation. However, when cells transform into cancerous forms, this regulatory system becomes dysregulated, departing from normal cellular control (Paterson et al., 1995).

Cyclin-dependent kinase inhibitors, such as Palbociclib, target CDK4 and CDK6 CDKs to modulate CDK-RB-E2F pathway. Palbociclib finds application in metastatic hormone-positive breast cancer cure, where its addition to hormonal therapy resulted in a notable 10-month enhancement in PFS, albeit without significant overall survival benefits. Hematological side effects, primarily neutropenia, constitute the primary adverse effects of this drug but they are generally manageable and well tolerated (Finn et al., 2015).

11.2.3 Avoiding Immune Destruction

Immune system serves as a crucial defender against cancer cells through various mechanisms. However, cancer cells can exploit this system to inflict harm on the host (Mittal et al., 2014). Immunotherapy represents a groundbreaking advancement in modern cancer treatment. These treatments leverage the body's immune system with few special effects in evaluation to chemotherapy. Primary outcomes have demonstrated an important reduction in tumor volume upon their practical application (Rosenberg et al., 2011). Programmed cell death-1 (PD-1) serves as a checkpoint protein expressed on activated T cells, working in conjunction with an additional protein known as cytotoxic T-lymphocyte-associated protein 4 (CTLA-4). CTLA-4, which is also found on T-cell surfaces and prevents these immune cells from attacking healthy tissues. Both PD-1 and CTLA-4 act as checkpoint receptors that dampen immune responses. Binding of PD-1 with programmed death-ligand 1 (PD-L1) triggers a series of signals inhibiting T-cell activation and slowing down the immune response. Cancer cells often express elevated levels of PD-L1 to evade the immune system. Targeting PD-1, CTLA-4, and PD-L1 acts as an attractive strategy, as blocking these pathways can hinder cancer cells from evading immune surveillance.

Human monoclonal antibody, that is, Ipilimumab obstructs CTLA-4 and boosts the immune response of the body against cancer cells. This medicine has brought about significant advancements in melanoma management. However, its usage is linked with immune-associated adverse effects, such as diarrhea and hepatic toxicity (Hodi et al., 2010; Fecher et al., 2013). Nivolumab and Pembrolizumab, monoclonal antibodies inhibiting the programmed death-1 checkpoint, have showcased substantial enhancements in survival rates and have obtained approval for various cancer types like non-small cell lung cancer cells (NSCLC), melanoma, Hodgkin's lymphomas, and renal cell carcinoma (Buchbinder and Desai, 2016; Ott et. al., 2013). Resistance to these drugs has prompted investigations into their combination with other therapies. Ipilimumab is frequently used alongside PD-L1 inhibitors, displaying superior responses compared to ipilimumab alone in melanoma (Callahan et al., 2015). Notably, these drugs can induce skin reactions, gastrointestinal issues, and lung inflammation, potentially manifesting as severe pneumonitis (Callahan et al., 2015).

11.2.4 Enabling Replicative Immortality

Telomerase—a unique reverse transcriptase enzyme that binds to chromosome ends—is responsible for preserving the integrity of these critical regions during normal cell division. It achieves this by de novo formation of telomeric DNA. Telomeres shorten with each cell division, which limits a cell's proliferative capacity. Dysregulation of telomerase function during cancer leads to abnormal cell growth, a hallmark of cancer cells; however, normal cells have a defined number of cell divisions during a basic cellular pathway, and uncontrolled or cancerous cells can bypass the current limit by re-expressing or over-expressing telomerase. Nearly 90% of cancer cells express telomerase, making it an attractive goal for treatment (Hahn, 2001; Shay et al., 2001). Various approaches, including small-molecule inhibitors, gene therapy, and immunotherapies, aim to target the telomerase enzyme. Cell proliferation can be halted by the inhibition of RNA template. HSP90 inhibitors and G-quadruplex stabilizers indirectly block the active subunit of the enzyme, leading to cell death (Jafri et al., 2016).

Tertomotide HCL, GV1001™, an inhibitory drug for telomerase, is used in clinical trials in Korea. It acts as a therapeutic vaccine which enhances immune system's recognition of telomerase in cancer-associated cells. This medicine has demonstrated durable T-cell memory associated with prolonged PFS in NSCLC and improved outcomes. However, it has not advanced beyond Phase 2 trials and requires further investigation for confirmation. GV1001 has also been tested in hepatocellular carcinoma and metastatic melanoma, with positive immune responses, although these clinical trials have reached only Phase 1 or 2. In pancreatic and breast cancers, GV1001 did not yield survival benefits. Resistance is less likely to occur with this approach due to the rare mechanism of alternative telomere lengthening observed in tumor-derived cell lines and human tumors (Ruden and Puri, 2013).

11.2.5 Tumor-Promoting Inflammation

Inflammatory cells are a prominent element of the tumor surroundings, a concept established by Rudolf Virchow as far back as 1863. It has been proposed by Hanahan and Weinberg (2011) that tumor-linked inflammation might augment the cancerous cells, aiding their survival along with progression. Extensive research has linked chronic inflammation to cancer development, as inflammatory cells provide growth factors and other mechanisms that facilitate hallmark features such as cytokines and proteases (Balkwill and Mantovani, 2001). Moreover, these inflammatory cells can release chemicals that induce genetic mutations in the tumor microenvironment, enabling malignant cells to proliferate, survive, and metastasize (Grivennikov and Karin, 2010).

Nonsteroidal anti-inflammatory drugs (NSAIDs) demonstrated their action in several cancer types, particularly in colorectal and breast cancers, even in lesser extent in lung cancer. Mainly, the study has shown a noteworthy reduction in the risk associated with the use of aspirin for the treatment of breast and colon cancer (Thun et al., 2002). These drugs are believed to exert indirect anticancer effects by inducing apoptosis and inhibiting spread. Celecoxib, a specific cyclooxygenae-2 (COX-2) inhibitor, blocks prostaglandin discharge. Polyps are identified risk factors for colon cancer, and various studies have explored the association between COX-2 inhibitors and polyps, especially premalignant ones. These drugs have substantially reduced the size and numbers of polyps in adenomatous polyposis syndrome and others colorectal diseases while administered with 400 mg dose twice a day for at least 6 months, resulting in a 31% decline in polyps numbers (Steinbach et al., 2000). Researchers have investigated the mixture of COX-2 inhibitors with hormonal therapy in patients with breast cancers, reporting promising results that indicate a 20% reduction in breast cancer risk among females who used celecoxib and aspirin (Agrawal and Fentiman, 2008). However, the significant heart adverse effects of COX-2 inhibitors have limited their clinical use.

11.2.6 Invasion Activation and Metastasis

Understanding how tumor cells cultivate and metastasize has been a long-standing challenge for scientists. Grasping this mechanism is vital for managing various cancers, as the ability of malignant cells to change shape, downregulate cell–cell adhesion molecules such as E-cadherin, and lose attachment within extracellular matrix enables them to invade associated tissues and potentially metastasize (Thiery, 2002). Cells undergoing epithelial to mesenchymal transition (EMT) exhibit altered morphology and functionality, enabling them to migrate, invade, and ultimately metastasize (Smith and Bhowmick, 2016). Hepatocyte growth factor (HGF)/c-Met mechanism plays a pivotal function in cancer invasion and metastasis, as demonstrated in animal experiments. Human cell lines exhibit overexpression of c-Met and HCG becomes cancerous and metastatic after transplanted into exposed mice (Mizuno and Nakamura, 2013). HGF/c-Met signaling dysregulation has aroused as a significant contributor to metastasis and invasion in human malignancies, and EMT is induced by HGF. The HGF/c-Met mechanism plays a crucial role during embryonic development, contributing to the development of organs such as the placenta and the central nervous system. In adult life, this pathway also participates in organ regeneration, notably in wound healing. Aberrant regulation of the Met pathway is observed in a broad spectrum of human tumors, including lung cancer (in almost 50% of cases), colorectal cancer, ovarian malignancies, breast cancer, pediatric cancerous-like lymphoma, medulloblastoma, papillary cell carcinoma as well as bone cancer medulloblastoma, lymphoma, and papillary cell carcinoma, as well as bone tumors where Met has shown expression in osteoblast cells (Sierra and Tsao, 2011; Sacco and Clague, 2015). Met signaling dysregulation can occur due to various molecular principles, including c-Met gene mutations, chromosomal rearrangements involving c-Met, and c-MET amplifications. Activation of the MET (c-Met) receptors in this pathway requires the ligand HGF. Additionally, this pathway is implicated in conferring radioresistance and chemoresistance in cancers like colorectal cancer, breast cancer and lung tumor (Salgia, 2009; Raghav et al., 2012).

Rilotumumab (AMG102) is a type of monoclonal antibody that inhibits HGF/C-Met by targeting HGF(IgG2), which impedes the attachment of HGF to its receptor, thereby promoting apoptosis in cells that express c-Met. Medical investigations on the current drug were finished due to increase in mortality with the combination by chemotheraphy and rilotumumab treatment as compared with chemotherapy alone (Cunningham et al., 2013). Multikinase inhibitor cabozantinib mainly targets VEGFR2, MET, and FLT3 by following different pathways and is approved for the treatment of medullar thyroid cancer but in renal cell carcinoma their role is not specific (Fallahi et al., 2015; Elisei et al., 2013; Viola et al., 2013; Choueiri et al., 2015). The drug has shown unlimited continued existence and progression-free endurance in hepatocellular carcinoma patients (Abou-Alfa et al., 2018).

11.2.7 Inducing Angiogenesis

Through angiogenesis mechanisms, malignant cells promote their growth and metastasis. Vascular endothelial growth factor (VEGF) family and its receptor plays the primary role in neoplastic angiogenesis and their appearance can be induced by hypoxia. The interaction between pro-angiogenic factors such as angiopoietins, basic fibroblast-like growth factor, VEGF and TGF-β, along with various anti-angiogenic factors such as thrombospondin-1, angiostatin, and endostatin, is alleged to be a critical determinant in this procedure (Bakhtiyari et al., 2023).

11.2.7.1 Inhibitors of Vascular Endothelial Growth Factor Signaling

Targeting this path is highly appealing because it disrupts the supply of blood and leads to the death of malignant cells. This kind of pathway did not rely on a specific histological type, unlike chemotherapy, and primarily affects tumor microvasculature with fewer side effects compared with chemotherapy.

Humanized monoclonal antibody, Bevacizumab (Avastin), inhibits the angiogenesis by targeting VEGF-A. It is employed in various cancers, including metastatic colorectal cancer, breast cancer, lung cancer, and brain tumors. Avastin is associated with improved PFS in patients with colorectal cancer, contributing an extension in overall survival (Ilic et al., 2016). However, its support in the National Institute for

Health and Care Excellence guidelines (the UK's cancer guidelines) was delayed for several years due to its limited cost-effectiveness. Furthermore, the choice of chemotherapy combination with Bevacizumab remains unclear, and trial results have been conflicting. Severe side effects such as increased anxiety and impaired wound healing are important concerns in individuals undergoing cancer treatment (Baliga, et al., 2023). In breast cancer, its indications have been recently modified, and its use has declined due to disappointing results regarding its safety and failure to improve the quality of life or extend survival in patients. In patients with lung cancer, the non-squamous drug is used, as the squamous type can lead to blood loss, and firm restrictions on its use exist, including any past hemorrhage and the patient's recital condition (Leal, et al., 2023). Survival benefits associated with this feature has been unsatisfactory in clinical practice, with drugs failing to demonstrate significant improvements in clinical data. Additionally, angiogenesis inhibitors often inadvertently promote another hallmark, namely, invasion.

11.2.8 Targeting Mutation anGenome Instability

Aneuploidy occurs when genomic instability (GI) results from DNA damage and chromosomal abnormalities, including rearrangements and gains or losses. GI is primarily characterized by DNA damage and chromosomal abnormalities, with aneuploidy being the most common form of GI (Sahgal, et al., 2023). Poly-ADP-ribose polymerase (PARP) inhibitors are involved in the repair of single-strand DNA damage and in stabilizing the genome. Inhibition of PARP prevents DNA repair, ultimately leading to cell death (Kanev et al., 2023).

Olaparib, an oral PARP inhibitor, is used for persistent ovarian tumors in BRCA mutations. It is also employed in breast tumors, particularly in the triple-time negative subtype and in cases of BRCA mutations, where PARP is highly upregulated due to its effectiveness (Andrade et al., 2023). Olaparib treatment in prostate cancer patients who no longer responded to ordinary therapies has shown a significant response rate, particularly in those with deficiencies in DNA repair mechanisms.

11.2.9 Cell Death Resisting

Apoptosis of cells are generally regulated by intrinsic (or mitochondrial) and extrinsic pathways, which is based on peripheral signals for activation. BCL-2 family proteins are serving as the primary regulators of apoptosis in intrinsic pathways. Apoptotic and pro-apoptotic are the members of BCL-2 family. These groups interact to control apoptosis (Czabotar and Garcia, 2023). BCL-2, for example, has four BH domains. BH3-only proteins, a subclass of the BCL-2 family, play a crucial role in transmitting death signals between the intrinsic and extrinsic pathways (Xiao et al., 2023). They function by disrupting the activity of anti-apoptotic Bcl-2 proteins or by activating pro-apoptotic proteins. In cases where cancer cells elevate the presence of anti-apoptotic BCL2 or reduce pro-apoptotic proteins, the equilibrium within the BCL2-family is disturbed, allowing the evasion of apoptosis (Wani et al., 2023). ABT-737, identified as a BH3 mimetic inhibitor, has been utilized in follicular lymphoma treatments and chronic lymphocytic leukemia (CLL); yet it has demonstrated inefficacy in treating solid tumors. Venetoclax (Abt-199) was specifically crafted as an orally available, highly selective small-molecule inhibitor targeting Bcl-2 family proteins (Aksoy et al., 2023). It exhibits a strong binding affinity for Bcl-2 and a comparatively weaker affinity for proteins of this family (Bcl-XL and Bcl), and it has recently gained consent for managing relapsed/refractory CLL (Xu et al., 2023).

11.2.10 Deregulating Cellular Energetics

Cancer cells primarily depend on anaerobic glycolysis to fulfill their elevated energy requirements for proliferation and expansion. This metabolic process yields little energy, compelling cancer cells to elevate the presence of glucose transporters, predominantly Glut-1, to augment the intake of glucose into the cytoplasm (Geng et al., 2023). Furthermore, cancer cells amplify the expression of numerous glycolytic enzymes. Additionally, oncogenes such as RAS, Myc, HIF-1alpha, and mutated P53 can also instigate glycolysis within these cells.

There is considerable interest in targeting the metabolic path because of its significant role in the development of tumors. Inhibitors of aerobic glycolysis, acting at various points along this pathway, have demonstrated hopeful outcomes. Notably, glutaminase 1 inhibition, which exhibits high upregulation in malignancies, has shown effectiveness. Targeting HIF, MCT1, and MCT4 can disrupt transfer of lactate, inducing the death of cells due to nutrient deprivation (Lobel et al., 2023). Another potential therapeutic strategy involves targeting mutant forms of isocitrate dehydrogenase 1 and 2. Metformin, a widely known antidiabetic medication, reduces liver mitochondrial ATP production, consequently lowering blood glucose levels. However, research in this domain is still in its nascent stages. This metabolic hallmark has significantly contributed to the advancement of positron emission tomography-computed tomography scans (18F-fluoro-2-deoxyglucose), enabling differentiation between malignant and normal cells based on their heightened glucose uptake (Aljizeeri and Mallah, 2023). In addition to the established cancer hallmarks, several emerging hallmarks and enabling characteristics have been proposed, including phenotypic plasticity unlocking, senescent cells, non-mutational epigenetic reprogramming and polymorphic microbiomes (Hanahan, 2022).

In addition to the aforementioned hallmarks of cancer, certain other hallmarks of cancer are also proposed as emerging hallmarks and enabling characteristics including unlocking phenotypic plasticity, non-mutational epigenetic reprogramming, polymorphic microbiomes, and senescent cells (Hanahan, 2022).

11.3 UNLOCKING PHENOTYPIC PLASTICITY

In the process of organogenesis, as cells develop and arrange themselves into tissues, they undergo terminal differentiation. This means that progenitor cells sometimes permanently cease their growth at the end of these processes. The ultimate outcome of cellular differentiation typically inhibits further cell division, which is essential to prevent the cancer development. A growing proof that suggests the ability to unlock the generally restricted potential for phenotype plasticity, allowing cells to run off from terminal differentiation state, is a crucial element in the development of cancer.

Phenotypic plasticity is a distinctive acquired capability that can lead to a variety of disruptions in cellular differentiation. These disruptions include reverting from mature to progenitor states (dedifferentiation), stops (terminal) differentiation from progenitor cell states, and transforming into diverse cell lineages (Hanahan, 2022).

These three types of mechanisms highlight cellular plasticity and selective regulators which are distinct at least in part from intrinsic tumorgenic drivers along with other hallmark capabilities. These three phenotypic plasticity subtypes, involving the reversal of full-grown cells into progenitor states (dedifferentiation), the prevention of differentiation to maintain a stem-like cell state, and the transformation into atypical cell lines, appear to play a role in various types of tumor throughout early cancer development, malignant advancement, and response to treatment.

Two key points need to be taken into account regarding these mechanisms. First, dedifferentiation and blocked differentiation often appear interconnected, making it difficult to differentiate them in many tumor types where it's unclear whether the cell-of-origin is a dividing cell. Second, when cancer cells acquire or maintain phenotypes resembling progenitor cells and lose their differentiated characteristics, this is typically a crude representation of the normal developmental stage. This process occurs alongside other changes that enable hallmark characteristics in cancer cells but are absent in naturally developing cells (Hanahan, 2022).

Similar to other hallmark features, cellular plasticity is not a recent development unique to cancer cells. Rather, it signifies the transformation of dormant but activatable capabilities found in various normal cells, which are utilized to maintain homeostasis, facilitate repair, and participate in regeneration. Consequently, the suggestion that harnessing cellular plasticity to induce disruptions in differentiation constitutes a distinct hallmark capability merits further exploration. This proposed capability exhibits unique regulatory attributes and cellular characteristics, distinguishing it from the firmly established core hallmarks of tumors.

11.3.1 Non-Mutational Epigenetic Reprogramming

DNA mutations and GI are main factors during the development of tumors. Several studies have demonstrated that mutations in genes responsible for organizing, modulating, and maintaining chromatin structure, which in turn globally control gene expression, are increasingly being identified and functionally linked to the hallmark characteristics of tumors (Flavahan et al., 2017; Jones et al., 2016). However, there is another scenario in which a seemingly independent process of genome reprogramming, driven solely by epigenetically regulated changes in gene expression, referred to as non-mutational epigenetic reprogramming, may possibly also contribute to the initiation of cancer. The conversation regarding cancer evolution without mutations and the epigenetic shaping of cancer hallmark traits began within the scientific community nearly a decade ago (Feng et al., 2021; Nam et al., 2021).

The concept of non-mutational control over gene expression is widely accepted as the primary method guiding embryonic maturity, differentiation, and organ formation (Bitman and Orian, 2021). In adults, non-mutational regulations typically function through long-lasting memory, involving alterations in histone and gene modifications, changes in chromosome structure, along with the activation of gene expression controls that are regulated over time through encouraging and unconstructive feedback loops (Hegde and Smith, 2019).

Furthermore, research has shown that non-mutational epigenetic regulation encompasses several specific mechanisms, including micro-environmental mechanisms for epigenetic reprogramming, the heterogeneity of epigenetic regulation, and the epigenetic control of the various stromal cell types present in the micro surroundings of tumors (Hanahan, 2022). Additionally, several pieces of evidence suggest that similar epigenetic mutations may also play a function in acquiring hallmark traits during tumor development and malignant progression.

11.3.2 Polymorphic Microbiomes

There is a growing concern regarding the influence of resident microbiomes, which include bacteria and fungi, on health and disease (Thomas et al., 2017). Concerning cancer, there is mounting evidence suggesting that the variations in microbiomes among individuals within a population can have a significant impact on cancer characteristics (Dzutsev et al., 2017).

Studies linking human data with investigational manipulations in tumor mouse models are revealing specific microbes, primarily but not entirely bacteria, that may have either defensive or harmful effects on the progression of cancer, its progression, and its response to treatment. The link between individuals' microbiomes and cancer is established through the gut microbiome's modulating effects on cancer and the influence of microbiota within tumors on cancer outcomes.

There is evidence suggesting that confident microbial (bacterial) species are able to straightforwardly trigger the proliferative signaling hallmark, as seen in the colon epithelium (Dzutsev et al., 2017). They can also influence the suppression of growth by modifying the activity of tumor suppressors in different sections of the intestine (Kadosh et al., 2020). However, their direct effects on other hallmark traits, like preventing cell mortality, promoting angiogenesis, and encouraging metastasis and invasion, remain unclear. It's also uncertain whether these observations can be applied broadly to various types of cancers in humans. Nevertheless, a progressively persuasive case can be constructed asserting that the diverse variations in microbiomes present in the intestine and other organs constitute a distinctive facilitator for acquiring hallmark traits. This factor converges with and complements those associated with genome instability, mutations, and inflammation that promotes tumor development.

11.3.3 Senescent Cells

Cellular senescence embodies a generally irreversible form of cell division arrest, likely evolving as a protective mechanism to maintain tissue equilibrium. It operates as a complementary process to programmed cell death, aiming to deactivate and ultimately eliminate unhealthy, dysfunctional, or unnecessary cells. Various factors can trigger senescence in cells, including environmental stresses such as nutrient deficiency and DNA damage, as well as harm to cell structures and imbalances in cellular communication networks (Birch and Gil, 2020). These factors have been associated with the observed increase in the number of senescent cells in different organs during the aging process (He and Sharpless, 2017; Kowald et al., 2020). For a considerable period, cellular senescence has been perceived as a defense against cancer, inducing cancerous cells to undergo senescence (Lee and Schmitt, 2019). However, mounting evidence suggests the opposite: under certain circumstances, senescent cells can play diverse roles in promoting tumor development and the progression of malignancies (Kowald et al., 2020; Wang et al., 2020).

The primary mechanism by which senescent cells enhance tumor traits is believed to be through the senescence-associated secretory phenotype (SASP). This phenotype is capable of transmitting signaling molecules and proteases to nearby viable cancer cells and other cells within the tumor microenvironment (TME) in a paracrine manner. This transmission serves to convey hallmark characteristics. Consequently, senescent cancer cells are considered contributors to proliferative signaling, evasion of apoptosis, promotion of angiogenesis, encouragement of invasion and metastasis, and suppression of the immune response against tumors (He and Sharpless, 2017; Kowald et al., 2020; Wang et al., 2020). The notion that tumors comprise genetically transformed cancer cells interacting with and benefiting from the recruitment of epigenetically or phenotypically altered accessory (stromal) cells is crucial to understanding cancer development. Furthermore, acknowledging the significance of senescent cells advocates for an additional objective: therapeutically targeting tumor-promoting senescent cells of all types. This can be achieved through pharmacological or immunological removal or by reprogramming the SASP into variants that counteract the tumor (Birch and Gil, 2020; Wang et al., 2020; Amor et al., 2020).

11.4 CONCLUSION

The significant strides made in comprehending cancer biology have translated into tangible clinical advantages, with new drugs enhancing both survival and response rates in numerous cases. Despite some setbacks, the future holds promise with a deepening understanding of the proteins involved in these pathways. Customizing treatment based on specific biomarkers or receptors presents an avenue to surmount various challenges. Resistance remains a formidable obstacle, and a potential solution lies in combining multiple targeted pathway inhibitors in the future. Moreover, uncovering the intricate relationships between various pathways and proteins within the hallmarks, such as P53's multifaceted role, may enable the targeting of multiple pathways by focusing on a single protein. Additional features of cancer, such as the activation of phenotypic plasticity, non-mutational epigenetic reprogramming, the influence of polymorphic microbiomes, and the presence of senescent cells, have been proposed as emerging hallmarks and enabling characteristics.

REFERENCES

Abou-Alfa, G. K., Meyer, T., Cheng, A. L., El-Khoueiry, A. B., Rimassa, L., Ryoo, B. Y., … & Kelley, R. K. (2018). Cabozantinib in patients with advanced and progressing hepatocellular carcinoma. *New England Journal of Medicine*, *379*(1), 54–63.

Agrawal, A., & Fentiman, I. S. (2008). NSAIDs and breast cancer: A possible prevention and treatment strategy. *International Journal of Clinical Practice*, *62*(3), 444–449.

Aksoy, O., Lind, J., Sunder-Plaßmann, V., Vallet, S., & Podar, K. (2023). Bone marrow microenvironment-induced regulation of Bcl-2 family members in multiple myeloma (MM): Therapeutic implications. *Cytokine*, *161*, 156062.

Aljizeeri, A., & Al-Mallah, M. H. (2023). The role of noninvasive cardiac imaging in the management of diseases of the cardiovascular system. In: A. Gholamrezanezhad, M. Assadi, H. Jadvar (Eds.), *Radiology-Nuclear Medicine Diagnostic Imaging: A Correlative Approach* (pp. 257–284). John Wiley & Sons Ltd.

Amor, C., Feucht, J., Leibold, J., Ho, Y. J., Zhu, C., Alonso-Curbelo, D., … & Lowe, S. W. (2020). Senolytic CAR T cells reverse senescence-associated pathologies. *Nature*, *583*(7814), 127–132.

Andrade, J. R., Gallagher, A. D., Maharaj, J., & McClelland, S. E. (2023). Disentangling the roles of aneuploidy, chromosomal instability and tumour heterogeneity in developing resistance to cancer therapies. *Chromosome Research*, *31*(4), 28.

Bakhtiyari, M., Liaghat, M., Aziziyan, F., Shapourian, H., Yahyazadeh, S., Alipour, M., … & Pornour, M. (2023). The role of bone marrow microenvironment (BMM) cells in acute myeloid leukemia (AML) progression: Immune checkpoints, metabolic checkpoints, and signaling pathways. *Cell Communication and Signaling*, *21*(1), 252.

Baliga, M. S., Lasrado, S., Krishna, A., George, T., Madathil, L. P., D'souza, R. F., & Palatty, P. L. (2023). Social, ethical and treatment related problems faced by healthcare workers in the care of head and neck cancer patients: A narrative review from the bioethics consortium from India. *Indian Journal of Otolaryngology and Head & Neck Surgery*, *75*(4), 4137–4147.

Balkwill, F., & Mantovani, A. (2001). Inflammation and cancer: Back to Virchow? *The Lancet*, *357*(9255), 539–545.

Birch, J., & Gil, J. (2020). Senescence and the SASP: Many therapeutic avenues. *Genes & Development*, *34*(23–24), 1565–1576.

Bitman-Lotan, E., & Orian, A. (2021). Nuclear organization and regulation of the differentiated state. *Cellular and Molecular Life Sciences*, *78*, 3141–3158.

Brose, M. S., Cabanillas, M. E., Cohen, E. E., Wirth, L. J., Riehl, T., Yue, H., … & Sherman, E. J. (2016). Vemurafenib in patients with BRAFV600E-positive metastatic or unresectable papillary thyroid cancer refractory to radioactive iodine: A non-randomised, multicentre, open-label, phase 2 trial. *The Lancet Oncology*, *17*(9), 1272–1282.

Buchbinder, E. I., & Desai, A. (2016). CTLA-4 and PD-1 pathways: Similarities, differences, and implications of their inhibition. *American Journal of Clinical Oncology*, *39*(1), 98–106.

Callahan, M. K., Postow, M. A., & Wolchok, J. D. (2015). CTLA-4 and PD-1 pathway blockade: Combinations in the clinic. *Frontiers in Oncology*, *4*, 385.

Choueiri, T. K., Escudier, B., Powles, T., Mainwaring, P. N., Rini, B. I., Donskov, F., … & Motzer, R. J. (2015). Cabozantinib versus everolimus in advanced renal-cell carcinoma. *New England Journal of Medicine*, *373*(19), 1814–1823.

Clarke, C. N., & Kopetz, E. S. (2015). BRAF mutant colorectal cancer as a distinct subset of colorectal cancer: Clinical characteristics, clinical behavior, and response to targeted therapies. *Journal of Gastrointestinal Oncology*, *6*(6), 660.

Cunningham, D., Al-Batran, S. E., Davidenko, I., Ilson, D. H., Murad, A. M., Tebbutt, N. C., … & RILOMET-1 investigators. (2013). RILOMET-1: An international phase III multicenter, randomized, double-blind, placebo-controlled trial of rilotumumab plus epirubicin, cisplatin, and capecitabine (ECX) as first-line therapy in patients with advanced MET-positive gastric or gastroesophageal junction (G/GEJ) adenocarcinoma. *Annals of Oncology*, *25*(4), iv–250.

Czabotar, P. E., & Garcia-Saez, A. J. (2023). Mechanisms of BCL-2 family proteins in mitochondrial apoptosis. *Nature Reviews Molecular Cell Biology*, *24*(10), 732–748.

Davies, H., Bignell, G. R., Cox, C., Stephens, P., Edkins, S., Clegg, S., … & Futreal, P. A. (2002). Mutations of the BRAF gene in human cancer. *Nature*, *417*(6892), 949–954.

Dick, F. A., & Rubin, S. M. (2013). Molecular mechanisms underlying RB protein function. *Nature Reviews Molecular Cell Biology*, *14*(5), 297–306.

Dzutsev, A., Badger, J. H., Perez-Chanona, E., Roy, S., Salcedo, R., Smith, C. K., & Trinchieri, G. (2017). Microbes and cancer. *Annual Review of Immunology*, *35*, 199–228.

Elisei, R., Schlumberger, M. J., Müller, S. P., Schöffski, P., Brose, M. S., Shah, M. H., … & Sherman, S. I. (2013). Cabozantinib in progressive medullary thyroid cancer. *Journal of Clinical Oncology*, *31*(29), 3639.

Fallahi, P., M Ferrari, S., Di Bari, F., Materazzi, G., Benvenga, S., Miccoli, P., & Antonelli, A. (2015). Cabozantinib in thyroid cancer. *Recent Patents on Anti-Cancer Drug Discovery*, *10*(3), 259–269.

Fecher, L. A., Agarwala, S. S., Hodi, F. S., & Weber, J. S. (2013). Ipilimumab and its toxicities: A multidisciplinary approach. *The Oncologist*, *18*(6), 733–743.

Feng, Y., Liu, X., & Pauklin, S. (2021). 3D chromatin architecture and epigenetic regulation in cancer stem cells. *Protein & Cell*, *12*(6), 440–454.

Finn, R. S., Crown, J. P., Lang, I., Boer, K., Bondarenko, I. M., Kulyk, S. O., … & Slamon, D. J. (2015). The cyclin-dependent kinase 4/6 inhibitor palbociclib in combination with letrozole versus letrozole alone as first-line treatment of oestrogen receptor-positive, HER2-negative, advanced breast cancer (PALOMA-1/TRIO-18): A randomised phase 2 study. *The Lancet Oncology*, *16*(1), 25–35.

Flaherty, K. T., Robert, C., Hersey, P., Nathan, P., Garbe, C., Milhem, M., … & Schadendorf, D. (2012). Improved survival with MEK inhibition in BRAF-mutated melanoma. *New England Journal of Medicine*, *367*(2), 107–114.

Flavahan, W. A., Gaskell, E., & Bernstein, B. E. (2017). Epigenetic plasticity and the hallmarks of cancer. *Science*, *357*(6348), eaal2380.

Geng, C., Pang, S., Ye, R., Shi, J., Yang, Q., Chen, C., & Wang, W. (2023). Glycolysis-based drug delivery nanosystems for therapeutic use in tumors and applications. *Biomedicine & Pharmacotherapy*, *165*, 115009.

Grivennikov, S. I., & Karin, M. (2010). Inflammation and oncogenesis: A vicious connection. *Current Opinion in Genetics & Development*, *20*(1), 65–71.

Hahn, W. C. (2001). Telomerase and cancer: Where and when? *Clinical Cancer Research*, *7*(10), 2953–2954.

Hanahan, D. (2022). Hallmarks of cancer: New dimensions. *Cancer Discovery*, *12*(1), 31–46.

Hanahan, D., & Weinberg, R. A. (2000). The hallmarks of cancer. *Cell*, *100*(1), 57–70.

Hanahan, D., & Weinberg, R. A. (2011). Hallmarks of cancer: The next generation. *Cell*, *144*(5), 646–674.

Hauschild, A., Grob, J. J., Demidov, L. V., Jouary, T., Gutzmer, R., Millward, M., … & Chapman, P. B. (2012). Dabrafenib in BRAF-mutated metastatic melanoma: A multicentre, open-label, phase 3 randomised controlled trial. *The Lancet*, *380* (9839), 358–365.

He, S., & Sharpless, N. E. (2017). Senescence in health and disease. *Cell*, *169*(6), 1000–1011.

Hegde, A. N., & Smith, S. G. (2019). Recent developments in transcriptional and translational regulation underlying long-term synaptic plasticity and memory. *Learning & Memory*, *26*(9), 307–317.

Hodi, F. S., O'day, S. J., McDermott, D. F., Weber, R. W., Sosman, J. A., Haanen, J. B., … & Urba, W. J. (2010). Improved survival with ipilimumab in patients with metastatic melanoma. *New England Journal of Medicine*, *363*(8), 711–723.

Holderfield, M., Deuker, M. M., McCormick, F., & McMahon, M. (2014). Targeting RAF kinases for cancer therapy: BRAF-mutated melanoma and beyond. *Nature Reviews Cancer*, *14*(7), 455–467.

Ilic, I., Jankovic, S., & Ilic, M. (2016). Bevacizumab combined with chemotherapy improves survival for patients with metastatic colorectal cancer: Evidence from meta-analysis. *PLoS One*, *11*(8), e0161912.

Jafri, M. A., Ansari, S. A., Alqahtani, M. H., & Shay, J. W. (2016). Roles of telomeres and telomerase in cancer, and advances in telomerase-targeted therapies. *Genome Medicine*, *8*, 1–18.

Jones, P. A., Issa, J. P. J., & Baylin, S. (2016). Targeting the cancer epigenome for therapy. *Nature Reviews Genetics*, *17*(10), 630–641.

Kadosh, E., Snir-Alkalay, I., Venkatachalam, A., May, S., Lasry, A., Elyada, E., … & Ben-Neriah, Y. (2020). The gut microbiome switches mutant p53 from tumour-suppressive to oncogenic. *Nature*, *586*(7827), 133–138.

Kanev, P. B., Atemin, A., Stoynov, S., & Aleksandrov, R. (2023, September). PARP1 roles in DNA repair and DNA replication: The basi(c) s of PARP inhibitor efficacy and resistance. *Seminars in Oncology*.

Kowald, A., Passos, J. F., & Kirkwood, T. B. (2020). On the evolution of cellular senescence. Aging Cell, *19*(12), e13270.

Leal, T., Kotecha, R., Ramlau, R., Zhang, L., Milanowski, J., Cobo, M., … & Langer, C. (2023). Tumor treating fields therapy with standard systemic therapy versus standard systemic therapy alone in metastatic non-small-cell lung cancer following progression on or after platinum-based therapy (LUNAR): A randomised, open-label, pivotal phase 3 study. *The Lancet Oncology*, *24*(9), 1002–1017.

Lee, S., & Schmitt, C. A. (2019). The dynamic nature of senescence in cancer. *Nature cell Biology*, *21*(1), 94–101.

Lobel, G. P., Jiang, Y., & Simon, M. C. (2023). Tumor microenvironmental nutrients, cellular responses, and cancer. *Cell Chemical Biology*, *30*(9), 1015–1032.

Marani, A., Gioacchini, H., Paolinelli, M., Offidani, A., & Campanati, A. (2023). Potential drug–drug interactions with mitogen-activated protein kinase (MEK) inhibitors used to treat melanoma. *Expert Opinion on Drug Metabolism & Toxicology*, *19*(8), 555–567.

Mittal, D., Gubin, M. M., Schreiber, R. D., & Smyth, M. J. (2014). New insights into cancer immunoediting and its three component phases—Elimination, equilibrium and escape. *Current Opinion in Immunology*, *27*, 16–25.

Mizuno, S., & Nakamura, T. (2013). HGF–MET cascade, a key target for inhibiting cancer metastasis: The impact of NK4 discovery on cancer biology and therapeutics. *International Journal of Molecular Sciences*, *14*(1), 888–919.

Mughal, T. I., & Goldman, J. M. (2006). Molecularly targeted treatment of chronic myeloid leukemia: Beyond the imatinib era. *Front Bioscience*, *11*, 209–220.

Nam, A. S., Chaligne, R., & Landau, D. A. (2021). Integrating genetic and non-genetic determinants of cancer evolution by single-cell multi-omics. *Nature Reviews Genetics*, *22*(1), 3–18.

Nevins, J. R. (2001). The Rb/E2F pathway and cancer. *Human Molecular Genetics*, *10*(7), 699–703.

Ott, P. A., Hodi, F. S., & Robert, C. (2013). CTLA-4 and PD-1/PD-L1 blockade: New immunotherapeutic modalities with durable clinical benefit in melanoma patients. *Clinical Cancer Research*, *19*(19), 5300–5309.

Paterson, I. C., Patel, V., Sandy, J. R., Prime, S. S., & Yeudall, W. A. (1995). Effects of transforming growth factor beta-1 on growth-regulatory genes in tumour-derived human oral keratinocytes. *British Journal of Cancer*, *72*(4), 922–927.

Raghav, K. P., Wang, W., Liu, S., Chavez-MacGregor, M., Meng, X., Hortobagyi, G. N., … & Gonzalez-Angulo, A. M. (2012). cMET and phospho-cMET protein levels in breast cancers and survival outcomes. *Clinical Cancer Research*, *18*(8), 2269–2277.

Rosenberg, S. A., Yang, J. C., Sherry, R. M., Kammula, U. S., Hughes, M. S., Phan, G. Q., … & Dudley, M. E. (2011). Durable complete responses in heavily pretreated patients with metastatic melanoma using T-cell transfer immunotherapy. *Clinical Cancer Research*, *17*(13), 4550–4557.

Ruden, M., & Puri, N. (2013). Novel anticancer therapeutics targeting telomerase. *Cancer Treatment Reviews*, *39*(5), 444–456.

Sacco, J. J., & Clague, M. J. (2015). Dysregulation of the Met pathway in non-small cell lung cancer: Implications for drug targeting and resistance. *Translational Lung Cancer Research*, *4*(3), 242.

Sahgal, P., Patil, D. T., Bala, P., Sztupinszki, Z. M., Tisza, V., Spisak, S., … & Sethi, N. S. (2023). Replicative stress in gastroesophageal cancer is associated with chromosomal instability and sensitivity to DNA damage response inhibitors. *Iscience*, *26*, 11.

Salgia, R. (2009, April). Role of c-Met in cancer: Emphasis on lung cancer. *Seminars in Oncology*, *36*, S52–S58.

Shay, J. W., Zou, Y., Hiyama, E., & Wright, W. E. (2001). Telomerase and cancer. *Human Molecular Genetics*, *10*(7), 677–685.

Sherr, C. J., & McCormick, F. (2002). The RB and p53 pathways in cancer. *Cancer Cell*, *2*(2), 103–112.

Sierra, J. R., & Tsao, M. S. (2011). c-MET as a potential therapeutic target and biomarker in cancer. *Therapeutic Advances in Medical Oncology*, *3*(1_suppl), S21–S35.

Smith, B. N., & Bhowmick, N. A. (2016). Role of EMT in metastasis and therapy resistance. *Journal of Clinical Medicine*, *5*(2), 17.

Steinbach, G., Lynch, P. M., Phillips, R. K., Wallace, M. H., Hawk, E., Gordon, G. B., … & Kelloff, G. (2000). The effect of celecoxib, a cyclooxygenase-2 inhibitor, in familial adenomatous polyposis. *New England Journal of Medicine*, *342*(26), 1946–1952.

Sullivan, R. J., & Flaherty, K. T. (2013). Resistance to BRAF-targeted therapy in melanoma. *European Journal of Cancer*, *49*(6), 1297–1304.

Thiery, J. P. (2002). Epithelial–mesenchymal transitions in tumour progression. *Nature Reviews Cancer*, *2*(6), 442–454.

Thomas, S., Izard, J., Walsh, E., Batich, K., Chongsathidkiet, P., Clarke, G., … & Prendergast, G. C. (2017). The host microbiome regulates and maintains human health: A primer and perspective for non-microbiologists. *Cancer Research*, *77*(8), 1783–1812.

Thun, M. J., Henley, S. J., & Patrono, C. (2002). Nonsteroidal anti-inflammatory drugs as anti-cancer agents: Mechanistic, pharmacologic, and clinical issues. *Journal of the National Cancer Institute*, *94*(4), 252–266.

Tol, J., Nagtegaal, I. D., & Punt, C. J. (2009). BRAF mutation in metastatic colorectal cancer. *New England Journal of Medicine*, *361*(1), 98–99.

Villanueva, J., Vultur, A., & Herlyn, M. (2011). Resistance to BRAF inhibitors: Unraveling mechanisms and future treatment options. *Cancer Research*, *71*(23), 7137–7140.

Viola, D., Cappagli, V., & Elisei, R. (2013). Cabozantinib (XL184) for the treatment of locally advanced or metastatic progressive medullary thyroid cancer. *Future Oncology*, *9*(8), 1083–1092.

Wang, B., Kohli, J., & Demaria, M. (2020). Senescent cells in cancer therapy: Friends or foes? *Trends in Cancer*, *6*(10), 838–857.

Wani, A. K., Akhtar, N., Mir, T. U. G., Singh, R., Jha, P. K., Mallik, S. K., … & Prakash, A. (2023). Targeting apoptotic pathway of cancer cells with phytochemicals and plant-based nanomaterials. *Biomolecules*, *13*(2), 194.

Xiao, S. Q., Cheng, M., Wang, L., Cao, J., Fang, L., Zhou, X. P., … & Hu, Y. F. (2023). The role of apoptosis in the pathogenesis of osteoarthritis. *International Orthopaedics*, *47*(8), 1895–1919.

Xu, J., Dong, X., Huang, D. C., Xu, P., Zhao, Q., & Chen, B. (2023). Current advances and future strategies for BCL-2 inhibitors: Potent weapons against cancers. *Cancers*, *15*(20), 4957.

Zhang, W., & Liu, H. T. (2002). MAPK signal pathways in the regulation of cell proliferation in mammalian cells. *Cell Research*, *12*(1), 9–18.

Zhang, Z., & Zhu, Q. (2023). WD repeat and HMG Box DNA binding protein 1: An oncoprotein at the hub of tumorigenesis and a novel therapeutic target. *International Journal of Molecular Sciences*, *24*(15), 12494.

Part III

Environmental Biotechnology

12 Polyhydroxyalkanoate *A Greener Approach towards Environmental Sustainability*

Sunena Dhania, Ruma Rani, Manju Bernela, and Rajesh Thakur

12.1 INTRODUCTION

Plastics have become increasingly important in our daily lives over the past few decades due to their versatility, durability, and affordability (Sabapathy et al., 2020). Synthetic polymers which are the building material of plastics have attracted substantial interest across all industries to the point where we can no longer conceive a product without them (Luo et al., 2016; Yadav et al., 2020). The production of synthetic plastics is largely dependent on fossil fuels, which are limited and non-renewable in nature, and have been overused as a result of industrialization and population growth. A paradigm shift toward sustainable energy resource alternatives has been triggered by the rising global demand and depleting fossil fuel supplies. Currently, there are significant concerns with plastic pollution also known as "white pollution" and "microplastic" pollution due to the vast buildup of petrochemical-derived compounds in the ecosystem (AI Battashi et al., 2021). It is predicted that the amount of plastic waste entering the oceans from land will dramatically increase by 2025, from 4.8 to 12.7 million tons annually, if a better waste management is not planned. These synthetic plastic materials have terrible side effects that include pollution, food chain contamination, threat to biodiversity, increased carbon footprint, and financial loss. Thus, looking toward the future, achieving a goal to adequate environmental safety and resources for all is a necessity (Sabapathy et al., 2020), and therefore there is an urgent need for switching to alternate environmentally benign plastic materials or biodegradable "green plastics" (Liu et al., 2021; AI Battashi et al., 2021).

Organic wastes from all types of daily operations, from domestic to industry, end up in landfills where their decomposition leads to the emission of greenhouse gases. For a sustainable future, it is important that these organic wastes should be used in the manufacturing of bio-based goods including biofuel, biochemicals, and biopolymers (Liu et al., 2021). Both the above concerns of plastic pollution and biological waste management can be addressed by utilizing the organic wastes from different sources for the production of biopolymers that could be subsequently used for manufacturing bioplastics for making diverse types of goods that exhibit sustainability (Banu et al., 2019), environment friendly manufacturing, biocompatibility for medical applications, and ease of waste management after end of a product's useful

DOI: 10.1201/9781032642567-15

life (Meeks et al., 2015). These polymer materials should totally break down into harmless products, when subjected to microbial action in a suitable environmental condition. In summary these can be defined as biologically synthesized biodegradable harmless plastics. Biopolymers are bio-based polymeric substances produced by microbes growing upon biomass and contain covalent bonds between monomers to form macromolecules (Kannah et al., 2022) fit this description. Only a few polymers have demonstrated these much-desired qualities; they include polyhydroxyalkanoates (PHAs), polylactic acid (PLAs), and poly(glycolic acid) (PGA), as well as a number of their blends (mechanical mixtures of different plastics) registered by various companies (Adeleye et al., 2020). PHAs are a class of biodegradable polymers that are produced by a variety of microorganisms as a means of storing carbon and energy. Their properties, such as biodegradability, biocompatibility, and physical qualities, have resulted in attracting substantial attention as environmentally-friendly alternatives to traditional petroleum-based plastics (Sabapathy et al., 2020; Sohn et al., 2020; Saratale, 2021). This chapter focuses on elucidating the current status on utilization of different industrial and agricultural wastes for the production of PHAs and on the use of PHAs for various agricultural, construction, and industrial applications.

12.2 BRIEF DESCRIPTION ABOUT POLYHYDROXYALKANOATES (PHAs)

Polyhydroxyalkanoates (PHAs) are a class of macromolecular bio-polyesters which exist widely in the microorganisms and are mainly used as a carbon source and for energy storage and accumulation (Liu et al., 2021). Due to their mechanical properties similar to polyethylene and polypropylene, PHAs are considered a potential rcplacement for these petroleum-based plastics (Lackner, 2015). PHAs are long, chain-like molecules made up of smaller molecules called hydroxyl acid monomers. These monomers are linked together by ester bonds. PHAs are produced by a variety of microorganisms as a means of storing carbon and energy. They have been studied extensively since they were first discovered in 1926 by Lemoigne. PHAs are more biodegradable than other biopolymers (Guo et al., 2012), meaning they can break down more easily in the environment. This makes them a good alternative to synthetic plastics, which are not biodegradable and can pollute the environment. The development of PHAs as substitutes for synthetic plastics contributes to the Sustainable Development Goals (SDGs) in three ways (Novelli et al., 2021):

- Water sanitation: PHAs can be used to make biodegradable packaging and other products, which can help to reduce plastic pollution in waterways and oceans.
- Innovation (to meet environmental objectives): The development of new PHA production methods and applications is an example of innovation that is helping to address environmental challenges.
- Responsible production: PHAs can be produced from renewable resources, such as agricultural waste, which helps to reduce the use of fossil fuels and greenhouse gas emissions.

Due to their superior characteristics, PHAs have attracted commercial interest and are considered suitable alternatives to traditional plastic materials, making them appropriate in the medical sector in vivo. These include biodegradability through biological activity composting without toxic products; biosynthetic in nature and renewability of their precursors, including alcohols, lipids, carbohydrates, CO_2 or methane. The global market for PHAs is expected to grow significantly in the coming years because of their advantages over conventional plastics with a wide range of potential applications in various industries, including biomedical devices, electronics, construction, automotive, packaging, and agriculture. (Sirohi et al., 2020; Anjum et al., 2016). Due to these properties, PHAs have since long been used for everyday packaging and for making articles like plastic bags, medical surgical garments, feminine hygiene products, shampoo bottles (Wella AG), carpets and upholstery (P&G, Biomer, Metabolix, and several other companies) (Chen, 2009). Table 12.1 lists some commercial materials based on PHAs, including their trade names, producers, countries of licensing and the product (Dietrich et al., 2017; Verlinden et al., 2007).

TABLE 12.1
A List of Some Commercial PHAs, Their Trade Names, Producers, the Countries of Licensing and the Product

Commercial Name	Product Name	Country of Origin	Producer	Product
Biocycle	Biocycle 1000 Biocycle 1189D-1 Biocycle 18BC-1 Biocycle 189C-1	Brazil	PHB Industrial	Plastic sheet extrusions, coating paper, injections
Ecogen	ENMAT Y1000 ENMAT Y1000P ENMAT Y3000 ENMAT Y3000P	China	TianAn biological material: PolyOne	Thermoplastics: extrusions, injections, molding, blown films; fiber and nonwovens, thermoforming, Denitrification: water treatment
Metabolix	Mvera™ B5010 Mvera™ B5011	USA	Telles Group LLC	Compostable resins, compostable bags, films and food waste compost bags
Minerv-PHB	MinervPHA™	Italy	BIO ON	Automotive, electronics, packaging
Mirel	Mirel P4001 Mirel P4010 Mirel P2100 Mirel P4100 Mirel P2200 Mirel P5001 Mirel P5004	USA	Metabolix	Additives for PLA and PVC, injection molding, coating of cardboard and paper, micro powders, Denitrification for aquariums
Nodax	Nodax™	USA and Japan	P&G Chemicals	Bottles, mulch, bags Hygiene
Tepha	P(4HB)	USA	Tepha, Inc.	Surgical films, mesh, suture

Although using PHA can successfully minimize the environmental risks associated with petrochemical plastics, PHA still only has a modest market share (Aeschelmann and Carus, 2015). PHAs have many excellent properties and environmental advantages, but their high production costs are preventing them from being used more widely. PHA production costs are high due to the cost of raw materials, energy, water, equipment, and the complexity of the process. The use of expensive fermentation carbon sources is a major contributor to these high costs, accounting for up to 50% of the total production cost (Amaro et al., 2019). By using waste and renewable carbon sources, researchers hope to reduce the cost of PHA production to make it more competitive with petroleum-based plastics. (Nielsen et al., 2017; Koller et al., 2017; Amaro et al., 2019). This would make PHA a more sustainable and environmentally-friendly alternative to petroleum-based plastics.

Development of competent technologies for the production of PHAs begins by understanding the microbes that produce them. Various gram-negative and gram-positive bacteria that can store PHA include *Alcaligenes latus*, *Bacillus cereus* and *Bacillus megaterium*, *Azospirillum rubrum*, *Azotobacter vinelandii*, *Leptothrix* sp., *Enterobacter* sp., *Methylocystis* sp., *Rhizobium* sp., *Pseudomonas* sp., *Rhodobacter sphaeroides*, and *Ralstonia eutropha* (Abd El-malek et al., 2020). Different types of substrates have been used by these organisms for the production of PHAs, such as agricultural leftovers, fatty acids, and industrial byproducts (Dai et al., 2015), fermented molasses, olive oil pomace (Waller et al., 2012), paper mill wastewater (Jiang et al., 2012), and palm oil mill effluent (Lee et al., 2015). These bacteria can store up to 40–70 wt% biodegradable PHAs. Industrial wastes including pulp and paper, sugar, fruit and food processing, slaughterhouses, poultries, and dairies can serve as excellent starting materials for the manufacture of PHA (Pakalapati et al., 2018). As a rcsult, scientists have designed many studies to find cheaper alternatives. (Abd El-malek et al., 2020). With the use of anaerobic effluents from waste paper, a number of pretreatment procedures, expensive enzymatic hydrolysis, and detoxification were no longer necessary for the synthesis of PHAs. Volatile fatty acids (VFAs), which are formed from waste paper, are one of the important components of municipal solid waste that can be used as a feedstock for the manufacture of PHAs (AI Battashi et al., 2021). Using waste materials benefit the society in two ways: in waste management and in the conversion of waste to an environmentally-friendly commercial biopolymer, thereby improving the process's overall economics. For instance, after the processing of apple, the leftover residue known as apple pomace is rich in fibers, polyphenols, and carbohydrates. This pomace serves as a potential carbon source for PHA production. An economically viable method for producing PHA using waste from the apple industry will be developed by considering the economics of genetically modified strains, life cycle analysis, and techno-economic analysis (Liu et al., 2021). Recent advances in the use of waste streams for PHA production suggest that integrating industrial waste streams into biorefineries to co-generate PHAs with other value-added products, such as biohydrogen and biogas, could lead to more widespread industrial-scale PHA production (Novelli et al., 2021). In another study, untreated crude paper industry effluent (PIE) was utilized for the production of PHA as a carbon substrate. The team concluded that when the appropriate bacterial strain is used, crude PIE might be implemented for the sustainable production of

PHA (Tyagi and Sharma, 2021). Additionally, PHA can be made using various waste fluxes from the food, dairy, and sugar processing sectors. Valorisation of these waste fluxes makes the process affordable and practically applicable. Recent developments in waste treatment methods, fermentation techniques, and genetic engineering can provide researchers with information about the route that waste-to-bioplastics manufacturing will take in the future (Saratale et al., 2021).

12.3 APPLICATIONS OF PHA

PHAs have a lot of applications in different sectors, such as industrial, medical, agricultural, and so on (Figure 12.1). Some of the applications of PHAs are discussed in the following sections.

12.3.1 Agricultural Applications

PHAs have several applications in the area of agriculture. The shelf life and nutritional properties of the plants under stressful environmental conditions are improved by PHAs. Biodegradable plastic films for crop protection, encapsulation of fertilizer, seeds encapsulation, etc., are potential applications of PHAs. The controlled release of insecticides and pesticides is one of the specialized applications of P(3HB) in agriculture (Sharma et al., 2021). They can be used to form soil-friendly compostable greenhouse films, protection nets, and grow bags, which could replace the commercial plastics. Grow bags made up of PHA do not release destructive pollutants upon degradation, and they provide matrix for microbial growth, can enable water denitrification, and are root friendly (Amelia et al., 2019). PHA-based agricultural nets can be effectively used to protect crops from insects, birds, and natural climatic fluctuations while avoiding overheating, thereby enhancing crop growth (Guerrini et al., 2017).

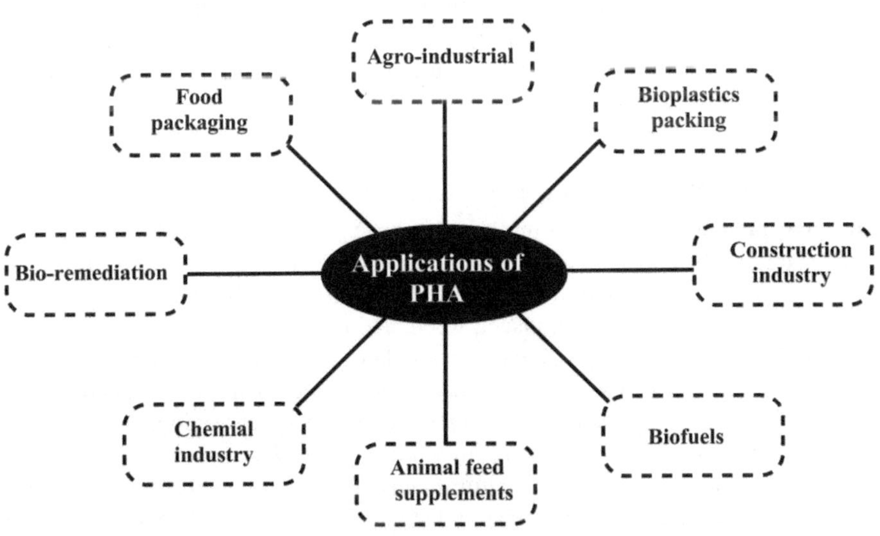

FIGURE 12.1 Different applications of PHA.

12.3.1.1 Mulch Films

Mulching is a technique for covering the soil's surface around crops to foster the best possible growing conditions for conducting controlled agriculture on a small amount of arable land with minimal irrigation. Normally mulch comes in two varieties: natural mulch and synthetic mulch. Synthetic mulch is preferable for use as compared to natural mulch on a broad scale in agriculture due to the latter being less effective in weed control and often inadequacy in material requirements for the production of natural mulch. Plastic mulch is a synthetic mulch that is utilized a lot, along with paper and other materials. Environmental issues have been raised by the disposal of used plastic mulch because high-density polyethylene (HDPE), low-density polyethylene (LDPE) and linear LDPE—materials that take a long time to decompose in the environment—are predominantly used to make plastic mulch. Used plastic mulch frequently ends up in landfills or is burnt, both of which cause severe pollution (Sintim and Flury, 2017).

Utilizing biodegradable mulch, films have gained much popularity as a sustainable mulching method because they self-degrade at the conclusion of the cycle, and it does not require the removal and disposal. Without harming the environment, biodegradable mulch films assist in improving crop quality and enhancing agricultural productivity by reducing soil erosion, supplying nutrients, raising soil temperature, and controlling the growth of weeds, among others (Kasirajan and Ngouajio, 2012; Rydz et al., 2014).

PHB polymers are a prime option to replace plastic as mulch films due to their advantageous properties like structural flexibility, raw material accessibility, and biodegradability. Moreover, mechanical strength can also be improved by blending various PHAs or with other polymers. PHA-based biodegradable mulch named Nodex™ has been designed and is patented by Danimer Scientific (Danimer Scientific 2018a; Chen 2009). The PHAs' ability to release agrochemicals under controlled conditions is a characteristic that makes mulch film application even more advantageous. PHB blends have been reported to be effective as mulch films for the gradual and controlled release of insecticides and fungicides (Chen et al., 2021). PHB blends' biodegradation has been investigated under a variety of circumstances, including a controlled laboratory setting and natural habitats such as various types of soil, river water, activated sludge, and compost. PHB is known to break down quickly in aerobic and anaerobic conditions, making it simple to dispose of without harming the environment.

12.3.1.2 Grow Bags/Seedling Bags/Planter Bags

Grow bags, planter bags or seedling bags improve sapling growth and management as they stabilize the soil temperature, retain moisture of soil, and ensure the definitive use of growing media (Rajdeep and Naithani, 2013). Additionally, the impact of external factors such as drought, insect, weed, and pH on seedling soil is reduced by using plastic grow bags. They make it convenient to isolate individual plants in order to lessen competition and root disturbance, thereby increasing crop survival rates. Additionally, polymer grow bags are lightweight and space-saving, making them effective for moving and transporting immature crops. The commonest polyethylene (PE) material engaged in horticultural, agricultural, and packaging purposes is low-density polyethylene (LDPE) (Haghighat and Torabi, 2004). The high durability of LDPE grow bags has an adverse effect on the environment, young plants, and the agricultural industry. Non-biodegradable planter bags and other identical plastic articles have necessitated several governments to impose tax and outright proscription on their use around the globe. With this prohibition, agriculturalists and forestry professionals around the world are looking for a more environmentally-friendly, sustainable alternative for the propagation of seedlings and crops (Adeleye et al., 2020). In this context, the scope for replacing LDPE seeding bags with PHA bags cannot be overemphasized as they can reduce soil, water, and air pollution.

12.3.1.3 Potential of PHAs in the Application of Slow-Release Fertilizer

In addition to causing serious environmental damage, the overuse of fertilizers can also lead to significant economic losses (Hafshejani et al., 2013). The use of fertilizers can be substantially minimized by using slow-release fertilizers, which are essential for the longevity of the end product while efficiently releasing nutrients for prolonged absorption by the plants, and is recognized as the best method (Harmaen et al., 2016). For decades, bio-plastic fertilizers have been used as slow-release fertilizers in the agricultural and horticultural industries. Slow-release fertilizers have been made using nano-composites of PHA-based bioplastics derived from bacterial biomass. The fertilizers are embedded in their granules, films, or bars. The use of poly(3-hydroxybutyrate), or P(3HB), as a slow-release matrix for urea fertilizer, insecticides, and herbicides has been demonstrated (Djamaan et al., 2015). This biodegradable covering does not contaminate the soil and gradually releases the stored nutrients. Compared with traditional water-soluble fertilizers, it releases the nutrients at a slower rate. However, the pace, pattern, and duration of release are difficult to control because they depend on the microbial activity, which is driven by soil moisture and temperature conditions. One example of a timely-released fertilizer is plastic-embedded fertilizer, which has been used successfully in oil palm plantations. This fertilizer releases nutrients gradually over a period of 0.5–2 years as the plastic degrades (Ivanov et al., 2015).

12.3.2 Industrial Applications

PHA bioplastics have features that make them acceptable substitutes for many of the applications presently provided by petroleum-based plastics, thus opening up a huge market potential, with a predicted CAGR of 15.9%. The PHA market is expected to grow in value from USD 93 million in 2023 to USD 195 million by 2028 (https://www.marketsandmarkets.com/Market-Reports/pha-market-395.html). Numerous variables contribute to this expectation of good growth. The increasing focus on minimizing the use of non-biodegradable plastics, which has resulted in the introduction of stringent government rules governing the use of petroleum-based plastics, is one important driving force. The PHA market is also growing as a result of the rising demand for environmentally-friendly products. PHA presents an economical option because it may be made with a variety of biomaterials and production techniques. Consequently, the global demand for PHA, particularly in applications such as packing and food services, biomedical and healthcare, is expected to witness significant growth in years to come (Lee et al., 2008; Nielsen et al., 2017).

In the industrial sector, PHAs have a wide range of uses ranging from manufacturing ubiquitous goods like packaging material, plastic cutlery such as cups, spoons, and plates to specialized industrial use such as manufacture of biofuels and industrial chemicals (Muneer et al., 2020). Altogether increased use of PHAs instead of synthetic plastics can help the UN accomplish its sustainable development goals while also reducing plastic pollution. PHA-based bioplastics may also be converted into hybrid biomaterials for a variety of sustainable usages employing chemical modifications such as epoxidation (Bassas-Galià et al., 2015).

12.3.3 Biofuels and Fine Chemicals

The production of biofuels using PHAs is an important research area to produce biofuels as it holds the potential for producing renewable fuel sources to meet the global demand and to reduce greenhouse gas emissions. Two main methods employed for the production of biofuel from PHA include PHA hydrolysis and direct microbial conversion. In the former approach, PHA can be enzymatically hydrolyzed into its monomers [hydroxyalkanoates (HAs)], which can then be further processed into biofuels such as biodiesel or bio-jet fuel. Through procedures like transesterification or microbial conversion, these HAs can be used as a feedstock for microbial fermentation to make biofuels. The latter method uses microorganisms that have been specifically designed to convert PHA into biofuels. PHA can be metabolized by some bacteria, resulting in the production of biofuels including bioethanol, biobutanol, and hydrogen gas. These microorganisms can degrade PHA and use the carbon that is stored there to produce molecules that are rich in energy. In 2009, the first study on biofuel synthesis using PHA-based materials was published. PHAs with short or medium chains can be esterified by using methanol to make HA methyl esters (Zhang et al., 2009). In another instance, the PHB synthesis route was used to create butanol (an important biofuel) from glucose (Bond-Watts et al., 2011). PHAs can be synthesized from activated sludge and nutrient-rich wastewater from a variety of sources, thereby making them a very sustainable source of biofuel (Johnson et al., 2009). The production cost of biofuel through PHA synthesis from activated sludge should be approximately US$1,200 per tonne (Chen, 2011). Apart from being a source of biofuel, PHAs may be utilized to produce industrially significant chemicals such as HAs, which are precursors for the production of aromatics, antibiotics, and pheromones (Ren et al., 2010).

12.3.4 Packaging

According to a recent report by United Nation Environment Programme on plastic waste, packaging contributes to 42% of the total global plastic waste generated. A significant portion of this waste comes from single-use plastics. Several European countries have already taken steps to ban the use of single-use plastics, thus setting the stage for PHAs to gain traction in the packaging and food service sector. PHAs are popular biodegradable packaging materials that may be found in films, foils, and other forms (Basumatary et al., 2022). They have qualities comparable to polypropylene and can be used to wrap food (Guillard et al., 2018). They are ideal for packaging items such as diapers, sanitary wipes, disposable feminine hygiene products, shampoo bottles, razors, and milk packs because of their gas barrier and water resistance properties (Singh and Kaur, 2015). PHAs have excellent moisture barrier properties, ten times stronger oxygen gas barrier properties, and a higher crystallinity degree than PLA and PCL (Keskin et al., 2017). PHAs are also used to make shopping bags, paper coatings, cups, and carpets, among other things. Many items, such as PHA-coated boxes, paper, sheets, and boards, are available now (Saxena and Tiwari, 2011). Barrier coatings on food packaging can keep moisture and oils inside the package. Food packaging materials are being developed by combining traditional packaging materials with nanoparticles by many companies. They have antibacterial qualities and can improve their mechanical performance. Castro-Mayorga and colleagues created an antibacterial PHA with in situ-stabilized silver nanoparticles that reduced *Salmonella enterica* growth. They claimed that it can be used as an antibacterial alternative to polyethylene in food packaging (Castro-Mayorga et al., 2016). PHAs are insoluble in water and resistant to hydrolytic breakdown due to their high hydrophobicity; therefore, they are also used as coatings in the paper industry to replace synthetic coatings. They provide a protective barrier against oxygen and UV radiation due to its thin, film-forming characteristics (Rastogi and Samyn, 2015). Commercial PHA, such as Tianan's Enmat Y1000P and Telles' Mirel F1006 and Mirel F3002, has intermediate water vapor barrier qualities compared with commercial PP and outperforms PLA (Corre et al., 2012). A recently developed PHA material that has been phloretin-functionalized has the potential to be used in bioplastic formulation, particularly in relation to food packaging, as it has acquired antioxidant activity and has inhibited the growth of food-borne pathogens (*Listeria monocytogenes*) (Mirpoor et al., 2023).

12.3.5 Environment Remediation by Removal of Heavy Metals

Apart from its various sustainable applications, PHAs may be used for environmental restoration, as they can also be utilized in bioremediation, a procedure where hazardous substances are converted into nontoxic forms by microbes or plants. PHAs are effective adsorbents for the removal of heavy metals, and they have been used to activate the adsorption of heavy metal ions in biochar. Removal of heavy metals from aqueous solutions by PHA composites such as PHB-polyethylene glycol (PEG) and PHB-starch composites was demonstrated by Hungund and coworkers in 2018. The adsorption of Pb^{2+} ions was effective in these composites, and the lead adsorption capacity of PHB was found to be 20 g/mL, which was greater than that of its composites, while that of PHB-PEG and PHB-starch had adsorption capacities of 13 and 19 g/mL, respectively (Hungund et al., 2018).

12.3.6 Bio-plastic-PHAs in the Construction Industry

When buildings or other structures are demolished, the construction industry generates a lot of hazardous and non-biodegradable waste from plastic foams and foam insulators. To minimize environmental damage from these, the demand for biodegradable materials and biopolymers is on the rise in the construction sector. Researchers are developing composite materials from biobased polymers and natural plant fibers for use in construction applications (Adeleye et al., 2020). Researchers have developed composites of hemp with cellulose acetate (HCA) or polyhydroxybutyrate (HPHB) that are stronger than plywood and comparable to structural timber in terms of strength. Cellulose acetate and PHB were chosen as matrix materials for their composite materials because they are both biodegradable and can break down quickly in the absence of oxygen. The HCA and HPHB panels are elastic up to a maximum stress of 6.4 and 6.5 MPa, respectively. This means that they will return to their original shape after being subjected to a load below this stress. This example shows how biocomposites could potentially replace plywood in formwork applications, such as slab forms.

Foam made of PHA-based bioplastic is sturdy and can be used to build non-structural (interior) elements, such as walls and partitions, and insulating matrix. At the end of their life, they can be disintegrated quickly and effectively. Crude nano-composites made from bacterial biomass containing PHAs can also be used to make disposable or compostable construction silt and dust fences. The renewable availability of PHA-based bioplastics and their end-of-life biodegradability in landfills are the factors establishing them as sustainable building materials that can help to reduce and manage the environmental impact of construction wastes. Additionally, because it doesn't need to be burned after demolition, this new material shall be extremely acceptable among all stakeholders (Ivanov et al., 2015).

Christian and Billington (2011) tested two biobased composites in which hemp fiber fabric was used as reinforcement either in a poly(hydroxybutyrate) (PHB) matrix produced by bacterial fermentation or in a cellulose acetate matrix made from plants. Their mechanical behavior was examined to determine whether construction applications would be appropriate for them. Both of the biobased composite materials under study exhibited flexural and shear strengths comparable with those of wood and wood-based building materials. Therefore, the use of engineered-wood for making objects such as furniture, shelves, pallets, crates, and formwork may be directly replaced with these biobased composites.

12.4 FUTURE OUTLOOK

This review aimed to highlight as to how PHA or PHA-based polymers are among the most suitable material for replacement of synthetic plastics in a myriad of applications in order to ensure an environmentally sustainable developmental approach. Findings from various studies have lent strong support to this idea and have paved the way for rapid conversion of this know-how into commercialization ventures and let the world enjoy the convenience of using plastics sans the concern for its disastrous impact on the environment and biodiversity. Materials based on PHAs are also likely to resolve concerns relating to food security and health by serving as much safer materials for achieving all the desired objectives. Fresh insights for cheaper production and novel applications of PHA in agriculture, industry, packaging etc. would fuel the rapid growth of commercial ventures. Thus, it is highly anticipated that the market size of PHA-based commercial ventures would grow rapidly in the future if cost-effective raw materials and economical production techniques are adopted. This will help achieve a world where plastics are no more dangerous for the environment and its biodiversity.

REFERENCES

Abd El-malek, F., Khairy, H., Farag, A., & Omar, S. (2020). The sustainability of microbial bioplastics, production and applications. *International Journal of Biological Macromolecules*, *157*, 319–328.

Adeleye, A. T., Odoh, C. K., Enudi, O. C., Banjoko, O. O., Osiboye, O. O., Odediran, E. T., & Louis, H. (2020). Sustainable synthesis and applications of polyhydroxyalkanoates (PHAs) from biomass. *Process Biochemistry*, *96*, 174–193.

Aeschelmann, F., & Carus, M. (2015). Biobased building blocks and polymers in the world: Capacities, production, and applications–Status quo and trends towards 2020. *Industrial Biotechnology*, *11*(3), 154–159.

Al Battashi, H., Al-Kindi, S., Gupta, V. K., & Sivakumar, N. (2021). Polyhydroxyalkanoate (PHA) production using volatile fatty acids derived from the anaerobic digestion of waste paper. *Journal of Polymers and the Environment*, *29*, 250–259.

Amaro, T. M., Comi, G., & Iacumin, L. (2019). Prospects for the use of whey for polyhydroxyalkanoate (PHA) production. *Frontiers in Microbiology*, *10*, 443843.

Amelia, T. S. M., Govindasamy, S., Tamothran, A. M., Vigneswari, S., & Bhubalan, K. (2019). Applications of PHA in agriculture. In: V. C. Kalia (Ed.), *Biotechnological Applications of Polyhydroxyalkanoates*. Springer Singapore, 347–361.

Anjum, A., Zuber, M., Zia, K. M., Noreen, A., Anjum, M. N., & Tabasum, S. (2016). Microbial production of polyhydroxyalkanoates (PHAs) and its copolymers: A review of recent advancements. *International Journal of Biological Macromolecules*, *89*, 161–174.

Banu, J. R., Kumar, M. D., Gunasekaran, M., & Kumar, G. (2019). Biopolymer production in bio electrochemical system: Literature survey. *Bioresource Technology Reports*, *7*, 100283.

Bassas-Galià, M., Gonzalez, A., Micaux, F., Gaillard, V., Piantini, U., Schintke, S., & Mathieu, M. (2015). Chemical modification of polyhydroxyalkanoates (PHAs) for the preparation of hybrid biomaterials: FH-HES. *Chimia*, *69*(10), 627–627.

Basumatary, I. B., Mukherjee, A., Katiyar, V., & Kumar, S. (2022). Biopolymer-based nanocomposite films and coatings: Recent advances in shelf-life improvement of fruits and vegetables. *Critical Reviews in Food Science and Nutrition*, *62*(7), 1912–1935.

Bond-Watts, B. B., Bellerose, R. J., & Chang, M. C. (2011). Enzyme mechanism as a kinetic control element for designing synthetic biofuel pathways. *Nature Chemical Biology*, *7*(4), 222–227.

Castro-Mayorga, J. L., Fabra, M. J., Cabedo, L., & Lagaron, J. M. (2016). On the use of the electrospinning coating technique to produce antimicrobial polyhydroxyalkanoate materials containing in situ-stabilized silver nanoparticles. *Nanomaterials*, *7*(1), 4.

Chen, G. Q. (2009). A microbial polyhydroxyalkanoates (PHA) based bio-and materials industry. *Chemical Society Reviews*, *38*(8), 2434–2446.

Chen, G. Q. (2011). Biofunctionalization of polymers and their applications. In: G. S. Nyanhongo, W. Steiner, & G. Gübitz (Eds.), *Biofunctionalization of Polymers and their Applications*, 29–45.

Chen, G., Cao, L., Cao, C., Zhao, P., Li, F., Xu, B., & Huang, Q. (2021). Effective and sustained control of soil-borne plant diseases by biodegradable polyhydroxybutyrate mulch films embedded with fungicide of prothioconazole. *Molecules*, *26*(3), 762. Berlin, Heidelberg: Springer.

Christian, S. J., & Billington, S. L. (2011). Mechanical response of PHB-and cellulose acetate natural fiber-reinforced composites for construction applications. *Composites Part B: Engineering*, *42*(7), 1920–1928.

Corre, Y. M., Bruzaud, S., Audic, J. L., & Grohens, Y. (2012). Morphology and functional properties of commercial polyhydroxyalkanoates: A comprehensive and comparative study. *Polymer Testing*, *31*(2), 226–235.

Dai, J., Gliniewicz, K., Settles, M. L., Coats, E. R., & McDonald, A. G. (2015). Influence of organic loading rate and solid retention time on polyhydroxybutyrate production from hybrid poplar hydrolysates using mixed microbial cultures. *Bioresource Technology*, *175*, 23–33.

Danimer Scientific. (2018a) Danimer scientific film resins. https://danimerscientific.com/compostable-solutions/a-family-of-biopolymers/film-resins/ (accessed on 15 August 2023).

Dietrich, K., Dumont, M. J., Del Rio, L. F., & Orsat, V. (2017). Producing PHAs in the bioeconomy: Towards a sustainable bioplastic. *Sustainable Production and Consumption*, *9*, 58–70.

Djamaan, A., Monica, R., Asiska, P. D., Suardi, M., Ben, E. S., & Erizal, Z. (2015). The use of biopolymer of poly (3-hydroxybutyrate) as matrix of urea slow release fertilizer. *Journal of Chemical and Pharmaceutical Research*, *7*(7), 558–563.

Guerrini, S, Borreani, G., & Voojis, H. (2017). Biodegradable materials in agriculture: Case histories and perspectives. In: M. Malinconico (Ed.), Soil Degradable Bioplastics for a Sustainable Modern Agriculture. Springer, Cham, 35–65.

Guillard, V., Gaucel, S., Fornaciari, C., Angellier-Coussy, H., Buche, P., & Gontard, N. (2018). The next generation of sustainable food packaging to preserve our environment in a circular economy context. *Frontiers in Nutrition*, *5*, 121.

Guo, W., Tao, J., Yang, C., Song, C., Geng, W., Li, Q., ... Wang, S. (2012). Introduction of environmentally degradable parameters to evaluate the biodegradability of biodegradable polymers. *PLoS One*, *7*(5), e38341.

Hafshejani, M. K., Khandani, F., Heidarpour, R., Arad, A., Choopani, S., & Branch, A. (2013). Study the sources of mercury vapor in atmosphere as a threatening factor for human health and bio-filtering methods for removal of toxic pollution. *Life Science Journal*, *10*(1), 293–296.

Haghighat, P. H., & Torabi, A. M. (2004). Preparation of biodegradable low density polyethylene by starch-urea composition for agricultural applications. *Iranian Journal of Chemistry and Chemical Engineering (IJCCE)*, *23*(1), 7–11.

Harmaen, A. S., Khalina, A., Ali, H. M., & Azowa, I. (2016). Thermal, morphological, and biodegradability properties of bioplastic fertilizer composites made of oil palm biomass, fertilizer, and poly (hydroxybutyrate-co-valerate). *International Journal of Polymer Science*, 1–8.

Polyhydroxyalkanoate (PHA) Market, Top Market Reports, Marketsandmarkets. https://www.marketsandmarkets.com/Market-Reports/pha-market-395.html.

Hungund, B. S., Umloti, S. G., Upadhyaya, K. P., Manjanna, J., Yallappa, S., & Ayachit, N. H. (2018). Development and characterization of polyhydroxybutyrate biocomposites and their application in the removal of heavy metals. *Materials Today: Proceedings*, *5*(10), 21023–21029.

Ivanov, V., Stabnikov, V., Ahmed, Z., Dobrenko, S., & Saliuk, A. (2015). Production and applications of crude polyhydroxyalkanoate-containing bioplastic from the organic fraction of municipal solid waste. *International Journal of Environmental Science and Technology*, *12*, 725–738.

Jiang, Y., Marang, L., Tamis, J., van Loosdrecht, M. C., Dijkman, H., & Kleerebezem, R. (2012). Waste to resource: Converting paper mill wastewater to bioplastic. *Water Research*, *46*(17), 5517–5530.

Johnson, K., Jiang, Y., Kleerebezem, R., Muyzer, G., & Van Loosdrecht, M. C. (2009). Enrichment of a mixed bacterial culture with a high polyhydroxyalkanoate storage capacity. *Biomacromolecules*, *10*(4), 670–676.

Kannah, R. Y., Kumar, M. D., Kavitha, S., Banu, J. R., Tyagi, V. K., Rajaguru, P., & Kumar, G. (2022). Production and recovery of polyhydroxyalkanoates (PHA) from waste streams: A review. *Bioresource Technology*, *366*, 128203.

Kasirajan, S., & Ngouajio, M. (2012). Polyethylene and biodegradable mulches for agricultural applications: A review. *Agronomy for Sustainable Development*, *32*, 501–529.

Keskin, G., Kızıl, G., Bechelany, M., Pochat-Bohatier, C., & Öner, M. (2017). Potential of polyhydroxyalkanoate (PHA) polymers family as substitutes of petroleum based polymers for packaging applications and solutions brought by their composites to form barrier materials. *Pure and Applied Chemistry*, *89*(12), 1841–1848.

Koller, M., Maršálek, L., de Sousa Dias, M. M., & Braunegg, G. (2017). Producing microbial polyhydroxyalkanoate (PHA) biopolyesters in a sustainable manner. *New Biotechnology*, *37*, 24–38.

Lackner, M. (2015). Bioplastics-biobased plastics as renewable and/or biodegradable alternatives to petroplastics. In: K. Othmer (Ed.) *Encyclopedia of Chemical Technology*, 6th ed., John Wiley & Sons, New York, 1–43.

Lee, W. H., Loo, C. Y., Nomura, C. T., & Sudesh, K. (2008). Biosynthesis of polyhydroxyalkanoate copolymers from mixtures of plant oils and 3-hydroxyvalerate precursors. *Bioresource Technology*, *99*(15), 6844–6851.

Lee, W. S., Chua, A. S. M., Yeoh, H. K., Nittami, T., & Ngoh, G. C. (2015). Strategy for the biotransformation of fermented palm oil mill effluent into biodegradable polyhydroxyalkanoates by activated sludge. *Chemical Engineering Journal*, *269*, 288–297.

Lemoigne, M. (1926). de l'acide 3-oxybutyrique. *Bulletin de la Société de chimie biologique (Paris)*, *8*, 770.

Liu, H., Kumar, V., Jia, L., Sarsaiya, S., Kumar, D., Juneja, A., ... Awasthi, M. K. (2021). Biopolymer poly-hydroxyalkanoates (PHA) production from apple industrial waste residues: A review. *Chemosphere*, *284*, 131427.

Luo, X., Ge, X., Cui, S., & Li, Y. (2016). Value-added processing of crude glycerol into chemicals and polymers. *Bioresource Technology*, *215*, 144–154.

Meeks, D., Hottle, T., Bilec, M. M., & Landis, A. E. (2015). Compostable biopolymer use in the real world: Stakeholder interviews to better understand the motivations and realities of use and disposal in the US. *Resources, Conservation and Recycling*, *105*, 134–142.

Mirpoor, S. F., Patanè, G. T., Corrado, I., Giosafatto, C. V. L., Ginestra, G., Nostro, A., ... Pezzella, C. (2023). Functionalization of polyhydroxyalkanoates (PHA)-based bioplastic with phloretin for active food packaging: Characterization of its mechanical, antioxidant, and antimicrobial activities. *International Journal of Molecular Sciences*, *24*(14), 11628.

Muneer, F., Rasul, I., Azeem, F., Siddique, M. H., Zubair, M., & Nadeem, H. (2020). Microbial polyhydroxyalkanoates (PHAs): Efficient replacement of synthetic polymers. *Journal of Polymers and the Environment*, *28*, 2301–2323.

Nielsen, C., Rahman, A., Rehman, A. U., Walsh, M. K., & Miller, C. D. (2017). Food waste conversion to microbial polyhydroxyalkanoates. *Microbial Biotechnology*, *10*(6), 1338–1352.

Novelli, L. D. D., Sayavedra, S. M., & Rene, E. R. (2021). Polyhydroxyalkanoate (PHA) production via resource recovery from industrial waste streams: A review of techniques and perspectives. *Bioresource Technology*, *331*, 124985.

Pakalapati, H., Chang, C. K., Show, P. L., Arumugasamy, S. K., & Lan, J. C. W. (2018). Development of polyhydroxyalkanoates production from waste feedstocks and applications. *Journal of Bioscience and Bioengineering*, *126*(3), 282–292.

Rajdeep, S. P., & Naithani, S. (2013). Scope of natural and synthetic biodegradable materials in development of ecofriendly substitutes for nursery polybags: A review. *Applied Forest Ecology*, *1*, 35–44.

Rastogi, V. K., & Samyn, P. (2015). Bio-based coatings for paper applications. *Coatings*, *5*(4), 887–930.

Ren, Q., Ruth, K., Thöny-Meyer, L., & Zinn, M. (2010). Enatiomerically pure hydroxycarboxylic acids: Current approaches and future perspectives. *Applied Microbiology and Biotechnology*, *87*, 41–52.

Rydz, J., Sikorska, W., Kyulavska, M., & Christova, D. (2014). Polyester-based (bio) degradable polymers as environmentally friendly materials for sustainable development. *International Journal of Molecular Sciences*, *16*(1), 564–596.

Sabapathy, P. C., Devaraj, S., Meixner, K., Anburajan, P., Kathirvel, P., Ravikumar, Y., ... Qi, X. (2020). Recent developments in polyhydroxyalkanoates (PHAs) production: A review. *Bioresource Technology*, *306*, 123132.

Saratale, R. G., Cho, S. K., Saratale, G. D., Kadam, A. A., Ghodake, G. S., Kumar, M., ... Shin, H. S. (2021). A comprehensive overview and recent advances on polyhydroxyalkanoates (PHA) production using various organic waste streams. *Bioresource Technology*, *325*, 124685.

Saxena, A., & Tiwari, A. (2011). Polyhydroxyalkonates: Green plastics of the future. *International Journal of Biomedical and Advance Research*, 2, 356–367.

Sharma, V., Sehgal, R., & Gupta, R. (2021). Polyhydroxyalkanoate (PHA): Properties and modifications. *Polymer*, *212*, 123161.

Singh, R. S., & Kaur, N. (2015). Microbial biopolymers for edible film and coating applications. *Advances in Biotechnology*, *12*, 187–216.

Sintim, H. Y., & Flury, M. (2017). Is biodegradable plastic mulch the solution to agriculture's plastic problem? *Environmental Science & Technology*, *51*(3), 1068–1069.

Sirohi, R., Pandey, J. P., Gaur, V. K., Gnansounou, E., & Sindhu, R. (2020). Critical overview of biomass feedstocks as sustainable substrates for the production of polyhydroxybutyrate (PHB). *Bioresource Technology*, *311*, 123536.

Sohn, Y. J., Kim, H. T., Baritugo, K. A., Jo, S. Y., Song, H. M., Park, S. Y., Park, S. K., Pyo, J., Cha, H. G., Kim, H., Na, J-G., Park, C., Choi, J., Joo, J. C., & Park, S. J. (2020). Recent advances in sustainable plastic upcycling and biopolymers. *Biotechnology Journal*, *15*(6), 1900489.

Tyagi, P., & Sharma, A. (2021). Utilization of crude paper industry effluent for polyhydroxyalkanoate (PHA) production. *Environmental Technology & Innovation*, *23*, 101692.

Verlinden, R. A., Hill, D. J., Kenward, M. A., Williams, C. D., & Radecka, I. (2007). Bacterial synthesis of biodegradable polyhydroxyalkanoates. *Journal of Applied Microbiology*, *102*(6), 1437–1449.

Waller, J. L., Green, P. G., & Loge, F. J. (2012). Mixed-culture polyhydroxyalkanoate production from olive oil mill pomace. *Bioresource Technology*, *120*, 285–289.

Yadav, B., Pandey, A., Kumar, L. R., & Tyagi, R. D. (2020). Bioconversion of waste (water)/residues to bioplastics – A circular bioeconomy approach. *Bioresource Technology*, *298*, 122584.

Zhang, X., Luo, R., Wang, Z., Deng, Y., & Chen, G. Q. (2009). Application of (R)-3-hydroxyalkanoate methyl esters derived from microbial polyhydroxyalkanoates as novel biofuels. *Biomacromolecules*, *10*(4), 707–711.

13 Environmental and Health Risk Assessment of Emerging Organic Pollutants

Ritu Rani and Dharmender Kumar

13.1 INTRODUCTION

Water is necessary for life and occupies nearly about 70% of the surface of Earth. Only 2.5% of groundwater is suitable for consumption. Water consumption is increasing because of population expansion, industrial demand, and living circumstances. As per the European Environmental Agency, only approximately 40% of surface water bodies (rivers, water bodies, and transition and coastline waters) are considered to be in excellent potential conditions ecologically, and 38% are in acceptable chemical conditions (Kristensen et al., 2018). The extensive implementation of chemicals in daily life activities, along with the unfettered utilization of medications, has resulted in increasing garbage production and an intense release of traditional and novel organic pollutants viz. pesticides, insecticides (Rani and Kumar, 2024), and even heavy metals as well as microplastics (Rani et al., 2024) into the surrounding atmosphere. In recent decades, newly formed chemicals known as emerging organic contaminants (EOCs) have become increasingly visible (Karpińska and Kotowska, 2017). Persistent organic pollutants, also known as POPs, can persist in the environment for an extended period. Due to their physical and chemical properties, POPs tend to move long distances and resist breakdown biologically and chemically. This makes it possible for POPs to bioaccumulate to a deeper level through biomagnification, and exposure to them can cause severe damage to health and the environment. Moreover, their presence in marine waters threatens the aquatic ecosystem affecting habitats, biodiversity, and human health. Several organic compounds have been found in dissolved form in water, including humic acid and fulvic compounds, halogenated aromatic and hydrocarbons with aliphatic structures. These pollutants may naturally degrade into nontoxic or less-toxic forms in freshwater ecosystems, persist for an extended period, get ingested into the diet or food chain, and are amplified via bioaccumulation, having a cascade effect on terrestrial and aquatic ecosystems. Biological food chains acquire these compounds via remobilization, changes in form, or speciation from these reservoirs. This has led to the bioaccumulation and biomagnification of these contaminants and the development of chronic and acute diseases due to their mutagenicity, carcinogenicity, and teratogenicity.

DOI: 10.1201/9781032642567-16

To control organic compounds' appearance and undesirable transformation into the ecosystem, we need to understand the group of organic substances, their sources, and the processes by which they are transformed in the environment. Additionally, we must be able to detect contaminants in freshwater readily and easily while understanding their chemistry and their impact on the surrounding environment and human health. We inevitably need to better understand these pollutants to develop better techniques for purifying water of persistent organic contaminants. Both the point and nonpoint sources of these contaminants enter the freshwater system. A point source can be defined spatially as a result of the chemical residues they contribute from organic pollutants and by the epidemiological characteristics resulting from them, such as morbidity, mortality, and disruption of communities (Kronimus et al., 2004).

13.2 RELEASE OF POLLUTANTS FROM DIFFERENT SITES AND THEIR ENVIRONMENTAL EFFECTS

Many studies on POPs' dependence on industrial and agricultural sectors have been observed in recent decades. Many were extensively employed commercially during the postwar industrial output boom, resulting in widespread geographical dispersion (Shumbula et al., 2021).

They enter the environment through various industrial processes, including power plants, heating plants, incinerators, domestic fireplaces, transport, agricultural aerosols, and soil. POP may be formed inadvertently in incinerations, biochemical facilities, wildfires, putrefying, and polychlorinated biphenyl (PCB)-containing debris.

This makes it possible for them to bioaccumulate to a deeper level through biomagnification, and exposure to them can cause severe damage to health and the environment (Göktaş et al., 2016). POPs are bioaccumulate in living organisms and they are semi-volatile, which can evaporate or turn into a gas and stick to tiny particles in the air. Considering the dangers to the environment, the Environmental Protection Agency (EPA) and the International Agency for Research on Cancer (IARC) have suggested reducing the release of these chemicals into the environment. Figure 13.1 describes the various sources and routes of the organic pollutants in the environment.

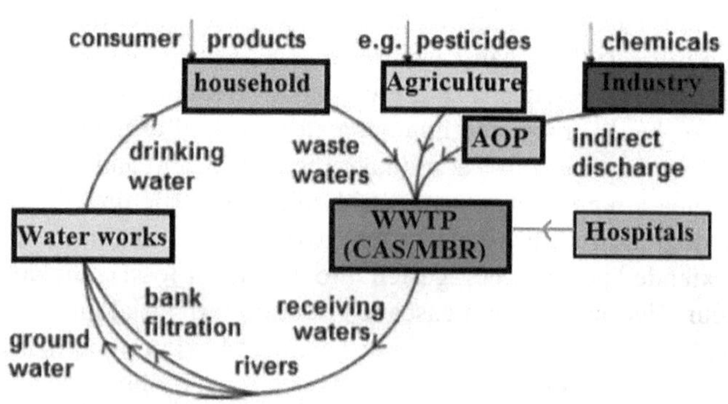

FIGURE 13.1 Major source of organic pollutants in the environment.

13.2.1 Hospitals

The levels of biological oxygen demand (BOD), the chemical demand for oxygen, and all organic compounds concentrations in hospital wastewater (HWW) are high, as are the concentrations of ammonia, and nitrogen compounds such as nitrites, nitrates, and phosphorus (Majumder et al., 2021). According to many studies, these characteristics are more excellent in hospital wastewater than municipal sewage wastewater (Carraro et al., 2016). Hazardous and nonbiodegradable pollutants such as pharmaceutically active substances (PhACs), X-ray contrast media, and surfactants contribute to the low BOD/COD ratio of HWW. Most of these pollutants, known as emerging contaminants (ECs), may be harmful to people along with other aquatic inhabitants at low doses (g/L to ng/L) (Meo et al., 2014).

13.2.2 Household

Organic contaminants commonly found in house dust fall into two main categories: (i) Flame-retardant substances such as polybrominated diphenyl ethers (PBDEs), novel bromine-containing flame retardants (NBFRs), and organic phosphate flame retardants (OPFRs) and (ii) the chemicals PCBs, organochlorine pesticides (OCPs), and plasticizing agents (phthalates, phenol) and products for consumers (Besis et al., 2021). Because most of these contaminants are semi-volatile organic substances (SVOSs), they can be found in the form of gases, suspended in the air, settled dust, and exposed surfaces. Furthermore, several compounds in subcategory (i) exhibit considerable biological magnification, persistence, and accumulation in food chains in humans and animals (Wei et al., 2018).

13.2.3 Industrial Sources

Industries are one of the most critical ways that water systems get dirty. Depending on the type of production process being used, the quantities and chemical properties of the wastewater components are different. Because many production processes stop and start, there is a lot of variation in the wastewater's quality, making it hard to give a clear scientific description (Dsikowitzky and Schwarzbauer, 2013).

Some of the most common pollutants made by industry include polychlorinated naphthalenes (PCNs), polymerized biphenyls , and hexachlorobenzenes (HCBs), and so on. (Mishra et al., 2022). Aromatic hydrocarbons, which include toluene, benzene, ethyl benzene, and xylene (BTEX) are persistent organic pollutants, also known as POPs, discharged into the environment as a result of petroleum drilling (Varjani et al., 2017). POP source of release and their uses and contamination pathway are described in Figure 13.2.

13.3 EFFECTS OF ORGANIC POLLUTANTS ON THE ENVIRONMENT

Air, water, animals, and humans are connected to the environment. As a result of pollution from POPs, the ecological balance is disturbed, threatening all living organisms as well as the environment. Environmental effects, such as shifts in gender ratios, cancer, and impaired fertility, can cause other physical abnormalities.

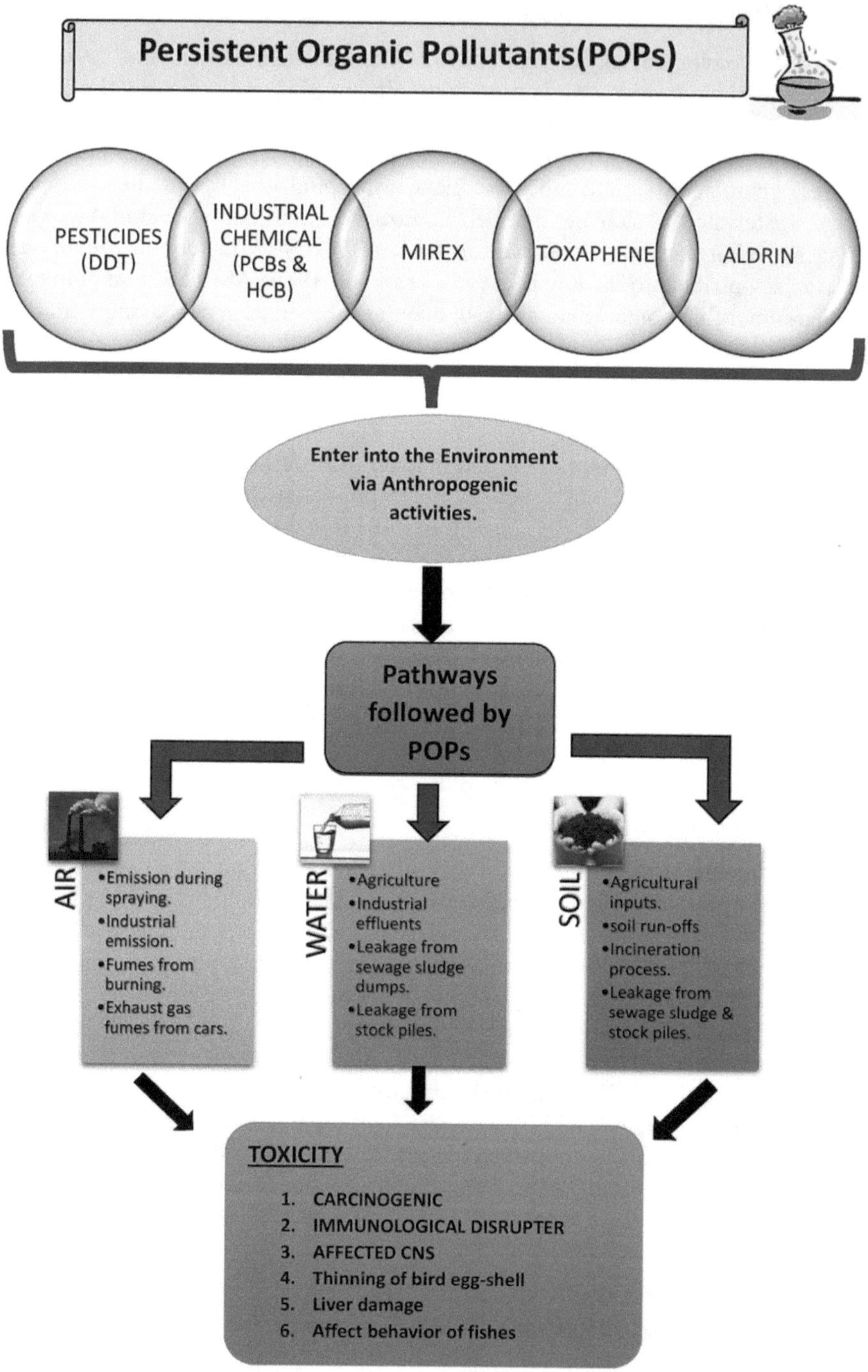

FIGURE 13.2 Uses, toxicity, and pathways of contamination by various POPs.

13.3.1 Organic Pollutants Risk Assessment of Aquatic Organisms

Zhou et al. (2018) investigated the occurrence, congener arrangement, and trophic distribution of 14 marine species' short-chain chlorine-based paraffin (CP) compounds in the aquatic food chain. The findings showed that there were extremely high levels of CPs, reaching as much as 1,300 g of fatty weight, with the benthic invertebrates having the greatest concentrations and nonbenthic invertebrates having the lowest amounts. Most of the PCBs present in aquatic water were between 0.032 and 1.4 ng/g by wet weight. Compared with *Cyprinus carpio*, *Oncorhynchus mykiss*, and *Eriocheir sienensis* had many more PCBs in their bodies. Haegerbaeumer et al. (2018) examined how exposure to polycyclic aromatic hydrocarbon (PAH) affected worm populations in freshwater microcosms. The findings revealed that the organism (*Caenorhabditis elegans*) was susceptible to the metabolic strain of pollution, particularly biomass. The study ultimately proved that nematodes (*C. elegans*) are the best way to determine how many PAHs are in freshwater communities and that they are essential species needed for risk assessment studies. Xiong et al. (2018) looked at how *Daphnia magna* biomarkers changed after it was exposed to two different types of PBDEs. The data showed that the flame hindering stopped the activity that made the *D. magna* stay still and made its heartbeat faster. A study found that fish and *D. magna* were most sensitive to chlorpyrifos, with LC50 values of 0.08 and 0.001 mol/L. Carbaryl and diuron have LC_{50} values for fish that are 43.19 and 43.48 mol/L. *D. magna* was also very sensitive to carbaryl, which had an LC_{50} of 0.031 mol/L, while diuron had an LC50 of 32.11 mol/L. The time–response relationships were almost the same, and the LT_{50} values, ranging from 60.86 to 68.04 hours, were very close, showing that the same exposure time was needed to cause toxicity. The toxicology tests showed that *D. magna* was more sensitive to organic pollution than fish and mixtures were harmful to fish and *D. magna* (Yasser et al., 2015).

13.3.2 Risk Assessment of Organic Pollutants on Flora

Green flora with lipid-rich cuticles covers over 80% of the Earth's surface. OPs like PAHs, BPA, and medicines aren't very volatile, so plants mostly absorb them in through their roots. Plants can take up pollutants in several ways. It includes (i) getting it from the Earth, moving it to the roots, and then moving it through the xylem and (ii) transporting it through the phloem after receiving it from the atmosphere on the surface of the leaves or the stomata.

Several studies look at the effects of OPs on plants worldwide (Kreslavski et al., 2017). Most of the new research on how OPs affect plant cells focuses on how they hurt the process of photosynthesis (Tomar and Jajoo, 2013; Sharma et al., 2018). OPs adversely affect plants by stopping biological processes like mitosis, germination, enzyme function, hormone production, development of leaves, photosynthesis proteins or DNA, breakdown of cell walls, etc.

Also, some OPs, like PAHs, breakdown when they are exposed to UV light. The photoproducts hurt plants more than the chemicals (Tomar and Jajoo, 2015). Damage to DNA and RNA, peroxidation of membrane lipids, and oxidation of cellular lipids (LPO) are all caused by the buildup of reactive oxygen species (ROS) triggered by oxidant stress from OP. In a variety of parts of *Bruguiera gymnorrhiza* (L.) Savigny, such as the roots, stems, and leaves, researchers examined the influence of PAHs (pyrene) on the enzymes ascorbate peroxidase (APX), super oxygen dismutase (SOD), malondialdehyde (MDA), and the amino acid proline. Many studies have shown how different stresses affect creatures that use light to make food, significantly higher plants. Much research has been done on environmental stressors like lack of water, excessive temperature, and high salt (Oukarroum et al., 2012).

13.3.3 Risk Assessment of Organic Pollutants on Human

Pollutants of various forms can be found in many locations. Some resist environmental deterioration (chemical, biological, and photolytic processes) and can survive long. POPs have been present in our environment for a long time. In addition to being present for an extended period, these pollutants are transported across and even over national borders. The bioaccumulation of these pollutants into food chains (animals and humans) causes a variety of infamous health hazards and environmental problems. The impact of these pollutants is very much in the spotlight among scientists, governments, and nongovernmental organizations (NGOs). All age groups are affected by POPs, with elderly populations having the highest concentrations. Many serious health problems may result from exposure to these pollutants, including hormonal disturbance, cancer, coronary artery disease, weight gain, nervous system disorders, learning difficulties, diabetes, and also cause inadequacy in women's embryos. According to the World Health Organization, the most toxic organic pollutant is 2,3,7,8-tetrachlorodibenzo-*p*-dioxin (2,3,7,8-TCDD). Dioxins are known to cause cancer, according to the EPA. 2,3,7,8-TCDD is classified as an active carcinogenic agent by the IARC. Disruptors of endocrine function include chemicals that interfere with the nervous, immune, and reproductive systems. There are several types of endocrine-disrupting pesticides, the most common pesticides made from organo-chlorine compounds (DDT, chlordane, dieldrin, endosulfan, toxaphene, mirex, etc.). POPs are associated with various illnesses, such as obesity, cancer, reproductive problems, and other health problems.

13.4 BIOSENSOR-BASED ASSESSMENT OF ORGANIC POLLUTANTS

Biosensor is a biotechnology-based analysis method for pollutants that comprises signal transducer components that produce observable or measurable signals while detecting contaminants (Dhull et al., 2013). Biosensors may be cell-free or whole-cell based, general or particular in terms of biological constituents and sensing capabilities (Kaur et al., 2015). The important variables for separating certain pollutants from others are those regulatory systems that interact with environmental contaminants (Huang et al., 2023). Recently, biosensors earmarked heavy metals, organo-pesticides, or medicines demonstrate biosensor benefits such as mobility, simplicity of use, and time savings (Chadha et al., 2022). Recent research has emphasized advancements in sensitivity, stability, and selectivity, as well as their importance to environmental monitoring (Voon et al., 2022). Some of the studies are shown in Table 13.1.

TABLE 13.1
Detection of Organic Pollutants Using Biosensor

Pollutants	Class	Sensing Element of Biosensor	Detection Range	Inference	References
Bisphenol A	EDC	Enzyme (tyrosinase)	0.01 nM	Promising Reliable	Zehani et al. (2015)
		Labeled antibody	0.014 µg/L	Satisfactory Repeatability Good agreement	Rodriguez-Mozaz et al. (2005)
		Sol-gel films	66 nM	High sensitivity Good precision	Kochana et al. (2015)
Estrogen Estradiaol Phenol Nonylphenol	EDCs	Enzyme (tyrosinase)	4.4, 4.06, 0.94, 2.2, 5.2 mg/L	Fast good Sensitive compound	Dempsey et al. (2004)
Estrone	EDC	Labeled antibody	0.2 ng/L	Detection a powerful tool in aquatic analytics	Tschmelak et al. (2005)
Triclosan	EDCs	Polymers with molecular imprinting	0.017 µg/L	Magnificent performance, fast, sensitive and simple nano-sensor	Atar et al. (2015)
Atrazine Bsphenol AestroneIsoproturon	Herbicide EDCE DCHerbicide	Labeled antibodies	0.002, 0.005, 0.019, 0.016 µg/L	Early warning application at a low cost	Tschmelak et al. (2004)
Acetamiprid	Pesticide	Electrochemical DNA Aptasensor	3.3×10^{-7}nM	Excellent sensitivity	Chen et al. (2019)
Organophosphorus (OPs)	Pesticide	A portable LIG-based electrochemical biosensor	1.2 µg/L	Excellent performance in applications of pesticides for the environment and the safety of food	Liu et al. (2022)
Pyrethroid	Pesticide	Whole cell and antibodies	3 µg/L	Simple and accessible colorimetric implementation on-site. In excess of 90 days after the lyophilization it is functional	Riangrungroj et al. (2019)

(*Continued*)

TABLE 13.1 (*Continued*)
Detection of Organic Pollutants Using Biosensor

Pollutants	Class	Sensing Element of Biosensor	Detection Range	Inference	References
Naphthalene Phenanthrene	PAHs	DNA/Sol-gel array	1–10 mg/L	Effective detection of PAHs within water and samples of biology	Doong (2005)
Poly-brominated diphenyl ether (PBDE)	POP	Whole cell	10 nM	Novel PBDE-specific sensing element	Chen et al. (2022)
PCB-28PCB-101	PCB	Peroxidase enzyme/ immobilized poly aniline augmented Pt electrodes	0.016 μg/L 0.019 μg/L	Useful for pollutants detection in leachates from landfills	Nomngongo et al. (2012)
TPH	TPH	Enzymes and antibodies	25 mg/kg	Simple on-site application Determine TPH levels in soil and water	Lambert et al. (2001)

13.5 GENETICALLY MODIFIED ORGANISMS FOR ORGANIC POLLUTANTS

Recombinant genetic material (DNA) and other methods from molecular biology have made it possible to (i) amplify, disrupt, and/or modify specific genes that code for enzymes in metabolic processes, (ii) remove bottlenecks in the processes, (iii) improve the redox and production of energy, and (iv) recruit heterologous genetic material to give rise to new traits (Liu et al., 2006). Different genetic methods have been developed and used to enhance degradable enzymes, metabolic processes, and microorganisms (Pieper and Reineke, 2000). The additional genes or enzymes must be absorbed into the metabolic and regulatory networks even in a single altered organism for proper expression (Cases and Lorenzo, 2005).

Bio-augmentation, which uses genetically modified microbial strains to break down trash and pollutants, has become a popular way to combat the issue of garbage and pollutants decomposing slowly and ineffectively (Peng et al., 2020). Aside from bacterial engineering, there are several additional ways to modify plants', animals', and microorganisms' genetic sequences, each with a unique objective in mind.

The term "genetically engineered microorganisms" (GEMs) is used about genetically modified organisms, or GMOs explicitly refers to microorganisms (bacterial species, fungi, yeast, and many others) that have been converted by humans using molecular biology in vitro procedures (Saxena et al., 2020). The insertion of genes transforms a single microbe into numerous microorganisms. GEMs may also be used in bioremediation, which is a method that uses organisms or their enzyme systems to break down and remove toxins from the environment (Perera and Hemamali, 2022).

In the fields of phytoremediation and rhizoremediation, microorganisms that have been changed genetically are being used more and more to clean up polluted areas. For instance, it has been said that toxins like trichloroethylenes and polychlorinated biphenyls break down. By cloning the phytochelatin synthetase gene from *Astragalus sinicus* into *Mesorhizobium huakuii* subsp. *rengei strain B3*, a symbiotic relationship between the two, has been created to gather cadmium metals. A hybrid rhizospheric was made by cloning the *o*-mono-oxygenase genes of *Burkholderia cepacia* into the *Pseudomonas* fluorescens 2–79 strain. This rhizosphere was able to break down trichloroethylene by 63% compared to the engineered host, which was only able to break it down by 9%. Organophosphorus toxins in the soil are broken down by the enzyme organophosphorus hydrolase, which is made by the OPH gene. The issue was solved by fusing the *E. coli* OPH gene with the signal sequence, which enabled the OPH protein to enter the periplasm. The recombinant *E. coli* causes the cytosolic OPH activity to increase 1.8-fold when combined with a native strain that produces it (Kang et al., 2006).

13.6 NANOPARTICLES IN THE DETECTION OF ORGANIC POLLUTANTS

Nanoparticles have shown great promise in the detection of organic pollutants due to their different physio-chemical properties. They offer several advantages over traditional detection methods, including increased sensitivity, selectivity, and versatility (Malik and Kumar, 2023). Here are some ways nanoparticles are used in the detection of organic pollutants:

13.6.1 Surface-Enhanced Raman Scattering

Enhanced Raman scattering on the surface (SERS) effect is caused by the coupling of an electromagnetic field from an appropriate roughed surface, typically made of silver or gold, with the vibrational patterns of an analyte. It is an ultrasensitive, ultrafast, and global analytical method with a detection limit that can reach the level of a single molecule (Alvarez-Puebla and Liz-Marzán, 2012). It can significantly increase the scattering impact, overriding the usual Raman technique's limited sensitivity. The "Fingerprint-like" Raman spectra can be used to gather descriptive data on numerous chemical and biological variables in a complicated system (Xie et al., 2012). In Table 13.2, the implications of the nanoparticles have been described for the detection of organic pollutants.

TABLE 13.2
Role of Nanoparticles for Detection of Organic Pollutants

Nanoparticles	Organic Pollutant	Sensitivity	References
Silver nanoparticles in the form of dried films	Fonofos	10 ppm	Vongsvivut et al. (2010)
Aggregated Silver nanoparticles	Omethoate	2 ppm	Guerrini et al. (2011)
Gold nanoparticles	Furaltadone	5 ppm	Xie et al. (2012)
Silver nanoparticles on silver nanoflowers	rhodamine 6G	10^{-13} M	Barveen et al. (2020)
	Congo Red	10^{-8} M	
1-Hexanethiol functioned silver nanoparticles aggregates on copper foil	Fluoranthene	5 μg/L	Jiang et al. (2012)
Silver nanoparticles modified by L-cysteine methyl ester hydrochloride	2,4-dinitroanisole	20 μg/L	Xu et al. (2011)

13.6.2 Quantum Dots

Semiconducting nanocrystals having a diameter of 2–10 nm are known as quantum dots (QDs). These are used in many applications involving sensing and have strong fluorescence (Smith and Nie, 2010). Due to the quantum confinement effect, QDs have good fluorescence characteristics, which is favorable. The extraordinarily high quantum yields of QDs, which enable them to be utilized in tiny amounts to produce a discernible fluorescence signal, are another distinctive quality of these particles. Semiconducting QDs' fluorescence characteristics are greatly influenced by their surface (Chandra et al., 2020).

Nsibande et al. show how the QDs might be used for identifying the presence of pyrene, a typical PAH component. Due to aggregation, the QDs' interaction with ferric ions quenched the fluorescence; however, their subsequent engagement with pyrene via p–p stacking activated the fluorescence. Thus, the presence of pyrene was seen (Nsibande et al., 2020).

13.6.3 Nanocomposite

Due to their favorable physical and chemical characteristics, graphene and its analogs have received a lot of attention in a variety of scientific fields (Liu et al., 2016). Such carbon-based substances are quickly becoming viable foundations for creating nanocomposites for various polycyclic aromatic hydrocarbon (PAH) sensing applications (Pan et al., 2015). It is simple to blend semiconductor QDs with graphene, graphene oxide, and reduced graphene oxide to create composites that have improved sensitivity and selectivity for identifying the presence of organic contaminants. Due to their excellent adsorption capacities, graphene, GO, and composites made of them are being extensively utilized for wastewater treatment. Graphene, along with its derivatives are excellent regulators of QDs that facilitate fluorescence detection of π-electron-rich organic contaminants due to their π-electron system (Huang et al., 2015).

Mitra et al. employed dextran-fluorescein as a fluorescent probe whose fluorescence is quenched over the surface of the graphene. Recovered fluorescence is used to detect organic contaminants as a positive signal. This method of detection is repeatable, highly sensitive, and only needs a modest volume of samples (Mitra and Saha, 2017). Adegoke et al. demonstrated the quantitative ability of a covalently connected core/shell/shell QD-GO nanocomposite material fluorescent probe made of CdSeTe, ZnSe, and ZnS for the detection of PAH compounds in aqueous solutions. Compared with unconjugated QDs, the QD-GO nanoconjugate had increased fluorescence (Adegoke and Forbes, 2016).

13.6.4 Metal Oxide Nanoparticle-Based Chemiresistor Sensors

Various oxides such as tungsten, titanium dioxide, tin oxide, and others are deposited over two interdigitated gold or platinum electrodes. The target analyte's presence causes a shift in the balance state of its surface oxygen reactions, which is how the sensor works. By their higher surface-to-bulk ratios compared with microcrystalline materials, nanoparticulate metal oxides enhance the sensor's sensitivity, response, and recovery times. This is because the surface oxygen chemisorption process dictates the biosensor responses. The selectivity is increased by doping the metal oxide nanoparticles with the appropriate metals, which generate catalytically active sites for a specific analyte (Khanna, 2008).

13.6.5 Nanostructured Electrochemical Sensors

The advantages of electrochemical sensors over conventional analytical techniques based on optical and mass spectra are low cost, ease of miniaturization, sensitivity, selectivity, speed, and ease of use (Gan and Zhao, 2019). Electrochemical sensors have been built utilizing nanostructured ZrO_2, TiO_2, and layered double hydroxides because of their naturally large surface areas and excellent adsorptive. These nanostructured metal oxides/hydroxides can be utilized to enrich and detect organic chemicals such as electroactive insecticides, herbicides, and phenolic compounds with superior selectivity and sensitivity due to their abundance of adsorption sites and great affinity for organic molecules. Thus, nitroaromatic organophosphates such as fenitrothion, methyl parathion, and paraoxon might show high redox activities at the electrode surface (Yu et al., 2014).

13.7 CONCLUSION

The unrestricted release of contaminants and untreated wastewater into the atmosphere is a major worldwide problem. In industrial and farming operations, efforts should be made to reduce or replace toxic compounds with less-toxic ones. Several conventional techniques for decontaminating multipollutants have been reported for the remediation process, including adsorption, precipitation via chemicals, photocatalytic process, oxido-reductase reaction, and others. The removal of all pollutants does not happen simultaneously, as some eco-friendly catalysts (bacteria/microbiological enzymes) may be effective in removing pollutants at the same time of duration in a contaminated site, whereas not all contaminants are released simultaneously. The use of nanoparticles in the detection of organic pollutants is an active area of research, and ongoing advancements in nanotechnology will likely lead to even more sensitive and efficient detection methods. However, it is essential to consider nanoparticles' potential environmental and health impacts when designing these detection systems.

Understanding how surface and ground waters are contaminated worldwide is crucial for addressing measures to stop the accumulation of appropriate organic micropollutants in the natural environment. Additionally, future monitoring campaigns should address some critical issues, such as determining some pollutants' precursors, metabolites, and transformation products, nonstandard techniques for ecotoxicological research, and analytic restrictions.

Furthermore, an enzyme's activity and stability are influenced by its surroundings. When utilized in bioremediation applications, enzymes are also efficient, quick, focused, incredibly adaptable, simple to operate, and require no nutrient supply.

REFERENCES

Abd-El-Haleem, D., Ripp, S., Scott, C., & Sayler, G. S. (2002). A luxCDABE-based bioluminescent bioreporter for the detection of phenol. *Journal of Industrial Microbiology and Biotechnology*, *29*(5), 233–237. https://doi.org/10.1038/sj.jim.7000309

Adegoke, O., & Forbes, P. B. C. (2016). l-cysteine-capped core/shell/shell quantum dot–graphene oxide nanocomposite fluorescence probe for polycyclic aromatic hydrocarbon detection. *Talanta*, *146*, 780–788. https://doi.org/10.1016/j.talanta.2015.06.023

Alvarez-Puebla, R. A., & Liz-Marzán, L. M. (2012). Traps and cages for universal SERS detection. *Chemical Society Reviews*, *41*(1), 43–51. https://doi.org/10.1039/c1cs15155j

Atar, N., Eren, T., Yola, M. L., & Wang, S. (2015). A sensitive molecular imprinted surface plasmon resonance nanosensor for selective determination of trace triclosan in wastewater. *Sensors and Actuators Part B*, *216*, 638–644. https://doi.org/10.1016/j.snb.2015.04.076

Barveen, N. R., Wang, T. J., & Chang, Y. H. (2020). In-situ deposition of silver nanoparticles on silver nanoflowers for ultrasensitive and simultaneous SERS detection of organic pollutants. *Microchemical Journal*, 159, 105520. https://doi.org/10.1016/j.microc.2020.105520

Besis, A., Botsaropoulou, E., Balla, D., Voutsa, D., & Samara, C. (2021). Toxic organic pollutants in Greek house dust: Implications for human exposure and health risk. *Chemosphere*, *284*, 131318. https://doi.org/10.1016/j.chemosphere.2021.131318

Carraro, E., Bonetta, S., Bertino, C., Lorenzi, E., Bonetta, S., & Gilli, G. (2016). Hospital effluents management: Chemical, physical, microbiological risks and legislation in different countries. *Journal of Environmental Management*, *168*, 185–199. https://doi.org/10.1016/j.jenvman.2015.11.021

Cases, I., & De Lorenzo, V. (2005). Promoters in the environment: Transcriptional regulation in its natural context. *Nature Reviews. Microbiology*, *3*(2), 105–118. https://doi.org/10.1038/nrmicro1084

Chadha, U., Bhardwaj, P., Agarwal, R., Rawat, P., Agarwal, R., Gupta, I., Panjwani, M., Singh, S., Ahuja, C., Selvaraj, S. K., Banavoth, M., Sonar, P., Badoni, B., & Chakravorty, A. (2022). Recent progress and growth in biosensors technology: A critical review. *Journal of Industrial and Engineering Chemistry*, *109*, 21–51. https://doi.org/10.1016/j.jiec.2022.02.010

Chandra, A., Prasad, S., Gigli, G., & del Mercato, L. L. (2020). Fluorescent nanoparticles for sensing. *Frontiers of Nanoscience*, *16*, 117–149. https://doi.org/10.1016/B978-0-08-102828-5.00006-1

Chen, P., Qiao, X., Liu, J., Xia, Fangquan, Tian, D., & Zhou, C. (2019). Dual-signaling amplification electrochemical aptasensor based on hollow polymeric nanospheres for acetamiprid detection. *ACS Applied Materials and Interfaces*, *11*(16), 14560–14566. https://doi.org/10.1021/acsami.9b00308

Chen, X., Yao, H., Song, D., Sun, G., & Xu, M. (2022). Extracellular chemoreceptor of deca-brominated diphenyl ether and its engineering in the hydrophobic chassis cell for organics biosensing. *Chemical Engineering Journal*, *433*, 133266. https://doi.org/10.1016/j.cej.2021.133266

Dempsey, E., Diamond, D., & Collier, A. (2004). Development of a biosensor for endocrine disrupting compounds based on tyrosinase entrapped within a poly(thionine) film. *Biosensors and Bioelectronics*, *20*(2), 367–377. https://doi.org/10.1016/j.bios.2004.02.007

Dhull, V. Gahlaut A., Gothwal, A. Duhan, J. S., & Hooda, V. (2013). Nitrite Biosensors: analytical tools for determination of toxicity due to presence of nitrite ions. In: R. K. Salar, S. K. Gahlawat, P. Siwach, & J. S. Duhan (eds), *Biotechnology: Prospects and Applications* (pp. 217–226). Springer, Springer.

Doong, R. (2005). Sol–gel-derived array DNA biosensor for the detection of polycyclic aromatic hydrocarbons in water and biological samples. *Sensors and Actuators Part B*, *111*, 323–330. https://doi.org/10.1016/j.snb.2005.06.065

Dsikowitzky, L., & Schwarzbauer, J. (2013). Organic contaminants from industrial wastewaters: Identification, toxicity and fate in the environment. *Pollutant Diseases, Remediation and Recycling*. https://doi.org/10.1007/978-3-319-02387-8_2

Focazio, M. J., Kolpin, D. W., Barnes, K. K., Furlong, E. T., Meyer, M. T., Zaugg, S. D., Barber, L. B., & Thurman, M. E. (2008). A national reconnaissance for pharmaceuticals and other organic wastewater contaminants in the United States—II Untreated drinking water sources. *Science of the Total Environment*, 2–3, 201–216.

Gan, X., & Zhao, H. (2019). Understanding signal amplification strategies of nanostructured electrochemical sensors for environmental pollutants. *Current Opinion in Electrochemistry*, *17*, 56–64. https://doi.org/10.1016/j.coelec.2019.04.016

Göktaş, R. K., & MacLeod, M. (2016). Remoteness from sources of persistent organic pollutants in the multi-media global environment. *Environmental Pollution*, *217*, 33–41. https://pubmed.ncbi.nlm.nih.gov/26775726/; https://doi.org/10.1016/j.envpol.2015.12.058

Guerrini, L., Sanchez-Cortes, S., Cruz, V. L., Martinez, S., Ristori, S., & Feis, A. (2011). Surface-enhanced Raman spectra of dimethoate and omethoate. *Journal of Raman Spectroscopy*, *42*(5), 980–985. https://doi.org/10.1002/jrs.2823

Haegerbaeumer, A., Höss, S., Heininger, P., & Traunspurger, W. (2018). Response of nematode communities to metals and PAHs in freshwater microcosms. *Ecotoxicology and Environmental Safety*, *148*, 244–253. https://pubmed.ncbi.nlm.nih.gov/29065374/. https://doi.org/10.1016/j.ecoenv.2017.10.030

Huang, C.-W., Lin, C., Nguyen, M. K., Hussain, A., Bui, X.-T., & Ngo, H. H. (2023). A review of biosensor for environmental monitoring: Principle, application, and corresponding achievement of sustainable development goals. *Bioengineered*, *14*(1), 58–80. https://doi.org/10.1080/21655979.2022.2095089

Huang, Y. L., Walker, A. S., & Miller, E. W. (2015). A photostable silicon rhodamine platform for optical voltage sensing. *Journal of the American Chemical Society*, *137*(33), 10767–10776. https://doi.org/10.1021/jacs.5b06644

Jiang, X., Lai, Y., Yang, M., Yang, H., Jiang, W., & Zhan, J. (2012). Silver nanoparticle aggregates on copper foil for reliable quantitative SERS analysis of polycyclic aromatic hydrocarbons with a portable Raman spectrometer. *Analyst*, *137*(17), 3995–4000. https://doi.org/10.1039/c2an35713e

Kang, D. G., Choi, S. S., & Cha, H. J. (2006). Enhanced biodegradation of toxic organophosphate compounds using recombinant Escherichia coli with sec pathway-driven periplasmic secretion of organophosphorus hydrolase. *Biotechnology Progress*, *22*(2), 406–410. https://doi.org/10.1021/bp050356k

Karpińska, J., & Kotowska, U. (2017). Removal of organic pollution in the water environment. *Water*, *11*(10). https://doi.org/10.3390/w11102017

Kaur, H., Kumar, R., Babu, J. N., & Mittal, S. (2015). Advances in arsenic biosensor development–a comprehensive review. *Biosensors and Bioelectronics*, *63*, 533–545. https://doi.org/10.1016/j.bios.2014.08.003

Khanna, V. K. (2008). Nanoparticle-based sensors. *Defence Science Journal*, *58*(5), 608–616.

Kochana, J., Wapiennik, K., Kozak, J., Knihnicki, P., Pollap, A., Woźniakiewicz, M., Nowak, J., & Kościelniak, P. (2015). Tyrosinase-based biosensor for determination of bisphenol A in a flow-batch system. *Talanta, 144*, 163–170. https://doi.org/10.1016/j.talanta.2015.05.078

Kreslavski, V. D., Brestic, M., Zharmukhamedov, S. K., Lyubimov, V. Y., Lankin, A. V., Jajoo, A., & Allakhverdiev, S. I. (2017). Mechanisms of inhibitory effects of polycyclic aromatic hydrocarbons in photosynthetic primary processes in pea leaves and thylakoid preparations. *Plant Biology, 19*(5), 683–688. https://doi.org/10.1111/plb.12598

Kristensen, P., Whalley, C., Zal, F. N. N., & Christiansen, T. (2018). *European Waters Assessment of Status and Pressures 2018 EEA Report* 7/2018. EEA.

Kronimus, A., Schwarzbauer, J., Dsikowitzky, L., Heim, S., & Littke, R. (2004). Anthropogenic organic contaminants in sediments of the Lippe River, Germany. *Water Research, 38*(16), 3473–3484. https://doi.org/10.1016/j.watres.2004.04.054

Lambert, P., Fingas, M., & Goldthorp, M. (2001). An evaluation of field total petroleum hydrocarbon (TPH) systems. *Journal of Hazardous Materials, 83*(1–2), 65–81. https://doi.org/10.1016/s0304-3894(00)00328-9

Li, Z., Young, R. J., Wang, R., Yang, F., Hao, L., Jiao, W., & Liu, W. (2013). The role of functional groups on graphene oxide in epoxy nanocomposites. *Polymer, 54*(21), 5821–5829. https://doi.org/10.1016/j.polymer.2013.08.026

Liu, X., Cheng, H., Zhao, Yuecan, Wang, Y., & Li, F. (2022). Portable electrochemical biosensor based on laser-induced graphene and MnO2 switch-bridged DNA signal amplification for sensitive detection of pesticide. *Biosensors and Bioelectronics, 199*, 113906. https://doi.org/10.1016/j.bios.2021.113906

Liu, Z., Hong, Q., Xu, J. H., Jun, W., & Li, S.-P. (2006). Construction of a genetically engineered microorganism for degrading organophosphate and carbamate pesticides. *International Biodeterioration and Biodegradation, 58*(2), 65–69. https://doi.org/10.1016/j.ibiod.2006.07.009

Liu, Z., Tian, C., Lu, Lehui, & Su, X. (2016). 'A novel aptamer-mediated CuInS 2 quantum dots@graphene oxide nanocomposites-based fluorescence "turn off–on' nanosensor for highly sensitive and selective detection of kanamycin. *RSC Advances, 6*(12), 10205–10214. https://doi.org/10.1039/C5RA22753D

Majumder, A., Gupta, A. K., Ghosal, P. S., & Varma, M. (2021). A review on hospital wastewater treatment: A special emphasis on occurrence and removal of pharmaceutically active compounds, resistant microorganisms, and SARS-CoV-2. *Journal of Environmental Chemical Engineering, 9*(2), 104812. https://doi.org/10.1016/j.jece.2020.104812

Malik, S., & Kumar, D. (2023). Perspectives of nanomaterials in microbial remediation of heavy metals and their environmental consequences: A review. *Biotechnology and Genetic Engineering Reviews*, 1–48. https://doi.org/10.1080/02648725.2023.2182546

Mauriz, E., Calle, A., Abad, A., Montoya, A., Hildebrandt, A., Barceló, D., & Lechuga, L. M. (2006). Determination of carbaryl in natural water samples by a surface plasmon resonance flow-through immunosensor. *Biosensors and Bioelectronics, 21*(11), 2129–2136. https://doi.org/10.1016/j.bios.2005.10.013

Meo, M. I., Haydar, S., Nadeem, O., Hussain, G., & Rashid, H. (2014). Characterization of hospital wastewater, risk waste generation and management practices in Lahore. *Proceedings of the Pakistan Academy of Sciences, 51*(4), 317–329.

Mishra, A., Kumari, M., Swati, R., Kumar, R., Iqbal, K., & Thakur, I. S. (2022). Persistent organic pollutants in the environment: Risk assessment, hazards, and mitigation strategies. *Bioresource Technology Reports, 19*, 101143. https://doi.org/10.1016/j.biteb.2022.101143

Mitra, R., & Saha, A. (2017). Reduced graphene oxide based "turn-on" fluorescence sensor for highly reproducible and sensitive detection of small organic pollutants. *ACS Sustainable Chemistry and Engineering, 5*(1), 604–615. https://doi.org/10.1021/acssuschemeng.6b01971

Nomngongo, P. N., Catherine Ngila, J. C., Msagati, T. A. M., Gumbi, B. P., & Iwuoha, E. I. (2012). Determination of selected persistent organic pollutants in wastewater from landfill leachates, using an amperometric biosensor. *Physics and Chemistry of the Earth, Parts A/B/C*, *50–52*, 252–261. https://doi.org/10.1016/j.pce.2012.08.001

Nsibande, S. A., & Forbes, P. B. (2020). Development of a turn-on graphene quantum dot-based fluorescent probe for sensing of pyrene in water. *RSC Advances*, 10(21), 12119–12128. https://doi.org/10.1039/c9ra10153e

Oukarroum, A., Perreault, F., & Popovic, R. (2012). Interactive effects of temperature and copper on photosystem II photochemistry in Chlorella vulgaris. *Journal of Photochemistry and Photobiology B Biology*, *110*, 9–14. https://doi.org/10.1016/j.jphotobiol.2012.02.003

Pan, Z., He, L., Qiu, L., Korayem, A. H., Li, G., Zhu, J. W., Collins, F., Li, D., Duan, W. H., & Wang, M. C. (2015). Mechanical properties and microstructure of a graphene oxide–cement composite. *Cement and Concrete Composites*, *58*, 140–147. https://doi.org/10.1016/j.cemconcomp.2015.02.001

Peng, Z., Liu, X., Zhang, W., Zeng, Zhuotong, Liu, Z., Zhang, C., Liu, Y., Shao, B., Liang, Q., Tang, W., & Yuan, X. (2020). Advances in the application, toxicity and degradation of carbon nanomaterials in environment: A review. *Environment International*, *134*, 105298. https://doi.org/10.1016/j.envint.2019.105298

Perera, I. C., & Hemamali, E. H. (2022). Genetically modified organisms for bioremediation: Current research and advancements. *Bioremediation of Environmental Pollutants: Emerging Trends and Strategies*, 163–186. https://doi.org/10.1007/978-3-030-86169-8_7

Pieper, D. H., & Reineke, W. (2000). Engineering bacteria for bioremediation. *Current Opinion in Biotechnology*, *11*(3), 262–270. https://doi.org/10.1016/S0958-1669(00)00094-X

Rani, R., & Kumar, D. (2024). Recent advances in degradation of *N,N*-diethyl-3-toluamide (DEET)—An emerging environmental contaminant: A review. *Environmental Monitoring and Assessment*, *196*, 238. https://doi.org/10.1007/s10661-024-12414-7

Rani, R., Malik, S., Kumar, D., Kumar, R., Mukherjee, S., Saharan, P. S., & Duhan, J. S. (2024). Advances in the role of microorganisms, waste management strategies and policies on microplastic abatement in the era of bio-circular economy. *Sustainable Chemistry and Pharmacy*, *39*, 101595.

Riangrungroj, P., Bever, C. S., Hammock, B. D., & Polizzi, K. M. (2019). A label-free optical whole-cell Escherichia coli biosensor for the detection of pyrethroid insecticide exposure. *Scientific Reports*, *9*(1), 12466. https://doi.org/10.1038/s41598-019-48907-6

Rodriguez-Mozaz, S., de Alda, M. L., & Barceló, D. (2005). Analysis of bisphenol A in natural waters by means of an optical immunosensor. *Water Research*, *39*(20), 5071–5079. https://doi.org/10.1016/j.watres.2005.09.023

Saxena, G., Kishor, R., Saratale, G. D., & Bharagava, R. N. (2020). Genetically modified organisms (GMOs) and their potential in environmental management: Constraints, prospects, and challenges. In: *Bioremediation of Industrial Waste for Environmental Safety: Volume II: Biological Agents and Methods for Industrial Waste Management* (pp. 1–19). https://link.springer.com/chapter/10.1007/978-981-13-3426-9_1

Shumbula, P., Maswanganyi, C., & Shumbula, N. (2021). Type, sources, methods and treatment of organic pollutants in wastewater. In: *Persistent Organic Pollutants (POPs)-Monitoring, Impact and Treatment* (pp. 1–20). https://www.intechopen.com/chapters/79526

Smith, A. M., & Nie, S. (2010). Semiconductor nanocrystals: Structure, properties, and band gap engineering. *Accounts of Chemical Research*, *43*(2), 190–200. https://doi.org/10.1021/ar9001069

Tomar, R. S., & Jajoo, A. (2013). A quick investigation of the detrimental effects of environmental pollutant polycyclic aromatic hydrocarbon fluoranthene on the photosynthetic efficiency of wheat (Triticum aestivum). *Ecotoxicology*, *22*(8), 1313–1318. https://doi.org/10.1007/s10646-013-1118-1

Tomar, R. S., & Jajoo, A. (2015). Photomodified fluoranthene exerts more harmful effects as compared to intact fluoranthene by inhibiting growth and photosynthetic processes in wheat. *Ecotoxicology and Environmental Safety*, *122*, 31–36. https://doi.org/10.1016/j.ecoenv.2015.07.002

Tschmelak, J., Proll, G., & Gauglitz, G. (2004). Verification of performance with the automated direct optical TIRF immunosensor (River Analyser) in single and multi-analyte assays with real water samples. *Biosensors and Bioelectronics*, *20*(4), 743–752. https://doi.org/10.1016/j.bios.2004.04.006

Tschmelak, J., Proll, G., & Gauglitz, G. (2005). Optical biosensor for pharmaceuticals, antibiotics, hormones, endocrine disrupting chemicals and pesticides in water: Assay optimization process for estrone as example. *Talanta*, *65*(2), 313–323. https://doi.org/10.1016/j.talanta.2004.07.011

Varjani, S. J., Gnansounou, E., & Pandey, A. (2017). Comprehensive review on toxicity of persistent organic pollutants from petroleum refinery waste and their degradation by microorganisms. *Chemosphere*, *188*, 280–291. https://doi.org/10.1016/j.chemosphere.2017.09.005

Vongsvivut, J., Robertson, E. G., & McNaughton, D. (2010). Surface-enhanced Raman spectroscopic analysis of fonofos pesticide adsorbed on silver and gold nanoparticles. *Journal of Raman Spectroscopy*, *41*(10), 1137–1148. https://doi.org/10.1002/jrs.2579

Voon, C. H., Yusop, N. M., & Khor, S. M. (2022). The state-of-the-art in bioluminescent whole-cell biosensor technology for detecting various organic compounds in oil and grease content in wastewater: From the lab to the field. *Talanta*, *241*, 123271. https://doi.org/10.1016/j.talanta.2022.123271

Wei, W., Mandin, C., & Ramalho, O. (2018). Influence of indoor environmental factors on mass transfer parameters and concentrations of semivolatile organic compounds. *Chemosphere*, *195*, 223–235. https://doi.org/10.1016/j.chemosphere.2017.12.072

Xie, Y., Zhu, X., Sun, Y., Wang, Heya, Qian, H., & Yao, W. (2012). Rapid detection method for nitrofuran antibiotic residues by surface-enhanced Raman Spectroscopy. *European Food Research and Technology*, *235*(3), 555–561. https://doi.org/10.1007/s00217-012-1752-5

Xiong, Q., Shi, Y., Lu, Y., Pan, K., Dakhil, M. A., Zhang, L., & Xiao, Y. (2018). Sublethal or not? Responses of multiple biomarkers in Daphnia magna to single and joint effects of BDE-47 and BDE-209. *Ecotoxicology and Environmental Safety*, *164*, 164–171. https://doi.org/10.1016/j.ecoenv.2018.08.007

Xu, Z., Hao, J., Braida, W., Strickland, D., Li, F., & Meng, X. (2011). Surface-enhanced Raman scattering spectroscopy of explosive 2, 4-dinitroanisole using modified silver nanoparticles. *Langmuir*, *27*(22), 13773–13779. https://doi.org/10.1021/la202560t

Yasser, E.-N., Shawkat, E.-N., & Samir, A. (2015). Impact of organic contamination on some aquatic organisms. *Toxicology International*, *22*(1), 45–53.

Yu, X.-Y., Liu, Z.-G., & Huang, X.-J. (2014). Nanostructured metal oxides/hydroxides-based electrochemical sensor for monitoring environmental micropollutants. *Trends in Environmental Analytical Chemistry*, *3–4*, 28–35. https://doi.org/10.1016/j.teac.2014.07.001

Zehani, N., Fortgang, P., Saddek Lachgar, M. S., Baraket, A., Arab, M., Dzyadevych, S. V., Kherrat, R., & Jaffrezic-Renault, N. (2015). Highly sensitive electrochemical biosensor for bisphenol A detection based on a diazonium-functionalized boron-doped diamond electrode modified with a multi-walled carbon nanotube-tyrosinase hybrid film. *Biosensors and Bioelectronics*, *74*, 830–835. https://doi.org/10.1016/j.bios.2015.07.051

Zhou, Y., Yin, G., Du, X., Xu, M., Qiu, Y., Ahlqvist, P., Chen, Q., & Zhao, J. (2018). Short-chain chlorinated paraffins (SCCPs) in a freshwater food web from Dianshan Lake: Occurrence level, congener pattern and trophic transfer. *Science of the Total Environment*, *615*, 1010–1018. https://doi.org/10.1016/j.scitotenv.2017.10.026

14 Herbal Nanoformulations for Control of Storage Grain Pests

Asha Choudhary, Rajesh Thakur, and Raj Kumar Salar

14.1 INTRODUCTION

Stored grains are the most significant agricultural product in the entire world and a key component of nutrition in underdeveloped and emerging nations. Numerous factors, such as the rapidly increasing population, development, increasing financial resources, and resultant modifications of eating habits, are putting strain on the agricultural resources' capacity to yield wholesome food and to bring further favorable existence possibilities in an ecologically sound means. In 2050, the world's population is expected to be approximately ten billion, and there will be an increasing need for a roadmap that helps the Food and Agriculture Organization of the United Nations (FAO) help nations and governments implement steps toward sustainability of food systems.

According to the Food Insecurity Experience Scale (FIES), 11.7% of the global population (924 million people) experienced extreme hunger in 2021. The levels and trends during the 2014–2021 timeframe are mostly in line with the PoU, demonstrating how well both work together to assess the severity of serious hunger.

As the population worldwide has increased, so has a need for food grains. Farm products such as cereals, pulses, and oil crops are a far more common variety of substances taken by people. These agricultural products serve as a crucial component of our diet since they are abundant in a variety of dietary supplements (Jasrotia et al., 2022; Murugesan et al., 2021).

Grains, which are a vital component of our diet and abundant in a variety of nutrients, are considerably more susceptible to insects when they are stored. Around 10%–40% of all crops in developing countries are lost each year due to insect attacks. Aside from making food threatening to consume, insect pests cause losses to crop quantity as well as quality which results in major loss of money. The major factor influencing grain storage insects is the availability of favorable circumstances for their development and enduring. Humidity and temperature levels are two primary variables influencing stored crop damage, followed by biotic and abiotic elements (Kumar and Kalita, 2017; Jasrotia et al., 2022; Murugesan et al., 2021).

Sitophilus oryzae, *Tribolium castaneum*, *Rhizopertha dominica*, *Callosobruchus chinensis*, and several other species tend to be the main destructive insects of crops and their commodities as well.

DOI: 10.1201/9781032642567-17

Currently, synthetic insecticides are widely utilized to control pest infestations on agricultural commodities. But there is growing worry about their major negative effect on nature, nontarget species, and human wellness, as well as their persistent toxicities and security challenges. As a result, investigators are focusing on producing environmentally conscious plant-based products that decrease insect pest infestations and help in current integrated pest control methods, which are extremely significant globally (Yadav et al., 2021).

The use of plants or biologically active substances in contemporary methods for managing pests and for the prevention of postharvest illness is crucial on a global scale (Das et al., 2019). They comprise various groups of commonly lipophilic substances that are primarily categorized as flavonoids, alkaloids, isoflavones, monoterpenes, phenolic substances, carotenoids, and aldehydes (Sundararajan et al., 2018). These biologically active compounds are employed for insect pest management because they cause minimal harm to the surroundings and nontarget organisms (Cao et al., 2018).

Meanwhile, despite their demonstrated ability to preserve food, the use of EOs and their biologically active elements in agriculture is still difficult because of major limitations like instability, insoluble in water, and volatile chemicals when exposed to unfavorable environmental conditions when food preservation and handling.

Encapsulating essential oils in polymers has emerged as an effective strategy to boost their capability to protect edible products during storage to conquer these limitations. Various cellulose and protein-rich polymers have been employed successfully to encapsulate since they are biodegradable and stable (Bagheri et al., 2021).

Numerous studies are currently being conducted to investigate the protective encapsulation of bioactive compounds and their possible use in the food and agricultural sectors (Falleh et al., 2021). The use of herbal nanoformulations, particularly the active components from medicinal and aromatic plants along with their limitations for the control of stored grain pests will be discussed in this chapter. Further, methods for the synthesis of polymeric nanoformulations, nanoemulsions, and their intended use in agriculture will also be discussed later in this chapter.

14.2 APPLICATIONS OF MEDICINAL AND AROMATIC PLANTS

Medicinal and aromatic plant species contribute significantly to the biological diversity of the world and provide benefits to the agricultural sector in terms of economic, artistic, societal, and environmentally sound elements. These species serve as resources of vital components used in the fields of medicine, nutrients, beauty products, superfoods, ayurveda, homeopathy, and health care sectors. It is stated that various equatorial spots have diverse native vegetation containing more than 6,000 types of medicinal herbs. More than 1,000 species are categorized as aromatic and have originated such ecological gem/solitaires (Tripathi, 2018; Gahukar, 2018).

Human beings have relied on herbal products extensively in conventional medical practices for a long time to cure every kind of illness and disorder. Numerous research studies have, thus far, been devoted to learning about the ingredient composition and therapeutic properties of herbal products, showing that aromatic herbs are significant producers of bioactive compounds. In addition to helping plants survive, bioactive compounds serve as a biological barrier to pathogens and pests (Aati et al., 2020).

Numerous medicinal and aromatic herbs have been studied as crop preservatives; the majority of these compounds were tested as powdered and nanoformulation, which combine with grain to combat harmful insect infestations of agricultural products (Kavallieratos et al., 2020).

14.3 STORED GRAINS AND INSECT INFESTATIONS

As populations worldwide are increasing, so is the need for food. Around 10%–40% of the crops in developing countries are lost each year due to insects that attack agricultural products (Murugesan et al., 2021).

By their continued use as the primary components for the nourishment of individuals across many nations, cereals, and legumes tend to play a significant role throughout the world in ensuring the availability of food. Numerous pests and other threats decimate these foods to the detriment of storage arrangements every year, destroying 5%–10% in warmer places and 20%–30% even in tropical areas (Nenaah, 2014; Matos et al., 2020).

Insect infestations are considered the most frequent causes of annual harm to crop yields, particularly the storage of grains. The major insects that destroy whole grain hulls ought to be separated from the secondary pests that may lead to the warmth of the grains and a decrease in the grain's sale price without harming the crop hulls. *Callosobruchus chinensis*, *C. maculatus* and *C. analis* (Matos et al., 2020; Banga et al., 2019; Murugesan et al., 2021), *Tribolium confusum*, *Oryzaephilus surinamensis*, *Plodia interpunctella*, *Sitophilus granaries* (Mantzoukas et al., 2020); (Ziaee et al., 2014; Mossa et al., 2017), *T. castaneum*, *Lasioderma serricorne*, *Sitophilus oryzae* and *Liposcelis bostrychophila* (Rajkumar et al., 2020b; Cao et al., 2018; Rajkumar et al., 2020a), *S. oryzae* (Mishra et al., 2012), *Trogoderma granarium* (Kavallieratos et al., 2016), and *Reticulitermes dabieshanensis* (Yang et al., 2021) have the major pests of stored grains, respectively (Mantzoukas et al., 2020; Korunic and Fields, 2020).

Food grains can be infested by a variety of insects, but beetles, or bruchids, are among the most hazardous. Ninety percent of infestations in legumes are caused by such pests. The entire core is consumed by bruchids, which leave merely the outermost layer behind, rendering seeds unusable. Insects in their larval phase do worse harm than the other phases (Banga et al., 2019).

Recently, essential oils from plants were suggested as promising substitutes for preventing infestations in agricultural products due to their strong insecticide properties and biologically active elements (Huang et al., 2020).

14.4 ESSENTIAL OIL'S BIOACTIVE COMPONENTS AND ITS PROPERTIES

Nowadays, the usage of natural pesticides, including plant compounds and essential oils (EOs), has increased as an effective substitute for chemical pesticides for insect management in agricultural and healthcare domains (Mohafrash et al., 2020).

EOs are highly volatile, biologically active molecules found in herbs that are known to have antibacterial and antioxidant properties. Some of these substances include aromatic compounds (ketones, esters, alcohols, etc.), phenolic chemicals, terpenes, and terpenoids (Choudhary et al., 2024; Asha et al., 2023). It may be retained in every plant's part, including flowers, stems and leaves, roots, fruit, aerial parts, bark, and seeds. These are mostly generated in larger plants, particularly volatile in nature (about 17,000 species) (Rajkumar et al., 2020a; Campolo et al., 2020). Several plant groups, including *Asteraceae*, *Geraniaceae*, *Lamiaceae*, *Verbenaceae*, *Annonaceae*, and *Meliaceae* (Rizvi et al., 2018), *Anacardiaceae*, *Araceae*, *Brassicaceae*, *Chemopodiaceae*, *Pinaceae*, *Liliaceae*, *Zingiberaceae*, and *Schisandraceae* (Matos et al., 2020). *Apiaceae*, *Myrtaceae*, *Lauraceae*, and *Rutaceae* generate large quantities of such oily substances, which are effective insecticides for plants. Phytochemical substances have been obtained by various distillation methods and have an elaborate combination of monoterpenes, phenols, and sesquiterpenes; carotenoids alkaloids, phenolic acids, flavonoids, isoflavones, and aldehydes (Nattudurai et al., 2017; Rehman et al., 2021). EOs are extracted from plant parts using hydro-distillation, steam-distillation, dry extraction, and mechanical extraction methods. The very first individuals who invented the method of extraction from plant organic compounds (such as steam and hydro-distillation) were the Arabs in the time of the Middle Ages (Campolo et al., 2020). These compounds, which are taken primarily from plants with aromas, constitute a diverse combination of relatively small aromatic secondary metabolic products (Hashem et al., 2018; Gomes et al., 2022).

Aromatic compounds demonstrated considerable anti-storage grains' insect properties, operating both orally and topically. They also have fumigant, repulsive; growth regulation ovulation deterrence (Rizwan-ul-Haq and Aljabr, 2015) and anti-feedant properties. Aromatic compounds can be used to control infectious bacterial and fungal diseases that affect plants, as well as communities of microbes found in stored goods (Hashem et al., 2018). Owing to their significant antimicrobial, antimycotoxigenic, insecticidal, and oxidative efficiency, EOs and their active xe4elements were strongly driven to be employed as innovative sustainable preservatives.

The efficacy of aromatic compounds for both people and nature, as well as their various mechanisms of effectiveness on pests, seems as the primary basis of their applications. Since EOs are generally recognized as safe (GRAS), they are used in hotels, healthcare facilities, educational institutions, as well as households. EOs exhibit deadly and harmful impacts on a variety of insects while being safe for mammals therefore, they exhibit the ability to create novel, safe pesticides. EOs have numerous applications in pharmaceuticals, skincare products, aroma, and drink flavoring due to their insecticide properties. Over 300 plant species have been used in industry all over the world (Rizvi et al., 2018).

14.4.1 Limitations with the Efficient Application of Essential Oils

Several issues relating to certain kinds of plant extracts, and various chemical and physical properties might be used to determine the primary reasons for the limited application in everyday use. Those problems are as follows.

Since EOs often have a low water solubility, it's difficult to disperse outdoors.

When applied directly to plants, many EOs are harmful; they may significantly decrease productivity and petrify plants, while they can also be used to suppress weeds.

The components in EOs are extremely biodegradable and volatile; as they might be capable of undergoing hydrolysis, react with oxygen, and get destroyed by heat.

Though the degradation of EOs may be seen optimistically because there are no residual traces in the goods and the atmosphere, however, the short-term retention of natural EOs may lessen their prolonged effectiveness against insects (Campolo et al., 2020).

14.5 NANOTECHNOLOGY IN DRUG DELIVERY

Nanotechnology is a cutting-edge domain in science that examines and uses nanostructures ranging in size from 1 to 1,000 nm (Chaudhary et al., 2019; Venkatesh et al., 2022). The use of nanotechnology is developing as a very promising area of study for achieving these objectives, including novel techniques for generating innovative biological agents having very small sizes and also for their preparation and delivery, all of that being known together as "nano pesticides." It is a comparatively recent area that may be characterized as the use of biotechnology for protecting crops (Athanassiou et al., 2018).

Nanotechnologies contribute to global sustainability and have numerous possibilities in the agricultural sector, pest management (Asha et al., 2022), food production, biological sciences, as well as healthcare sectors. Several other sectors, including fabric, paper products, nourishment, fertilizers, chemical pesticides, crop safety, food colors, agricultural biomass as fuel, biological composites, pharmaceutical sectors, and multiple useful substances, have used bio-nanotechnology to develop exclusive goods for a variety of uses. It has been employed in healthcare sectors for disease diagnostics, proteins and nucleic acids analysis, genetic engineering, and drug delivery and their production. It is helpful in agriculture because it can be used to create nanopesticides and nanoformulations that are potent at keeping infections and insects out of stored grains and harming plants (Sangeetha et al., 2019; Lade et al., 2019).

An interdisciplinary method known as nanotechnologies comprises the development and use of numerous mechanisms on a non-uniform basis (Sundararajan et al., 2018).

The use of nanotechnology offers the chance to deliver bioactive compounds for an extended period using innate attraction in systemic drug delivery. Some common nanoformulation structures utilized in delivering drugs include dendrimers, micelles, liposomes, and nanoemulsions. These nanocomplex structures aid in enhancing the biological dispersion of drugs and their capacity for solubilizing in water, along with their bioaccumulation, dosage reduction, improved therapeutic selection, and toxic impact reduction, among other things (Natrajan et al., 2015).

Over the past ten years, various polysaccharide-based nanoformulations have been created for usage in systems that deliver plant bioactive compounds or drugs. Thus, some approaches are described here that could be employed in agriculture, whereby a mixture of polymers may retain insecticides (plant bioactive compounds), increasing their impact at low levels (Abreu et al., 2012).

14.5.1 Polymeric Nanoformulation

Various types of nanoformulations are used to synthesize nanoparticles but among them, two main types are lipid-dependent substances (such as nanoemulsions, nanoliposomes, and nanomaterials with lipid carriers), and solid-lipid nanoparticles and biopolymeric nano- or microcapsules (capsules contained by cellulose, amino acids, and its combo). The aforementioned nanoformulations overcome the issues caused by simply applying essential oil to stored grains. (Hemmatkhah et al., 2020).

Organic polymers have been employed in the synthesis of polymeric nanoformulation, such as polymer nanospheres and nanocapsules (Malaikozhundan et al., 2019). Nanoformulation, also referred to as nanocapsules, are somewhat stable spheres (1–100 nm) that have three distinct regions: the outermost layer that might be coated with polymeric material and surfactant, a membrane covering that can include a binding substance and an inner-core layer that contains EOs or their biologically active components.

In Figure 14.1, a diagrammatic representation of polymeric nanoparticles/nanoformulation is depicted.

14.5.2 Methods for the Synthesis of Polymeric Nanoformulations

The majority of procedures used to synthesize nanoformulations are synthetic substance dependent and could cause adverse consequences. Consequently, we have a crucial demand for sustainable particles that have been produced using reliable and ecologically sound processes. The development of nanoformulation from a variety of living materials, like microbes, larger trees, etc., has been studied as a more environmentally friendly and secure method (Deshpande, 2019).

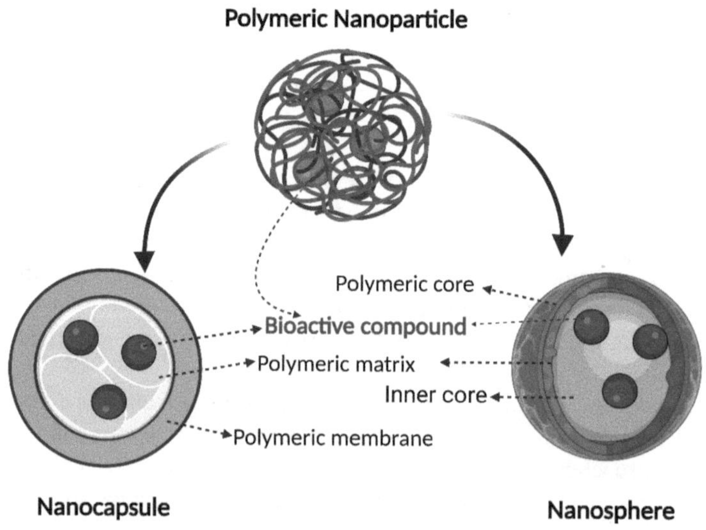

FIGURE 14.1 Diagrammatic illustration of polymeric nanoparticles.

The phrase "polymeric nanoformulation" refers to a variety of polymer-based nanoparticle shapes, while it is also simpler to define as either nanocapsules or nanospheres. Because of the cylindrical shape, nanocapsules work as therapeutic carriers as the bioactive compounds are kept within a water center that is deposited inside the particle chamber and encased through a hardened polymeric material crust. whereas a nanosphere is a complete polymeric cylindrical body wherein a chemical substance can either be confined in its center or adhered to the sphere's exterior (El-Say and El-Sawy, 2017).

The adsorption and entrapment within nanomaterials in terms of nanospheres and nanocapsules are two possible approaches to delivering bioactive insecticide compounds into nanoformulation (Figure 14.2).

Nanocapsules containing insecticides are created for use in farming operations to preserve the bioactive component from harmful ecological consequences, boost its toxic potential, and reduce treated dosage (Ziaee et al., 2014).

A variety of processes, including solvent extraction, salting-out, supercritical fluid technologies, and dialysis procedures, might be explored to create polymeric nanoparticles using polymers. In addition, several alternative polymerization processes, including controllable radical polymerization, emulsion, without surfactant emulsion, mini-emulsion, micro-emulsion, and interface polymerization, are also used (El-Say and El-Sawy, 2017).

Based on current investigations, nano- or microemulsions are effective substitutes for traditional formulations to encapsulate the EOs because they allow for the enhancement of physiochemical reliability, sustained release, and the accessibility of these bioactive compounds (Abreu et al., 2020).

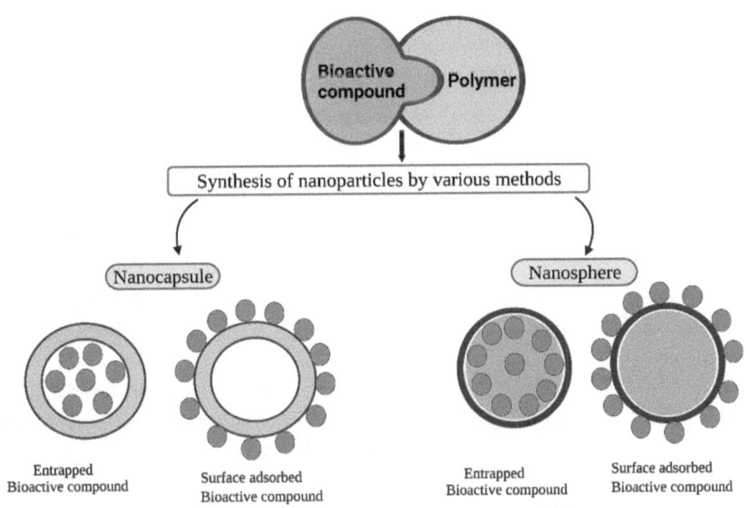

FIGURE 14.2 Several mechanisms that bioactive compounds can be linked between the nanosphere and nanocapsule.

14.6 NANOFORMULATION IN AGRICULTURE

Sustainable nanotechnology has become prevalent across many significant sectors, including medical science, agriculture technology, pharmaceutical science, parasitic biology, and the study of insects (Sundararajan et al., 2018).

It is feasible to employ EOs and their formulations in agricultural and crop after-harvest processes by integrating them and exploring antimicrobial properties in small amounts using synergistic acts (Nikkhah and Istúriz-Zapata, 2020).

The introduction of nanomaterials in the field of agriculture has not just aided in overcoming the aqueous insolubility of EOs. However, it has brought out significant advancements regarding the diagnosis of plant infection, preserving crops, preventing disorder, reducing loss of nutrients in implantation, and plant emergence and development. In this case, encapsulating EOs into a type of oil/water (O/W) nanoemulsions may be a viable strategy to address this issue. The primary benefits of nanoemulsions are decreased adverse consequences of artificial insecticides and controlling insects (Heydari et al., 2020).

The polymerization of insecticides in nanometer-sized structures such as agricultural products, biological fertilizers, and vital nutrients are all examples of useful ways of nanotechnologies in the agriculture sector. Nanoparticles may also be used for genetic modification, microscopic sensors for persistent insecticides, biological markers for the detection of infections in crops, and environmental preservation approaches (Christofoli et al., 2015).

TABLE 14.1
Effectiveness of Essential Oils and Nanoformulations against Insects

Plants	Parts	Test insects	References
Allium sativum, Aegle marmelos	Bulbs	*Sitophilus zeamais*	Chaubey (2017)
Artemisia annua	Leaves	*Tribolium casteneum*	Deb and Kumar (2020)
Azadirachta indica, Cannabis sativa	Essential oil	*Tribolium confusum, Oryzaephilus surinamensis,* and *Plodia interpunctella*	Mantzoukas et al. (2020)
Ocimum gratissimum and O. basilicum	Seed	*Tuta absoluta*	Yarou et al. (2018)
Carum copticum	Seed	*Tuta absoluta*	Piri et al. (2020)
Citronella		*Oryzaephilus surinamensis*	Gharsan et al. (2022)
Cymbopogon martinii, Cymbopogon flexuous, and *Lippia origanoides*	Aerial parts	*Tribolium castaneum*	Caballero-Gallardo et al. (2012)
Citrus	Leaves	*Monometallism tanajoa,*	Farias et al. (2020)
Cephalotaxus sinensis	Leaves	*Megoura japonica, Plutella xylostella,* and *Sitophilus zeamais*	Ma et al. (2020)
Cuminum cyminum,	Seeds	*Sitophilus granarius* and *Tribolium confusum*	Ziaee et al. (2014)

(*Continued*)

TABLE 14.1 (*Continued*)
Effectiveness of Essential Oils and Nanoformulations against Insects

Plants	Parts	Test insects	References
Citrus limonum	Leaves	*Sitophilus granarius*	Guettal et al. (2021)
Evodia lenticellata	Ripe fruits	*Tribolium castaneum*, *Lasioderma serricorne* and *Liposcelis bostrychophila*	Cao et al. (2018)
Eucalyptus resinifera	Leaves	*Rhyzopertha dominica*	Filomeno et al. (2020)
Eucalyptus globulus	Leaves	*Sitophilus granarius*	Mossa et al. (2017)
Foeniculum vulgare, *Mentha piperita*, and *Citrus sinensis*	Whole plant and peel	*Rhyzopetha dominica*	Giunti et al. (2021)
Illicium verum and *Eugenia caryophyllus*	Flower buds	*Callosobruchus maculatus*	Matos et al. (2020)
Myristica fragrans, *Bunium persicum*, *and Zanthoxylum alatum*	Aril, seeds, fruits	*Callosobruchus chinensis* and *Aspergillus flavus*	Yadav et al. (2021)
Mentha piperita	Leaves	*Tribolium castaneum* and *Sitophilus oryzae*	Rajkumar et al. (2020b)
Myristica fragrans	Seeds	*Aspergillus*, *Penicillium* and *Fusarium*	Das et al. (2019)
Manilkara subsericea	Fruits	*Dysdercus peruvianus*	Fernandes et al. (2014)
Mentha piperita, *Cymbopogon martinii*	Essential Oil	*Sitophilus oryzae* and *Lasioderma serricorne*	Yeguerman et al. (2022)
Ocimum basilicum and *Eucalyptus globulus*	Dry dates	*Oryzaephilus surinamensis*	Abd El-Salam et al. (2019)
Pelargonium Graveolens	Whole plant	*Tribolium castaneum* and *Rhyzopertha dominica*	Abouelatta et al. (2020)
Solanum torvum	Leaf	*Callosobruchus maculatus*	Murugesan et al. (2021)

14.6.1 Application of EO and Nanoformulation against Insects

Herbal nanoformulation is highly recommended for the future development of pesticides because many plant EOs fall inside the GRAS.

These include *Cinnamomum glaucescens*, *Artemisia scoparia*, *Illicium pachyphyllum*, *Lantana camara*, *Atalantiamonophylla*, *Mentha spicata*, *Laureliopsisphilippiana*,*Alpiniaconchigera*,*Menthapiperita*,*Cuminumcyminum*,andothers(Prakash et al., 2021).

Table 14.1 shows the insecticidal activity of plant essential oil on various insect pests. Some studies were conducted by multiple investigators.

Following Yeguerman et al. (2022), *Mentha piperita* and *Cymbopogon martinii* EOs and their polymeric nanoparticles were assessed against the adults of *Sitophilus oryzae* and *Lasioderma serricorne*. The findings showed that polymeric nanoparticles boosted the fatal efficiency of EOs, prolonged their repelling impacts for 3.5 days, and altered behaviors over the course of one day to *S. oryzae* and *L. serricorne*.

Rhyzopetha dominica was examined using a formulation based on EOs from *Foeniculum vulgare*, *Mentha piperita*, and *Citrus sinensis*. The findings revealed that habitual behavior might happen with repellant EO-based formulations; as a result, this behavior pattern can lessen the efficacy of such types of interventions for R. dominica and ought to be taken into account when developing efficient Integrated Pest Management programs toward insects that affect storage food production (Giunti et al., 2021).

In addition, following fumigating *Sitophilus granarius*, persistent lethality and repellent properties of *Citrus limonum* EO were reported. The efficacy of EO depended on the length of interaction, so their persistent action continued for as long as 30 hours after fatalities. The results demonstrated that the management of adult *S. granarius* had negative impacts on the reduction of their metabolic stores. The levels of lipids or carbohydrates contained in adults were decreased by *C. limonum* EO (Guettal et al., 2021).

Furthermore, contact and fumigation bioassays were used to investigate the effectiveness of *Eucalyptus resinifera* EO against *Rhyzopertha dominica*. The contact impact was brought by the interaction with the main elements from the EO of *E. resinifera*, whereas a fumigant impact was carried on by the terpene 1,8-cineole. Therefore, the EOs of *E. resinifera* possess the possibility of being employed as a substitute for conventional pesticides in the control of *R. dominica* (Filomeno et al., 2020).

14.6.2 Insecticidal Modes of Behavior of EO and Nanoformulation

The processes by which EO and nanoformulation act against insects are numerous, with effects often appearing through neurological structure, breathing, suppression of cuticle creation, or impeding genetic material synthesis. Several botanical products attack for various targeted reasons and the processes by which they work might be because they focus on various pest protection proteins, detoxing proteins, and act on another specific enzyme (Wang et al., 2021).

The biochemistry of agricultural materials, the kind of specific insects, or their life cycle all play a role in the intricate systems that underlie their toxic effects. The bioactive chemicals have a variety of modes of behavior, including antifeeding agents, causing metabolism disruption, inducing apoptosis, and inhibiting acetylcholine esterase (Piri et al., 2020). Mature are highly vulnerable compared with all stages of an insect's lifespan, including larval stages, pupa, and egg. The function of cytochrome P450-dependent monooxygenases, suppression of the activity of an enzyme called acetylcholinesterase enzyme, gamma-aminobutyric acid (GABA), and octopamine (neurotransmitters), and dysfunction in the protective antioxidant pathways are all factors that contribute to toxicities of agricultural substances. The most critical site for the activities of aromatic compounds on insects, such as EOs, is acetylcholine esterase activity (Prakash et al., 2021).

14.7 CONCLUSIONS AND PROSPECTIVE OUTCOMES

Nanotechnology has a huge range of possible uses and advantages that make it ideal for pest management. Plants and their bioactive chemicals have emerged as cost-effective and efficient possibilities for nanoformulation production. Plants and their bioactive components are becoming more and more integrated into agriculture on a global scale, improving the way they are delivered. A growing field of bioactive chemical delivery is the incorporation of nanotechnology in the agricultural sector. The most effective insect management choices are nano biopesticide formulations, and their continued implementation will open up opportunities for a wider range of solutions. These polymeric Nanoformulations have demonstrated excellent sustained release of bioactive substances, as well as being non-toxic to mammals, eco-friendly, biodegradable, and target selective.

Based on statistics, plant EOs are effective biological substitutes for chemical pesticides that provide long-term insect management of crops. The encapsulation method, which aims to enhance insoluble and bio-available active compounds, has recently come to prominence as a potential method for delivering and is currently effectively used for a range of pharmaceutical medications and agricultural sectors.

REFERENCES

Aati, H. Y., Perveen, S., Orfali, R., Al-Taweel, A. M., Aati, S., Wanner, J., … & Mehmood, R. (2020). Chemical composition and antimicrobial activity of the essential oils of Artemisia absinthium, Artemisia scoparia, and Artemisia sieberi grown in Saudi Arabia. *Arabian Journal of Chemistry*, *13*(11), 8209–8217.

Abd El-Salam, A. M. E., Salem, S. A. E., & Abdel-Rahman, R. S. (2019). Fumigant and toxic activity of some aromatic oils for protecting dry dates from *Oryzaephilus surinamensis* (L.) (Coleoptera: Silvanidae) in stores. *Bulletin of the National Research Centre*, *43*, 1–6.

Abouelatta, A. M., Keratum, A. Y., Ahmed, S. I., & El-Zun, H. M. (2020). Repellent, contact and fumigant activities of geranium (*Pelargonium graveolens* L.'Hér) essential oils against *Tribolium castaneum* (Herbst) and *Rhyzopertha dominica* (F.). *International Journal of Tropical Insect Science*, *40*, 1021–1030.

Abreu, F. O. M. D. S., Costa, E. F., Cardial, M. R. L., & André, W. P. P. (2020). Polymeric nanoemulsions enriched with *Eucalyptus citriodora* essential oil. *Polímeros*, *30*, e2020024.

Abreu, F. O., Oliveira, E. F., Paula, H. C., & de Paula, R. C. (2012). Chitosan/cashew gum nanogels for essential oil encapsulation. *Carbohydrate Polymers*, *89*(4), 1277–1282.

Asha, C., Usha, R., Kumar, S. R., & Rajesh, T. (2022). Synthesis, characterization and various in vitro activities of essential oil-loaded polymeric nanoformulations. In: Z. H. Khan, M. Jackson, & N. A. Salah (Eds.), *International Conference on Nanotechnology: Opportunities and Challenges* (pp. 31–41). Singapore: Springer Nature Singapore.

Asha, C., Usha, R., Kumar, S. R., & Rajesh, T. (2023). Synthesis, characterization and antibiotic capabilities of microspheres loaded with essential oils. *Indian Journal of Natural Products and Resources (IJNPR) [Formerly Natural Product Radiance (NPR)]*, *14*(1), 81–88.

Athanassiou, C. G., Kavallieratos, N. G., Benelli, G., Losic, D., Usha Rani, P., & Desneux, N. (2018). Nanoparticles for pest control: Current status and future perspectives. *Journal of Pest Science*, *91*, 1–15.

Bagheri, R., Ariaii, P., & Motamedzadegan, A. (2021). Characterization, antioxidant and antibacterial activities of chitosan nanoparticles loaded with nettle essential oil. *Journal of Food Measurement and Characterization*, *15*(2), 1395–1402.

Banga, K. S., Kotwaliwale, N., Mohapatra, D., Giri, S. K., & Babu, V. B. (2019). Bioacoustic detection of *Callosobruchus chinensis* and *Callosobruchus maculatus* in bulk stored chickpea (*Cicer arietinum*) and green gram (*Vigna radiata*). *Food Control, 104*, 278–287.

Caballero-Gallardo, K., Olivero-Verbel, J., & Stashenko, E. E. (2012). Repellency and toxicity of essential oils from *Cymbopogon martinii, Cymbopogon flexuosus* and *Lippia origanoides* cultivated in Colombia against *Tribolium castaneum. Journal of Stored Products Research, 50*, 62–65.

Campolo, O., Giunti, G., Laigle, M., Michel, T., & Palmeri, V. (2020). Essential oil-based nano-emulsions: Effect of different surfactants, sonication and plant species on physicochemical characteristics. *Industrial Crops and Products, 157*, 112935.

Cao, J. Q., Guo, S. S., Wang, Y., Pang, X., Geng, Z. F., & Du, S. S. (2018). Toxicity and repellency of essential oil from *Evodia lenticellata* Huang fruits and its major monoterpenes against three stored-product insects. *Ecotoxicology and Environmental Safety, 160*, 342–348.

Chaubey, M. K. (2017). Study of insecticidal properties of garlic, *Allium sativum* (Alliaceae) and Bel, *Aegle* mar*melos* (Rutaceae) essential oils against *Sitophilus zeamais* L.(Coleoptera: Curculionidae). *Journal of Entomology, 14*(5), 191–198.

Chaudhary, A., Kumar, N., Kumar, R., & Salar, R. K. (2019). Antimicrobial activity of zinc oxide nanoparticles synthesized from Aloe vera peel extract. *SN Applied Sciences, 1*, 1–9.

Choudhary, A., Salar, R. K., & Thakur, R. (2024). Synthesis, characterization and insecticidal activity of Mentha spicata essential oil loaded polymeric nanoparticles. *Biocatalysis and Agricultural Biotechnology, 55*, 102989.

Christofoli, M., Costa, E. C. C., Bicalho, K. U., de Cássia Domingues, V., Peixoto, M. F., Alves, C. C. F., ... & de Melo Cazal, C. (2015). Insecticidal effect of nanoencapsulated essential oils from *Zanthoxylum rhoifolium* (Rutaceae) in *Bemisia tabaci* populations. *Industrial Crops and Products, 70*, 301–308.

Das, S., Singh, V. K., Dwivedy, A. K., Chaudhari, A. K., Upadhyay, N., Singh, P., ... & Dubey, N. K. (2019). Encapsulation in chitosan-based nanomatrix as an efficient green technology to boost the antimicrobial, antioxidant and in situ efficacy of *Coriandrum sativum* essential oil. *International Journal of Biological Macromolecules, 133*, 294–305.

Deb, M., & Kumar, D. (2020). Bioactivity and efficacy of essential oils extracted from *Artemisia annua* against *Tribolium casteneum* (Herbst. 1797) (Coleoptera: Tenebrionidae): An eco-friendly approach. *Ecotoxicology and Environmental Safety, 189*, 109988.

Chaudhari, A. K., Singh, A., Das, S., & Dubey, N. K. (2021). Nanoencapsulated *Petroselinum crispum* essential oil: Characterization and practical efficacy against fungal and aflatoxin contamination of stored chia seeds. *Food Bioscience, 42*, 101117.

Deshpande, M. V. (2019). Nanobiopesticide perspectives for protection and nutrition of plants. In: O. Koul (Ed.), *Nano-Biopesticides Today and Future Perspectives* (pp. 47–68). Boca Raton, FL: Academic Press.

Falleh, H., Jemaa, M. B., Neves, M. A., Isoda, H., Nakajima, M., & Ksouri, R. (2021). Formulation, physicochemical characterization, and anti-E. coli activity of food-grade nanoemulsions incorporating clove, cinnamon, and lavender essential oils. *Food Chemistry, 359*, 129963.

Farias, A. P., dos Santos, M. C., Jumbo, L. O. V., Oliveira, E. E., de Lima Nogueira, P. C., de Sena Filho, J. G., & Teodoro, A. V. (2020). Citrus essential oils control the cassava green mite, *Mononychellus tanajoa*, and induce higher predatory responses by the lacewing *Ceraeochrysa caligata. Industrial Crops and Products, 145*, 112151.

Fernandes, C. P., de Almeida, F. B., Silveira, A. N., Gonzalez, M. S., Mello, C. B., Feder, D., … & Falcão, D. Q. (2014). Development of an insecticidal nanoemulsion with *Manilkara subsericea* (Sapotaceae) extract. *Journal of Nanobiotechnology, 12*, 1–9.

Filomeno, C. A., Barbosa, L. C. A., Teixeira, R. R., Pinheiro, A. L., de Sá Farias, E., Ferreira, J. S., & Picanço, M. C. (2020). Chemical diversity of essential oils of Myrtaceae species and their insecticidal activity against *Rhyzopertha dominica. Crop Protection, 137*, 105309.

Gahukar, R. T. (2018). Management of pests and diseases of important tropical/subtropical medicinal and aromatic plants: A review. *Journal of Applied Research on Medicinal and Aromatic Plants, 9*, 1–18.

Gharsan, F. N., Kamel, W. M., Alghamdi, T. S., Alghamdi, A. A., Althagafi, A. O., Aljassim, F. J., & Al-Ghamdi, S. N. (2022). Toxicity of citronella essential oil and its nanoemulsion against the sawtoothed grain beetle *Oryzaephilus surinamensis* (Coleoptera: Silvanidae). *Industrial Crops and Products, 184*, 115024.

Giunti, G., Campolo, O., Laudani, F., Zappalà, L., & Palmeri, V. (2021). Bioactivity of essential oil-based nano-biopesticides toward *Rhyzopertha dominica* (Coleoptera: Bostrichidae). *Industrial Crops and Products, 162*, 113257.

Gomes, D. S., da Costa, A., Pereira, A. M., Casal, M., & Machado, R. (2022). Biocomposites of silk-elastin and essential oil from mentha piperita display antibacterial activity. *ACS Omega, 7*(8), 6568–6578.

Guettal, S., Tine, S., Hamaidia, K., Tine-Djebbar, F., & Soltani, N. (2021). Effect of Citrus limonum essential oil against granary weevil, *Sitophilus granarius* and its chemical composition, biological activities and energy reserves. *International Journal of Tropical Insect Science, 41*, 1531–1541.

Hashem, A. S., Awadalla, S. S., Zayed, G. M., Maggi, F., & Benelli, G. (2018). *Pimpinella anisum* essential oil nanoemulsions against *Tribolium castaneum*—Insecticidal activity and mode of action. *Environmental Science and Pollution Research, 25*, 18802–18812.

Hemmatkhah, F., Zeynali, F., & Almasi, H. (2020). Encapsulated cumin seed essential oil-loaded active papers: Characterization and evaluation of the effect on quality attributes of beef hamburger. *Food and Bioprocess Technology, 13*, 533–547.

Heydari, M., Amirjani, A., Bagheri, M., Sharifian, I., & Sabahi, Q. (2020). Eco-friendly pesticide based on peppermint oil nanoemulsion: Preparation, physicochemical properties, and its aphicidal activity against cotton aphid. *Environmental Science and Pollution Research, 27*, 6667–6679.

Huang, Y., Liao, M., Yang, Q., Shi, S., Xiao, J., & Cao, H. (2020). Knockdown of NADPH-cytochrome P450 reductase and CYP6MS1 increases the susceptibility of *Sitophilus zeamais* to terpinen-4-ol. *Pesticide Biochemistry and Physiology, 162*, 15–22.

Jasrotia, P., Nagpal, M., Mishra, C. N., Sharma, A. K., Kumar, S., Kamble, U., … & Singh, G. P. (2022). Nanomaterials for postharvest management of insect pests: Current state and future perspectives. *Frontiers in Nanotechnology, 3*, 811056.

Kavallieratos, N. G., Athanassiou, C. G., Barda, M. S., & Boukouvala, M. C. (2016). Efficacy of five insecticides for the control of *Trogoderma granarium* Everts (Coleoptera: Dermestidae) larvae on concrete. *Journal of Stored Products Research, 66*, 18–24.

Kavallieratos, N. G., Boukouvala, M. C., Ntalli, N., Skourti, A., Karagianni, E. S., Nika, E. P., … & Benelli, G. (2020). Effectiveness of eight essential oils against two key stored-product beetles, *Prostephanus truncatus* (Horn) and *Trogoderma granarium* Everts. *Food and Chemical Toxicology, 139*, 111255.

Khalid, M., & El-Sawy, H. S. (2017). Polymeric nanoparticles: Promising platform for drug delivery. *International Journal of Pharmaceutics*, *528*(1–2), 675–691.

Korunic, Z., & Fields, P. G. (2020). Evaluation of three new insecticide formulations based on inert dusts and botanicals against four stored-grain beetles. *Journal of Stored Products Research*, *88*, 101633.

Kumar, D., & Kalita, P. (2017). Reducing postharvest losses during storage of grain crops to strengthen food security in developing countries. *Foods*, *6*(1), 8.

Kumar, M., Bishnoi, R. S., Shukla, A. K., & Jain, C. P. (2019). Techniques for formulation of nanoemulsion drug delivery system: A review. *Preventive Nutrition and Food Science*, *24*(3), 225.

Lade, B. D., Gogle, D. P., Lade, D. B., Moon, G. M., Nandeshwar, S. B., & Kumbhare, S. D. (2019). Nanobiopesticide formulations: Application strategies today and future perspectives. In: *Nano-Biopesticides Today and Future Perspectives* (pp. 179–206). Boca Raton, FL: Academic Press.

Ma, S., Jia, R., Guo, M., Qin, K., & Zhang, L. (2020). Insecticidal activity of essential oil from Cephalotaxus sinensis and its main components against various agricultural pests. *Industrial Crops and Products*, *150*, 112403.

Malaikozhundan, B., Vinodhini, J., & Vaseeharan, B. (2019). Nanopesticides for the management of insect pests of stored grains. In D. G. Panpatte, & Y. K. Jhala (Eds.), *Nanotechnology for Agriculture: Crop Production & Protection* (pp. 303–322). Singapore: Springer.

Mantzoukas, S., Ntoukas, A., Lagogiannis, I., Kalyvas, N., Eliopoulos, P., & Poulas, K. (2020). Larvicidal action of cannabidiol oil and neem oil against three stored product insect pests: Effect on survival time and in progeny. *Biology*, *9*(10), 321.

Matos, L. F., da Cruz Lima, E., de Andrade Dutra, K., Navarro, D. M. D. A. F., Alves, J. L. R., & Silva, G. N. (2020). Chemical composition and insecticidal effect of essential oils from Illicium verum and *Eugenia caryophyllus* on *Callosobruchus maculatus* in cowpea. *Industrial Crops and Products*, *145*, 112088.

Mishra, B. B., Tripathi, S. P., & Tripathi, C. P. M. (2012). Response of *Tribolium castaneum* (Coleoptera: Tenebrionidae) and *Sitophilus oryzae* (Coleoptera: Curculionidae) to potential insecticide derived from essential oil of Mentha arvensis leaves. *Biological Agriculture & Horticulture*, *28*(1), 34–40.

Mohafrash, S. M., Fallatah, S. A., Farag, S. M., & Mossa, A. T. H. (2020). Mentha spicata essential oil nanoformulation and its larvicidal application against *Culex pipiens* and *Musca domestica. Industrial Crops and Products*, *157*, 112944.

Mossa, A. T. H., Abdelfattah, N. A. H., & Mohafrash, S. M. M. (2017). Nanoemulsion of camphor (*Eucalyptus globulus*) essential oil, formulation, characterization and insecticidal activity against wheat weevil, *Sitophilus granarius. Asian Journal of Crop Science*, *9*(3), 50–62.

Murugesan, R., Vasuki, K., Kaleeswaran, B., Santhanam, P., Ravikumar, S., Alwahibi, M. S., … & Alkahtani, J. (2021). Insecticidal and repellent activities of *Solanum torvum* (Sw.) leaf extract against stored grain Pest, *Callosobruchus maculatus* (F.)(Coleoptera: Bruchidae). *Journal of King Saud University-Science*, *33*(3), 101390.

Natrajan, D., Srinivasan, S., Sundar, K., & Ravindran, A. (2015). Formulation of essential oil-loaded chitosan–alginate nanocapsules. *Journal of Food and Drug Analysis*, *23*(3), 560–568.

Nattudurai, G., Baskar, K., Paulraj, M. G., Islam, V. I. H., Ignacimuthu, S., & Duraipandiyan, V. (2017). Toxic effect of *Atalantia monophylla* essential oil on *Callosobruchus maculatus* and *Sitophilus oryzae*. *Environmental Science and Pollution Research*, *24*, 1619–1629.

Nenaah, G. E. (2014). Chemical composition, toxicity and growth inhibitory activities of essential oils of three Achillea species and their nano-emulsions against *Tribolium castaneum* (Herbst). *Industrial Crops and Products*, *53*, 252–260.

Nikkhah, M., & Hashemi, M. (2020). Boosting antifungal effect of essential oils using combination approach as an efficient strategy to control postharvest spoilage and preserving the jujube fruit quality. *Postharvest Biology and Technology*, *164*, 111159.

Piri, A., Sahebzadeh, N., Zibaee, A., Sendi, J. J., Shamakhi, L., & Shahriari, M. (2020). Toxicity and physiological effects of ajwain (*Carum copticum*, Apiaceae) essential oil and its major constituents against *Tuta absoluta* (Meyrick) (Lepidoptera: Gelechiidae). *Chemosphere*, *256*, 127103.

Prakash, B., Kumar, A., Singh, P. P., Das, S., & Dubey, N. K. (2021). Prospects of plant products in the management of insect pests of food grains: Current status and future perspectives. In R. P. Sinha, & D.-P. Häder (Eds.), *Natural Bioactive Compounds* (pp. 317–335). Academic Press.

Rajkumar, V., Gunasekaran, C., Dharmaraj, J., Chinnaraj, P., Paul, C. A., & Kanithachristy, I. (2020). Structural characterization of chitosan nanoparticle loaded with Piper nigrum essential oil for biological efficacy against the stored grain pest control. *Pesticide Biochemistry and Physiology*, *166*, 104566.

Rehman, A., Qunyi, T., Sharif, H. R., Korma, S. A., Karim, A., Manzoor, M. F., … & Mehmood, T. (2021). Biopolymer based nanoemulsion delivery system: An effective approach to boost the antioxidant potential of essential oil in food products. *Carbohydrate Polymer Technologies and Applications*, *2*, 100082.

Rizvi, S. A. H., Ling, S., Tian, F., Xie, F., & Zeng, X. (2018). Toxicity and enzyme inhibition activities of the essential oil and dominant constituents derived from *Artemisia absinthium* L. against adult Asian citrus psyllid Diaphorina citri Kuwayama (Hemiptera: Psyllidae). *Industrial Crops and Products*, *121*, 468–475.

Rizwan-ul-Haq, M., & Aljabr, A. M. (2015). *Rhynchophorus ferrugineus* midgut cell line to evaluate insecticidal potency of different plant essential oils. *In Vitro Cellular & Developmental Biology-Animal*, *51*, 281–286.

Sangeetha, J., Mundaragi, A., Thangadurai, D., Maxim, S. S., Pandhari, R. M., & Alabhai, J. M. (2019). Nanobiotechnology for agricultural productivity, food security and environmental sustainability. In D. G. Panpatte, & Y. K. Jhala (Eds.), Nanotechnology for Agriculture: Crop Production & Protection (pp. 1–23). Singapore: Springer Singapore.

Yearbook, F. S. (2013). *World Food and Agriculture*. Food and Agriculture Organization of the United Nations, Rome.

Sundararajan, B., Moola, A. K., Vivek, K., & Kumari, B. R. (2018). Formulation of nanoemulsion from leaves essential oil of *Ocimum basilicum* L. and its antibacterial, antioxidant and larvicidal activities (*Culex quinquefasciatus*). *Microbial Pathogenesis*, *125*, 475–485.

Tripathi, A. K. (2018). Pests of medicinal and aromatic plants. In: Omkar (Ed.), *Pests and Their Management* (pp. 719–822). Singapore: Springer.

Venkatesh, Y., Raju, P. S., & Reddy, P. Y. (2022, November). Synthesis of titania nanoparticles and assessment of antioxidant activity. In: International Conference on Nanotechnology: Opportunities and Challenges (pp. 7–14). Singapore: Springer Nature Singapore.

Wang, Z., Xie, Y., Sabier, M., Zhang, T., Deng, J., Song, X., … & Zhou, G. (2021). Trans-anethole is a potent toxic fumigant that partially inhibits rusty grain beetle (*Cryptolestes ferrugineus*) acetylcholinesterase activity. *Industrial Crops and Products*, *161*, 113207.

Yadav, A., Kumar, A., Singh, P. P., & Prakash, B. (2021). Pesticidal efficacy, mode of action and safety limits profile of essential oils based nanoformulation against *Callosobruchus chinensis* and *Aspergillus flavus*. *Pesticide Biochemistry and Physiology*, *175*, 104813.

Yang, X., Han, H., Li, B., Zhang, D., Zhang, Z., & Xie, Y. (2021). Fumigant toxicity and physiological effects of spearmint (*Mentha spicata*, Lamiaceae) essential oil and its major constituents against *Reticulitermes dabieshanensis*. *Industrial Crops and Products*, *171*, 113894.
Yarou, B. B., Bawin, T., Boullis, A., Heukin, S., Lognay, G., Verheggen, F. J., & Francis, F. (2018). Oviposition deterrent activity of basil plants and their essentials oils against *Tuta absoluta* (Lepidoptera: Gelechiidae). *Environmental Science and Pollution Research*, *25*, 29880–29888.
Yeguerman, C. A., Urrutia, R. I., Jesser, E. N., Massiris, M., Delrieux, C. A., Murray, A. P., & González, J. O. W. (2022). Essential oils loaded on polymeric nanoparticles: Bioefficacy against economic and medical insect pests and risk evaluation on terrestrial and aquatic non-target organisms. *Environmental Science and Pollution Research*, *29*(47), 71412–71426.
Ziaee, M., Moharramipour, S., & Mohsenifar, A. (2014). MA-chitosan nanogel loaded with *Cuminum cyminum* essential oil for efficient management of two stored product beetle pests. *Journal of Pest Science*, *87*, 691–699.

15 Duckweed Utilization for Environment Management

Present Understanding and Future Scope

Pawan Kumar Rose, Suman Rani, Meenakshi Suhag, Mohd. Kashif Kidwai, Sanju Bala Dhull, Jyoti Rani, and Manjeet Bansal

15.1 INTRODUCTION

Densely populated nations such as China and India have been encountering significant ecological and economic issues in the recent decades regarding water, food, unsustainable generation of energy, and waste (Rose et al., 2022). This has motivated researchers all over the globe to explore substitutes for nonrenewable resources and implement measures to reduce waste, thereby contributing to achieving Sustainable Development Goals. Recently, the concept of circular economy as a new development in waste management emphasizes the recovery of valuable compounds and the generation of energy from waste (Dhull et al., 2024). One of the most popular waste-treatment methodologies involves the utilization of microorganisms and plants (Rose et al., 2024; Zhao et al., 2014). Microbial techniques convert wastewater nutrients into gas (nitrogen) and sludge, making nutritional reuse difficult. Phytoremediation is established as a cost-effective and efficient wastewater treatment method that utilizes wastewater nutrients for the production of biomass (Shammas and Wang, 2010). Unlike terrestrial plants, aquatic plants such as water hyacinth, duckweed, water lettuce, pennywort, canna lily, reed, and so on absorb nutrients from wastewater without additional requirements of land (Cheng and Stomp, 2009). Selecting the most suitable aquatic plants for wastewater treatment depends on aspects such as biomass production, nutrient recovery efficiency, and biomass utilization potential (Rani et al., 2023; Zhao et al., 2014). Duckweed has greater protein (over 40%), starch (over 45%), biomass production (106 t/ha/year dry weight), and grows in a broader range of temperature, light, pH, and fertilizers than other aquatic plants, indicating its potential use (Femeena et al., 2023). Duckweeds are natural solutions to various issues and complement managing resources to achieve Sustainable Development Goals. Duckweed may outshine other aquatic plants due

DOI: 10.1201/9781032642567-18

to its regulated management and use of biomass to produce biofuels, animal feed, and human nourishment (Appenroth et al., 2018). Duckweed may replace or increase existing protein sources to fulfill the rising need for sustainable protein-rich feed and meals. It is also rich in antioxidants and healthy pigments and offers adequate essential amino acids, abiding by the guidelines of the World Health Organization (WHO) (Appenroth et al., 2017). Duckweeds are minuscule, rapidly proliferating, waterborne plants that can increase their bulk by two-fold every 16–24 hours. The yield of duckweed biomass may reach up to 106 t/ha/year (dry weight), significantly surpassing that of prominent starch crops such as maize (7.84 t/ha/year), wheat (3.15 t/ha/year), and barley (3.70 t/ha/year) (Sun et al., 2020). Duckweed starch is easily fermented into bioethanol for use in bioenergy production (Ge et al., 2012), with a yield of 6.42×10^3 L/ha, 50% greater than conventional maize-based bioethanol production (Xu et al., 2011). The commercial market is expected to develop primarily because of the increasing nutritional value and many uses. In 2023, the total value of the worldwide duckweed market was estimated at $76.0 million. Between 2023 and 2033, the sales of duckweed are expected to increase at a healthy CAGR of 10.8%, reaching a total of around US$ 195.4 million by the end of 2033 (https://www.futuremarketinsights.com/reports/duckweed market). This chapter is an attempt to discuss various aspects of duckweed and its applications in wastewater treatment, the production of potential protein-rich biomass used as animal feed, human nutrition, etc., and the use of starch-rich biomass as a renewable energy source.

15.2 DUCKWEEDS: INTRODUCTION

Duckweeds are members of the family Lemnaceae. Lemnaceae, also referred to as water lentils, water lens, or duckweed, is a subfamily of flowering aquatic plants from the family *Araceae*, which frequently appear on the surfaces of stagnant or slowly moving fresh water or wetlands, providing food and habitat for diverse organisms such as herbivore fish, macroinvertebrates, aquatic fowl, and so on (Lee et al., 2020; Sree et al., 2016). Species of duckweed range in size from 0.5 mm to less than 2 cm (Bog et al., 2019), but they form dense mats on the surface of water and generate enormous amounts of biomass. It is readily grown in a lab setting, develops rapidly by clonal development (mother/daughter fronds), and may quadruple its biomass in only 2–3 days (Zhao et al., 2014). Morphologically, duckweed species are without stems or true leaves and a few flat, oval-shaped fronds. This family has five genera as displayed in Figure 15.1 and a total of 36 species, further divided into two subfamilies, that is, *Lemnoidea* having 15 species and *Wolffioideae* having 21 species. However, ambiguity has been found in the number of species in these five genera as reported by different researchers, that is, Baek et al. (2021) (36 species) and Appenroth et al. (2013) (37 species). Duckweed species reproduce asexually by producing genetically identical clones by vegetative propagation, where daughter fronds emerge from the mother fronds. Despite some reports, duckweeds are not water-pollinated; instead, they reproduce sexually via tiny blooms that either self-pollinate or are pollinated by minute arthropod vectors like flies, mites, and spiders above the water's surface. The *Lemnoidea* family, which includes the genera *Landoltia* (dotted duck meat), *Lemna*, and *Spirodela* (duck meat), has two meristematic areas inside the mother frond. On

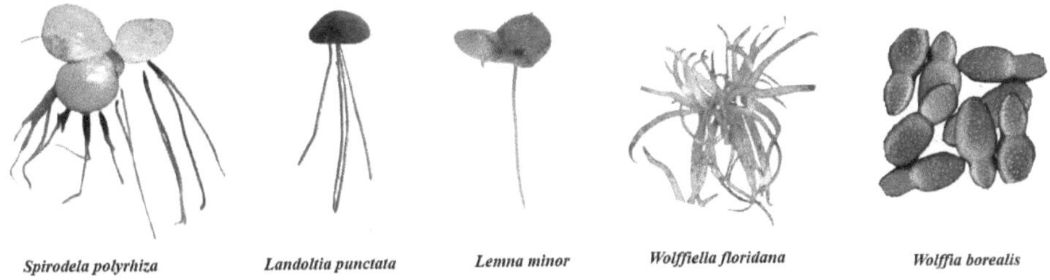

FIGURE 15.1 The structural characteristics of five selected species belonging to several duckweed genera. The five genera may be differentiated based on vegetative characteristics. For instance, *Landoltia* (dotted duck meat) and *Spirodela* (duck meat) species have comparatively large, floating fronds with many roots. *Lemna* (duckweed), often has floating fronds, except for the suspended *L. trisulca*, which has one root. *Wolffia* (watermeal) a species that has short and thick fronds that do not have roots. Similarly, *Wolffiella* (mud midget) plants lack roots and have small, elongated fronds.

the other hand, *Wolffioideae*, which includes the species *Wolffia* and *Wolffiella*, has just one meristematic pouch in the mother frond. *Landoltia*, formerly *Spirodela punctata*, is a kind of duckweed consisting of two fronds connected. *Lemna* species have a single root and often exhibit 2–4 fronds. Its maximum length has been documented as 6.0mm. *Spirodela* species is one of the most diminutive genomes among the duckweed family and has five substantial fronds ranging in length from 4 to 12mm. *Wolffia* species are the most advanced group, with a length that seldom goes above 0.5–1.2mm and a breadth of 0.4–1.0mm. *Wolffiella* species exhibits infrequent blooming, and its fronds typically measure between 1.0 and 5.0mm in length. Therefore, several kinds of duckweed exhibit distinct morphological traits (Baek et al., 2021).

15.3 SPECIES DIVERSITY AND TAXONOMY CLASSIFICATION

Duckweeds belong to the family *Lemnaceae* and are monophyletic, so they closely relate to the *Araceae* family (Bog et al., 2019; Choi et al., 2017). It does not necessarily mean that duckweeds must be integrated into the family of *Araceae* and can be treated by taxonomic rules as a plant family, Lemnaceae Dumort (Bog et al., 2019). Landolt (1980, 1986) divided the *Lemnaceae* family into two subfamilies: *Lemnoideae* (included the members with varying numbers of roots) and *Wolffioideae* (included the members with no roots). Later, Les and Crawford (1999) classified *Spirodela punctata* as a separate genus called Landoltia. Consequently, the only species within this genus was renamed *Landoltia punctata* (G.Mey.) Les and D.J. Crawford. Therefore, currently, this family has five genera, as shown in Figure 15.2 comprising 37 species, that is, *Spirodela* (2 species), *Landoltia* (1 species only), *Lemna* (13 species), *Wolffiella* (10 species), and *Wolffia* (11 species) (Appenroth et al., 2013).

Since the 1950s, Elias Landolt has collected various duckweeds across the globe, and for their identification, he published a "key of determination" of duckweeds based on morphology (Sree et al., 2016; Landolt, 1986). However, the taxonomic classification of duckweed has kept changing over the last three decades compared

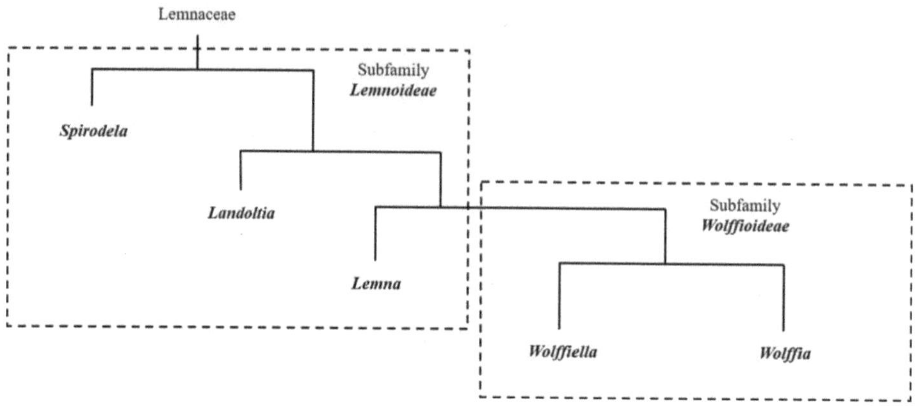

FIGURE 15.2 Generic organization within the Lemnaceae family. Adapted and modified from Bog et al. (2019), under Creative Commons Attribution License (CC-BY).

with the earlier published one. While molecular taxonomy and barcoding have been effective in identifying duckweed species, it is not practical or cost-effective to use amplified fragment length polymorphism or DNA marker sequencing to determine each collected duckweed clone. Therefore, it is essential to determine the morphology of duckweed species to identify them accurately (Sree et al., 2016).

15.4 DUCKWEED AND ENVIRONMENTAL MANAGEMENT

Duckweeds provide various ecosystem services, such as wastewater treatment, synthesis of phytopharmaceuticals, animal feed, human nutrition, generation of protein-rich biomass (Vu et al., 2020), and so on. Duckweeds are among the best at using nitrogen effectively, which helps them build biomass. These characteristics may be further improved by selecting the optimal species or ecotypes and optimizing the plant's growth factors, such as nutrition ratio, light intensity, frond density, and so on. When employed for rehabilitating agricultural, municipal, and even industrial wastewater streams, duckweeds exhibit relatively high tolerance to various water contaminants (organic and inorganic compounds) and other environmental pressures (Appenroth et al., 2015). In recent years, the corporate sector has shown a keen interest in application-based research on duckweeds due to various properties of *in situ* water remediation and quick biomass buildup.

15.4.1 Environmental Pollutant Removal

15.4.1.1 Organic Pollutant

Duckweed in water may interact with organic contaminants such as industrial organic compounds, medications, personal care items, and agricultural pesticides (Ekperusi et al., 2019). Norflurazon significantly decreased *L. minor* growth, although all norflurazon doses tested quickly recovered after removing the plant from the substrate (Wilson and Koch, 2013). *L. minor* efficiently degrades halogenated phenols when it covers the whole surface. Time consumed by *L. minor* to remove 2,4,5-trichlorophenol from the water phase is 16–22 hours (Tront et al., 2007). In *L. minor*, the main metabolic pathway for chlorinated phenols involves glycosidation, forming chlorinated malonyl-glucoside and apiosyl-glucoside conjugates stored in vacuoles and cell walls (Day and Saunders, 2004). *L. minor* has remarkable resilience, with a high capacity for recuperation even after repeated encounters with isoproturon (Varga et al., 2020) and terbuthylazine (Panfili et al., 2019). Matamoros et al. (2012) showed that in a microcosm wetland environment, three aquatic plants, including duckweed *L. minor*, were able to biodegrade and/or plant-absorb caffeine and ibuprofen at a rate of 83%–99%, respectively. Reinhold et al. (2010) reported that in all experimental settings using *L. minor*, the elimination of four of the eight pollutants, that is, triclosan (via absorption), 2,4-dichlorophenoxyacetic acid (absorption), fluoxetine (absorption), and ibuprofen (break down) were significant. Li et al. (2017) eliminated four emergent personal care product chemicals in a laboratory-scale artificial wetland using greater duckweed (*S. polyrhiza*). Duckweed's metabolism changes at high antibiotic concentrations, affecting its drug elimination potential (Gomes et al., 2017). Singh et al. (2018) investigated that amoxicillin posed adverse effects on *S. polyrhiza* even at low doses. In research, using methylene blue to evaluate *L. minor's* phytoremediation potential, Can-Terzi et al. (2021) observed the removal effectiveness of 98% in less than 24 hours. *Lemna gibba* phytoremediated 1,4 dioxane-rich wastewater, producing biomass with high protein and carbon contents but significantly influenced by reactor types. The baffled duckweed pond configuration removed 1,4 dioxane with 69.3% efficiency, leaving a residual concentration of 50.2 mg/L in the effluent. The traditional module had a removal effectiveness of 20.8% and a higher residual 1,4 dioxane concentration of 115.6 mg/L in the effluent. *L. gibba* absorbed 1,4-dioxane after decomposing it into glycolic and oxalic acids which are further degraded into phosphoglycerate, which plants easily take and store them as amino acids. The primary route by which 1,4 dioxane is eliminated via a duckweed pond is absorption, followed by a photo-degradation process. The economic study focuses primarily on the practicality of the suggested setup, which resulted in a net profit of $159.4 per year and a payback time of 37.7 years (Osama et al., 2020). Moreover, a laboratory-scale study with *Lemna* sp. exhibited impressive removal efficiency for numerous pharmaceutical residues such as carbamazepine (47.4%), ibuprofen (80%), caffeine (82.2%), paracetamol (100%), propranolol (100%), and clofibric acid (60.2%) (Maldonado et al., 2022).

15.4.1.2 Inorganic Pollutants

Duckweed species are reported to eliminate heavy metals from wastewater via adsorption and intracellular accumulation, chelation, and compartmentalizing in vacuoles to prevent toxicity. Duckweed co-cultivation enhances biomass, antioxidant enzyme activity, and heavy metal removal from contaminated water. Co-culturing (*L. punctata* and *L. minor*) was better than single cultures at low copper concentrations, whereas single cultures were better at higher copper concentrations (Zhao, 2015). Various species of duckweed, such as *L. minor*, *L. trisulca*, *S. polyrhiza*, *L. punctata*, *L. gibba*, and *Wolffia globosa* exhibited remediation potential for diverse heavy metals, such as arsenic, selenium, mercury, chromium, zinc, copper, iron, nickel, lead, cadmium, and so on from wastewater (Kidwai et al., 2023).

Specifically, *S. polyrhiza*, *L. gibba*, and *L. minor* effectively remediate boron, arsenic, uranium, and nickel. *L. gibba* was identified as the most effective duckweed for removing zinc, copper, and cadmium up to 5 mg/L with no negative impact on the duckweed biomass (Sharma and Lenaghan, 2022). Plants need boron, but its occurrence in high concentrations is an issue. Boron at high concentrations (0.5–37 mg/L) reduces normal frond production in *S. polyrhiza*. Salt stress substantially hinders *L. minor* growth and boron. Ohlendorf et al. (1986) reported that *L. minor* can tolerate NaCl up to 200 mM concentration, with an accumulation capacity of 7.9%–15.5% of boron. Duckweed species, including *L. gibba*, *L. minor*, *S. polyrhiza*, *W. globose*, *W. australiana*, and *L. valdiviana* remove arsenic from water. *L. gibba* accumulated ten times as much arsenic as background levels in tailing fluids from defunct uranium mines, reducing the concentrations by 40.3% (Mkandawire et al., 2005). *L. valdiviana*, when cultivated in a pH range of 6.3–7.0, absorbed 1190 mg/kg of arsenic (dry weight) (Marchand et al., 2011). The highest phytoremediation rate of chromium (VI) ion by *L. minor* was 64% at pH 3.8, while the lowest was 25% at pH 9.5 (Nassar and Ibrahim, 2021). Due to its antioxidant defense system, duckweeds can withstand extreme stress caused by heavy metals. Duckweeds use the formation of citrate and oxalate to sequester metal ions, enabling them to withstand heavy metals. Sequestration, compartmentalization, and exclusion help plants to overcome impairment due to heavy metals (Sharma and Lenaghan, 2022). The assessment of duckweed's efficacy in removing heavy metals aids in comprehending their biochemical reactions and their significance in evolutionary adaptations to stressors (Li et al., 2023).

15.4.2 Duckweed: Livestock Management and Feed

Duckweeds grow quickly in tiny ponds, ditches, and swamps where they take up nutrients and convert them into potential protein sources for humans and animals (Xu et al., 2023; Goopy and Murray, 2003), notably chicken and fish. Compared with other higher plants, duckweeds have a far lower energy cost per biomass unit produced (Gupta and Prakash, 2013). Under optimal nutrient availability, solar exposure, and water temperature, duckweed can double its bulk within a time frame ranging from 16 to 48 hours (Cheng and Stomp, 2009). Therefore, the plants expand exponentially until they become too large for their environments or exhaust their food supply. Traditionally, feeding duckweed to domesticated animals such as pigs, cattle, rams, sheep, horses, ducks, and fish is a common practice. This may be done by providing them temporary access to ponds where duckweed grows or by adding collected duckweed to their diet, either in fresh or dried form (Leng et al., 1995). Detailed reports on the use of *Wolffia arrhiza* meal as a soya replacement in the diet of Japanese quails, duckweed species in the feed of striped catfish, rohu and carp, broilers, and genetically modified *Lemna minor* to feed laying hens have been published in recent years (Appenroth et al., 2018). In many parts of the Asian subcontinent, people eat duckweed with traditional food. *Wolffia globosa* is widely available for purchase in many regions of Thailand's vegetable markets, often known as khai nam, kai-pum, and kai nhae. The protein, lipid, and carbohydrate levels, as well as the distribution of amino acids and fatty acids, in several species of duckweed from *Spirodela*,

Landoltia, *Lemna*, *Wolffiella*, and *Wolffia*, exhibited a protein composition range of 20%–35%, fat content ranging from 4% to 7%, and starch content ranging from 4% to 10% (dry weight) as shown in Figure 15.3. It's interesting to note that the percentages of various amino acids, such Lys (4.8%), Met (2.7%), Cys (0.7%), and Phe (0.7%), and Tyr (0.7%), are quite near to the WHO's recommended levels. In addition to having a n6/n3 ratio of 0.5 or below, the polyunsaturated fatty acid content was between 48% and 71%. Humans in Asia have traditionally used *Wolffia* species as a source of protein. Presently, duckweeds are mostly used as amino acid supplements. Parabel, Ltd., offers a range of duckweed plant powders as a substitute for high-protein diets. Commercial utilization of duckweeds is listed in Table 15.1. Replacing animal protein with plant protein is expected to reduce energy consumption and greenhouse gas emissions (Harwatt, 2019). The omega-3 to omega-6 fatty acid ratio found in *Landoltia punctata*, *L. gibba*, *L. minor*, *S. polyrrhiza*, *W. microscopica*, and

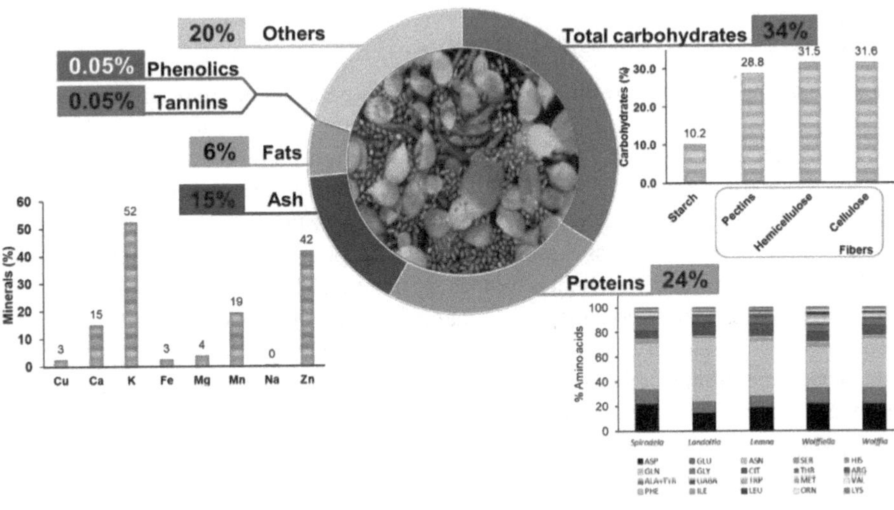

FIGURE 15.3 Nutritional profile of duckweed. Adapted from Pagliuso et al. (2022), under Creative Commons Attribution License (CC-BY).

TABLE 15.1
Commercial Application of Duckwood (Fourounjian et al., 2020).

Application	Company	Genera
Human food	Parabel, Hinoman, Green Onyx	*Wolffia*, *Lemna*
Protein	Plantible, Parabel, CAIS	*Lemna*
Livestock	Many small-scale farmers	*Lema*, *Spirodela*, others
Conversion chemicals	MamaGrande	*Lemna*
Wastewater treatment	MamaGrande, CAIS	*Mixture*
Space life support	Space Lab Technologies	*Wolffia*, *Lemna*
Biofuel or energy	Greenbelt resources	-

Wolffiella hyalina renders them appropriate for use as food and feed. Nevertheless, it is worth alarming that the consumption of duckweed species that contain high levels of oxalic acid may lead to the formation of kidney stones. Furthermore, using duckweed, which can adsorb toxic substances such as heavy metals, may result in enhancing their toxicity (Baek et al., 2021).

Duckweed species, such as *L. minor*, *L. trisulca*, and *S. polyrrhiza*, have been used as traditional remedies in China, Korea, and certain European nations. *L. minor* has antibacterial potential against both gram-negative and gram-positive bacteria. Consequently, it can serve as an alternative to conventional antibacterial medicines for treating diseases (Mane et al., 2017). *S. polyrrhiza* has shown inhibitory effects on seven gram-negative bacilli, one gram-positive bacterium, and two fungal pathogens. Additionally, flavonoids that reduce triglyceride levels were found to suppress adipogenesis (Das et al., 2012). Duckweeds' various pharmacological properties enable them as promising pharmaceutical candidates, but intensive studies and clinical trials are needed.

15.4.3 Duckweed: A Substrate for Biofuel

Duckweeds are ideal for producing bioenergy like ethanol, butanol, and biogas due to fast biomass growth, short reproductive cycle, buoyant water tolerance, and easy harvesting. Duckweed, having high starch content, may be used as a feedstock for ethanol production. Duckweed cultivated under mixotrophic conditions grew faster, producing the most biomass with high nutrient absorption. However, Mixotrophic duckweed has 2.06 times less starch than heterotrophic duckweed, indicating their combination to produce starch-rich biomass (Sun et al., 2020). Duckweed starch and other carbohydrates are converted to ethanol by two processes: initially, they are saccharified to create fermentable sugars (primarily glucose), followed by fermentation to produce ethanol. Yeasts or bacteria carry out the fermentation reaction, and enzymes catalyze the saccharification. Various enzymatic hydrolysis or saccharification methods have been studied to enhance the liberation of sugar, hence increasing the ethanol production from duckweed. Duckweed starch was hydrolyzed using α-amylase, pullulanase, and amyloglucosidase; the reduced sugar recovered was 96.8% of the theoretical value (Xu et al., 2011). Cellulase enzymes are very effective at saccharifying duckweed's cell walls but also costly. Pectinase hydrolysis was used to liberate glucose from *L. punctata* to enhance saccharification efficiency. The method yielded 218.63.1 mg/g of glucose from the dry matter, representing a 142% increase compared to the untreated mash (Chen et al., 2012). A preliminary study with *S. polyrhizaod* showed that the starch content of duckweed increased by 64.9% (from 18.8% to 31.0% on a dry basis) after transferring the plants from nutrient-rich wastewater to well water for 10 days. The study revealed that the starch-enriched duckweed cultivation could produce 9,420 kg starch/ha/year, which resulted in an ethanol yield of 6,420 L/ha/year, which is roughly 50% greater than that obtained using corn (Xu et al., 2011). Duckweed-based ethanol may produce an annual yield of 16.0×10^3 L/ha, around 2.6 times more than maize-based ethanol (Guo et al., 2023). The *Clostridium acetobutylicum* CICC 8012-based fermentation produced 68.46% butanol, 14.11% more than the amount in corn syrup-derived solvents, suggesting that duckweed is a superior feedstock for butanol production. Duckweed is an easily accessible source of biomass that can be supplied to anaerobic digesters.

Calicioglu et al. (2021) combined the biological processes of ethanol fermentation, acidogenic digestion, and methanogenic digestion to convert duckweed into bioenergy after treating wastewater, achieving the highest possible efficiency. Furthermore, using duckweed and microbial fermentation under anaerobic conditions resulted in a notable 51.2% rise in the generation of methane, surpassing the levels achieved with the exclusive fermentation of duckweed. Duckweeds significantly enhance biogas production in the anaerobic digester.

15.4.4 Duckweed: Reclamation Agent

Annually, 80 million tonnes of nitrogen-based fertilizers are used globally to enhance the agricultural productivity of various crops. Probably 40% of chemically synthesized nitrogenous fertilizers are used by crops, whereas the remaining gets carried in runoff water to freshwater bodies ultimately causing eutrophication. Ammonium, organic nitrogen, and phosphorus comprise most aquaculture wastewater pollutants. Fish and shrimp only absorb around 15%–25% of the nitrogen and phosphorus from the feed used in aquaculture. It has been estimated that duckweed biomass may hold up to 9.1 t/ha/year of total nitrogen and 0.8 t/ha/year of total phosphorus (Muradov et al., 2014). The primary nutrient concentrations (total nitrogen and total phosphorus) in local municipal wastewater decreased just after three days of *Lemna turionefera* incubation (Sun et al., 2021). After 10 days of mixotrophic culture, *S. polyrhiza* (L.) Schleid (5543) reduced phosphorus concentration to 0.20 mg/L (98.94% of total phosphorus) (Sun et al., 2020). Zhou et al. (2018) collected *Lemna aequinoctialis* matK duckweed from a nearby pond, which extracted nutrients (84% of NH_3-N and 98% of total phosphorus) from anaerobic-digested swine wastewater in 25 days. *L. minor* may thrive in ammonium ion solutions up to 84 mg/L concentration. Duckweeds prefer ammonium ion over nitrate for nitrogen. This preference was initially identified in dotted duckweed (*Landoltia punctata*) and verified in other genera also, that is, *Spirodela*, *Lemna*, and *Wolffia*. Duckweed produces high biomass with impressive growth in nutrient-rich effluent, making its production sustainable (Zhou et al., 2023).

15.5 CONCLUSION

Duckweeds are notable for their distinct physiology, fast growth rate, several services related to biomass, effortless maintenance, and uncomplicated harvesting. Duckweeds are widely accepted in scientific and commercial communities as potential organisms for bioremediation to mitigate the adverse effects of anthropogenic pollutants in aquatic environments and help prevent the eutrophication of water bodies. For more effective results of phytoremediation of various pollutants, exploring various species of duckweeds is still a significant and expanding topic of current research. The nutritional content and metabolite composition of duckweed substantially contributes to various sectors such as animal feed, aquaculture, health supplements, biofuels, and developing food items for the human population. Specific varieties of duckweed provide an adequate supply of vital nutrients, which include protein, dietary fiber, amino acids, micronutrients, and antioxidants, thereby complementing good health. Nevertheless, further investigations are necessarily required to have a better understanding of the chemical aspects, structure, functions, and health advantages of several constituents of duckweeds.

REFERENCES

Appenroth, K. J., Borisjuk, N., & Lam, E. (2013). Telling duckweed apart: Genotyping technologies for the Lemnaceae. *Chinese Journal of Applied and Environmental Biology, 19*(1), 1–10.

Appenroth, K. J., Sree, K. S., Bog, M., Ecker, J., Seeliger, C., Böhm, V., ... & Jahreis, G. (2018). Nutritional value of the duckweed species of the genus Wolffia (Lemnaceae) as human food. *Frontiers in Chemistry, 6*, 483.

Appenroth, K. J., Sree, K. S., Böhm, V., Hammann, S., Vetter, W., Leiterer, M., & Jahreis, G. (2017). Nutritional value of duckweeds (Lemnaceae) as human food. *Food Chemistry, 217*, 266–273.

Appenroth, K. J., Sree, K. S., Fakhoorian, T., & Lam, E. (2015). Resurgence of duckweed research and applications. *Plant Molecular Biology, 89*, 647–654.

Baek, G., Saeed, M., & Choi, H. K. (2021). Duckweeds: Their utilization, metabolites and cultivation. *Applied Biological Chemistry, 64*, 1–15.

Bog, M., Appenroth, K. J., & Sree, K. S. (2019). Duckweed (Lemnaceae): Its molecular taxonomy. *Frontiers in Sustainable Food Systems, 3*, 117.

Calicioglu, O., Richard, T. L., & Brennan, R. A. (2019). Anaerobic bioprocessing of wastewater-derived duckweed: Maximizing product yields in a biorefinery value cascade. *Bioresource Technology, 289*, 121716.

Can-Terzi, B., Goren, A. Y., Okten, H. E., & Sofuoglu, S. C. (2021). Biosorption of methylene blue from water by live Lemna minor. *Environmental Technology & Innovation, 22*, 101432.

Chen, Q., Jin, Y., Zhang, G., Fang, Y., Xiao, Y., & Zhao, H. (2012). Improving production of bioethanol from duckweed (*Landoltia punctata*) by pectinase pretreatment. *Energies, 5*(8), 3019–3032.

Cheng, J. J., & Stomp, A. M. (2009). Growing duckweed to recover nutrients from wastewaters and for production of fuel ethanol and animal feed. *Clean–Soil, Air, Water, 37*(1), 17–26.

Choi, K. S., Park, K. T., & Park, S. (2017). The chloroplast genome of *Symplocarpus renifolius*: A comparison of chloroplast genome structure in Araceae. *Genes, 8*(11), 324.

Das, B. K., Das, D. P., Jyotirmayee Pradhan, J. P., Barsha Priyadarshinee, B. P., Ipsita Sahu, I. S., Pragyan Roy, P. R., & Mishra, B. K. (2012). Evaluation of antimicrobial activity and phytochemical screening of ethanolic extract of greater duckweed, *Spirodela polyrrhiza*. *International Journal of Pharma and Bio Sciences, 3*(3), B-822-B-833 ref. 38 .

Day, J. A., & Saunders, F. M. (2004). Glycosidation of chlorophenols by Lemna minor. *Environmental Toxicology and Chemistry: An International Journal, 23*(3), 613–620.

Dhull, S. B., Rose, P. K., Rani, J., Goksen, G., & Bains, A. (2024). Food waste to hydrochar: A potential approach towards the sustainable development goals, carbon neutrality, and circular economy. *Chemical Engineering Journal, 490*, 151609.

Ekperusi, A. O., Sikoki, F. D., & Nwachukwu, E. O. (2019). Application of common duckweed (*Lemna minor*) in phytoremediation of chemicals in the environment: State and future perspective. *Chemosphere, 223*, 285–309.

Femeena, P. V., Roman, B., & Brennan, R. A. (2023). Maximizing duckweed biomass production for food security at low light intensities: Experimental results and an enhanced predictive model. *Environmental Challenges, 11*, 100709.

Fourounjian, P., Fakhoorian, T., & Cao, X. H. (2020). Importance of duckweeds in basic research and their industrial applications. In: X. H. Cao, P. Fourounjian, & W. Wang (Eds.), *The Duckweed Genomes* (pp. 1–17). Cham: Springer

Future Market Insights (2023). https://www.futuremarketinsights.com/reports/duckweed-market (accessed on 21 September 2023)

Ge, X., Zhang, N., Phillips, G. C., & Xu, J. (2012). Growing Lemna minor in agricultural wastewater and converting the duckweed biomass to ethanol. *Bioresource Technology*, *124*, 485–488.

Gomes, M. P., Gonçalves, C. A., de Brito, J. C. M., Souza, A. M., da Silva Cruz, F. V., Bicalho, E. M., ... & Garcia, Q. S. (2017). Ciprofloxacin induces oxidative stress in duckweed (Lemna minor L.): Implications for energy metabolism and antibiotic-uptake ability. *Journal of Hazardous Materials*, *328*, 140–149.

Goopy, J. P., & Murray, P. J. (2003). A review on the role of duckweed in nutrient reclamation and as a source of animal feed. *Asian-Australasian Journal of Animal Sciences*, *16*(2), 297–305.

Guo, L., Fang, Y., Jin, Y., He, K., & Zhao, H. (2023). High starch duckweed biomass production and its highly-efficient conversion to bioethanol. *Environmental Technology & Innovation*, *32*, 103296.

Gupta, C., & Prakash, D. (2013). Duckweed: An effective tool for phyto-remediation. *Toxicological & Environmental Chemistry*, *95*(8), 1256–1266.

Harwatt, H. (2019). Including animal to plant protein shifts in climate change mitigation policy: A proposed three-step strategy. *Climate Policy*, *19*(5), 533–541.

Kidwai, M. K., Malik, A., & Garg, V. K. (2023). Metal(loid)s sources, toxicity and bioremediation. In: A. Malik, M. K. Kidwai, & V. K. Garg (Eds.), *Bioremediation of Toxic Metal(loid)s* (pp. 3–32). CRC Press.

Landolt, E. (1980). Key to the determination of taxa within the family of Lemnaceae. *Veröffentlichungen des Geobotanischen Institutes der Eidg. Techn. Hochschule, Stiftung Rübel, Zürich*, *70*, 13–21.

Landolt, E. (1986). Biosystematic investigations in the family of duckweeds (Lemnaceae). II: The family of Lemnaceae: A monographic study. *Veröffentlichungen des Geobotanischen Institutes der ETH, Stiftung Rübel*, Zürich, 1, 71.

Lee, Y., Choi, H. J., & Shiga, T. (2020). Taxonomic Identity of *Landoltia punctata* (Araceae, Lemnoideae) in Korea. *Journal of Asia-Pacific Biodiversity*, *13*(3), 494–498.

Leng, R. A., Stambolie, J. H., & Bell, R. (1995). Duckweed-a potential high-protein feed resource for domestic animals and fish. *Livestock Research for Rural Development*, *7*(1), 36.

Les, D. H., & Crawford, D. J. (1999). Landoltia (Lemnaceae), a new genus of duckweeds. *Novon*, 530–533.

Li, C., Lin, Y., Li, X., Cheng, J. J., & Yang, C. (2023). Cupric ions inducing dynamic hormesis in duckweed systems for swine wastewater treatment: Quantification, modelling and mechanisms. *Science of the Total Environment*, *866*, 161411.

Li, J., Zhou, Q., & Campos, L. C. (2017). Removal of selected emerging PPCP compounds using greater duckweed (*Spirodela polyrhiza*) based lab-scale free water constructed wetland. *Water Research*, *126*, 252–261.

Maldonado, I., Terrazas, E. G. M., & Vilca, F. Z. (2022). Application of duckweed (Lemna sp.) and water fern (Azolla sp.) in the removal of pharmaceutical residues in water: State of art focus on antibiotics. *Science of the Total Environment*, *838*, 156565.

Mane, V. S., Gupta, A., Pendharkar, N., & Shinde, B. (2017). Exploration of primary metabolites from *Lemna minor* and determined its immunomodulatory and antimicrobial activity. *European Journal of Pharmaceutical and Medical Research*, *4*(4), 384–388.

Marchand, L., Mench, M., Marchand, C., Le Coustumer, P., Kolbas, A., & Maalouf, J. P. (2011). Phytotoxicity testing of lysimeter leachates from aided phytostabilized Cu-contaminated soils using duckweed (*Lemna minor* L.). *Science of the Total Environment*, *410*, 146–153.

Matamoros, V., Nguyen, L. X., Arias, C. A., Salvadó, V., & Brix, H. (2012). Evaluation of aquatic plants for removing polar microcontaminants: A microcosm experiment. *Chemosphere*, *88*(10), 1257–1264.

Mkandawire, M., & Dudel, E. G. (2005). Accumulation of arsenic in *Lemna gibba* L. (duckweed) in tailing waters of two abandoned uranium mining sites in Saxony, Germany. *Science of the Total Environment*, *336*(1–3), 81–89.

Muradov, N., Taha, M., Miranda, A. F., Kadali, K., Gujar, A., Rochfort, S., ... & Mouradov, A. (2014). Dual application of duckweed and azolla plants for wastewater treatment and renewable fuels and petrochemicals production. *Biotechnology for Biofuels*, *7*, 1–17.

Nassar, H. F., & Ibrahim, M. (2021). Duckweed-Lemna minor as green route for removal of chromium (VI) from aqueous solution. *International Journal of Environmental Research*, *15*(2), 275–284.

Ohlendorf, H. M., Lowe, R. W., Kelly, P. R., & Harvey, T. E. (1986). Selenium and heavy metals in San Francisco Bay diving ducks. *The Journal of Wildlife Management*, *50*, 64–70.

Osama, R., Awad, H. M., Ibrahim, M. G., & Tawfik, A. (2020). Mechanistic and economic assessment of polyester wastewater treatment via baffled duckweed pond. *Journal of Water Process Engineering*, *35*, 101179.

Pagliuso, D., Grandis, A., Fortirer, J. S., Camargo, P., Floh, E. I. S., & Buckeridge, M. S. (2022). Duckweeds as promising food feedstocks globally. *Agronomy*, *12*(4), 796.

Panfili, I., Bartucca, M. L., & Del Buono, D. (2019). The treatment of duckweed with a plant biostimulant or a safener improves the plant capacity to clean water polluted by terbuthylazine. *Science of the Total Environment*, *646*, 832–840.

Rani, P., Rose, P. K., Kidwai, M. K., & Meenakshi. (2023). *Brassica Juncea* L.: A potential crop for phytoremediation of various heavy metals. In: R. P. Singh, P. Singh, & A. Srivastava (Eds.), *Heavy Metal Toxicity: Environmental Concerns, Remediation and Opportunities* (pp. 285–311). Singapore: Springer Nature.

Reinhold, D., Vishwanathan, S., Park, J. J., Oh, D., & Saunders, F. M. (2010). Assessment of plant-driven removal of emerging organic pollutants by duckweed. *Chemosphere*, *80*(7), 687–692.

Rose, P. K., Kidwai, M. K., & Dhull, S. B. (2022). Food industry waste: Potential pollutants and their bioremediation strategies. In: S. B. Dhull, A. Singh, P. Kumar (Eds.), *Food Processing Waste and Utilization* (pp. 343–359). CRC Press.

Rose, P. K., Kidwai, M. K., & Kantiwal, P. (2024). Green approaches for the valorization of olive mill wastewater. In: V. K. Garg, A. Yadav, C. Mohan, S. Yadav, & N. Kumari (Eds.), *Green Chemistry Approaches to Environmental Sustainability* (pp. 313–336). Elsevier.

Shammas, N. K., & Wang, L. K. (2010). Natural environmental biotechnology. In L. K. Wang, V. Ivanov, & J.-H. Tay (Eds.), *Environmental Biotechnology*, 567–622. Humana Totowa, NJ

Sharma, R., & Lenaghan, S. C. (2022). Duckweed: A potential phytosensor for heavy metals. *Plant Cell Reports*, *41*(12), 2231–2243.

Singh, V., Pandey, B., & Suthar, S. (2018). Phytotoxicity of amoxicillin to the duckweed Spirodela polyrhiza: Growth, oxidative stress, biochemical traits and antibiotic degradation. *Chemosphere*, *201*, 492–502.

Sree, K. S., Bog, M., & Appenroth, K. J. (2016). Taxonomy of duckweeds (Lemnaceae), potential new crop plants. *Emirates Journal of Food and Agriculture*, *28*(5), 291–302.

Sun, F., Mu, Y., Leung, K. M., Su, H., Wu, F., & Chang, H. (2021). China is establishing its water quality standards for enhancing protection of aquatic life in freshwater ecosystems. *Environmental Science & Policy*, *124*, 413–422.

Sun, Z., Guo, W., Yang, J., Zhao, X., Chen, Y., Yao, L., & Hou, H. (2020). Enhanced biomass production and pollutant removal by duckweed in mixotrophic conditions. *Bioresource Technology*, *317*, 124029.

Tront, J. M., Reinhold, D. M., Bragg, A. W., & Saunders, F. M. (2007). Uptake of halogenated phenols by aquatic plants. *Journal of Environmental Engineering*, *133*(10), 955–961.

Varga, M., Žurga, P., Brusić, I., Horvatić, J., & Moslavac, M. (2020). Growth inhibition and recovery patterns of common duckweed *Lemna minor* L. after repeated exposure to isoproturon. *Ecotoxicology*, *29*(9), 1538–1551.

Vu, G. T., Fourounjian, P., Wang, W., & Cao, X. H. (2020). Future prospects of duckweed research and applications. In: X. H. Cao, P. Fourounjian, & W. Wang (Eds.), *The Duckweed Genomes* (pp. 179–185). Cham: Springer.

Wilson, P. C., & Koch, R. (2013). Influence of exposure concentration and duration on effects and recovery of *Lemna minor* exposed to the herbicide norflurazon. *Archives of Environmental Contamination and Toxicology*, *64*, 228–234.

Xu, J., Cui, W., Cheng, J. J., & Stomp, A. M. (2011). Production of high-starch duckweed and its conversion to bioethanol. *Biosystems Engineering*, *110*(2), 67–72.

Xu, J., Shen, Y., Zheng, Y., Smith, G., Sun, X. S., Wang, D., … & Li, Y. (2023). Duckweed (Lemnaceae) for potentially nutritious human food: A review. *Food Reviews International*, *39*(7), 3620–3634.

Zhao, Z., Shi, H., Duan, D., Li, H., Lei, T., Wang, M., … & Zhao, Y. (2015). The influence of duckweed species diversity on ecophysiological tolerance to copper exposure. *Aquatic Toxicology*, *164*, 92–98.

Zhao, Z., Shi, H., Liu, Y., Zhao, H., Su, H., Wang, M., & Zhao, Y. (2014). The influence of duckweed species diversity on biomass productivity and nutrient removal efficiency in swine wastewater. *Bioresource Technology*, *167*, 383–389.

Zhou, Q., Lin, Y., Li, X., Yang, C., Han, Z., Zeng, G., … & He, S. (2018). Effect of zinc ions on nutrient removal and growth of *Lemna aequinoctialis* from anaerobically digested swine wastewater. *Bioresource Technology*, *249*, 457–463.

Zhou, Y., Stepanenko, A., Kishchenko, O., Xu, J., & Borisjuk, N. (2023). Duckweeds for phytoremediation of polluted water. *Plants*, *12*(3), 589.

16 Exploiting Microbial Genome for Bioremediation of Toxic Pollutants

Recent Developments

Sachin Malik and Dharmender Kumar

16.1 INTRODUCTION

Bioremediation developed as an attractive method for detoxifying environmental contaminants. The effectiveness of bioremediation depends upon various elements, such as the chemical makeup and concentration of pollutants, the environment's physicochemical properties, and the contaminants' accessibility to microorganisms. With the help of bacteria, bioremediation has the potential for the removal, modification, and decontamination of different chemicals and physical pollutants from the environment. It ensures a low-cost, straightforward, and environmentally friendly cleaning method (Lovley, 2003) than the physicochemical remediation technologies. In addition, they own numerous intracellular and extracellular enzymes that convert complex toxicants into carbon and energy (Thakur et al., 2019). These microbial species provided certain biotechniques for enhanced and cost-effective degradation of heavy metal pollutants based on microbial metabolic capabilities, such as biosorption, bioextraction, and bioaccumulation (Kumar et al., 2023; Malik and Kumar, 2023). Therefore, successful biodegradation has an integrated avenue involving chemistry, ecology, engineering, geology, and microbiology. The fundamental benefit of bioremediation is that on-site clean-up frequently results in site preservation.

Today, heavy metal contamination is the foremost significant environmental issue. The daily activities of humans, such as agriculture, metalliferous smelting, mining, waste disposal, or industrial effluents, release varieties of metals such as arsenic (As), cadmium (Cd), chromium (Cr), cobalt (Co), mercury (Hg), nickel (Ni), lead (Pb), and strontium (Sn), etc. in the surrounding medium such as terrestrial and aquatic environment. Biotechnology uses genetic engineering to increase the effectiveness and lower the cost of using microorganisms and plants widely in the future to lessen

DOI: 10.1201/9781032642567-19

the load of hazardous pollutants on the environment (Rafeeq et al., 2023). The fact that over a long period, even without human involvement, nature eradicates both natural and the majority of fabricated contaminants through natural bioprocesses inspired bioremediation. Chemical precipitation, ultrafiltration, reverse osmosis, ion exchange, and electrodialysis are physicochemical techniques frequently used for the analysis of metal ions (Renu et al., 2017). These traditional physicochemical procedures for heavy metal removal are costly from an economic standpoint and have some drawbacks, such as insufficient metal removal, higher amount of reagent, requirements of energy, and formation of hazardous sludge. These methods do not apply to large contaminated areas. So, microbial bioremediation processes are the best choice for alleviating the toxic impacts of dangerous compounds from the environment. Various bacterial strains have been used for the degradation of toxic compounds, such as pesticides (Pannu and Kumar, 2023a, b), removal of colored organic waste degradation by nanoparticles (Saharan et al., 2023), paracetamol (Chopra and Kumar, 2022, 2023), ibuprofen (Chopra and Kumar, 2022a, b), and microplastics (Rani et al., 2024) This chapter, therefore, highlights the bioremediation potential of microorganisms integrating with the modern biotechnological perspectives.

16.2 TOXICITY ANALYSIS OF NOXIOUS COMPOUND IN THE ENVIRONMENT THROUGH BIOTECHNOLOGY

16.2.1 Microbial Biosensor

To quantify the presence of a target substrate, a microbial biosensor will generate a signal which may be measured electrically, physiologically, and biochemically. The biosensor shows the benefit of being inexpensive compared to others. Natural and genetically modified microorganisms, whether bacteria, fungi, algae, protozoa, or viruses, are used in microbial biosensors because of their ability to interact with various analytes and generate a signal that can be detected and quantified by a dedicated transducer. Different types of transducers based on electrical, like amperometric, conductometric, potentiometric, and optical, like colorimetric, luminescent, and fluorescent principles, have been merged with microorganisms (Gavrilaş et al., 2022). Biosensors from microbes are sensitive to environmental signals and can function in various environments. They have several benefits, including excellent selectivity, low detection limits, and high sensitivity. Some examples related to the microbial-based biosensor detect environmental pollutants, such as heavy metals (Cu, As, Cd, Hg, and Pb), pesticides, phenols, and other toxic substances that use terrestrial and aquatic biota (Table 16.1).

16.2.2 Bioindicator

"Biological Indicators" come from an extensive range of biotic and abiotic sources reacting to environmental changes over a while.

The moderate tolerance of bioindicator species to environmental variability allows them to efficiently indicate the current situation of the ecosystem. According to Holt and Miller (2010), using bioindicators isn't confined to a particular species with a

TABLE 16.1
Environmental Pollutants Detection Using Microbial Biosensors

Target Site	Microorganisms	Immobilized Method	Transducer	Analytes Used	LOD	References
Environmental monitoring	*E.Coli* DH5α	Microbial culture in a microfluidic device	Fluorescent	Pb^{2+}	–	Bae et al. (2018)
Waste water	*Saccharomyces cerevisiae* S288C	Physical adsorption on BND-chitosan hydrogel polymer on GCE	Amperometric	Cu2+, Cd2+, Ni2+, Pb2+	–	Gao et al. (2017)
Soil	Bacillus megaterium VR1	Entrapment in sol-gel matrix	Fluorescent	Cd^{2+}, Cu^{2+}, Zn^{2+}	1.42×10^{-4}, 3.16×10^{-4}, 2.42×10^{-4}	Rathnayake et al. (2021)
Water	*Chlamydomonas reinhardtii*	Ti/TiO2 ultramicroe-lectrodes in algal suspension	Chronoamperometric	Diuron (herbicide)	0.2 μM	Tsopela et al. (2016)
Environmental monitoring	Anabaena variabilis	Entrapment in alginate	Amperometric	Atrazine (herbicide)	0.07 μM	Tucci et al. (2019)

narrow range of environmental tolerances. Whole communities can function as bioindicators and function as many different sources of data to evaluate their ecological health in a "biotic index," as they span a wide range of environmental tolerances.

Living organisms, such as microorganisms to macroorganisms, lower plants to higher plants, and lower invertebrates to invertebrates are major sources for bioindication to identify various levels of contamination in the soil matrix (Gautam et al., 2022).

16.2.2.1 Microorganisms

Amoeba is one of the significant sources of the bioindicator of arsenic contamination, whereas *Oscillatoria tenuis and Phormedium bohneri* are two examples of cyanobacteria that are potential bioindicators of chromium bioaccumulation. Heavy metal resistance in certain bacteria is a result of plasmid-borne resistant genes which help be employed as latent bioindicators (Thriene et al., 2012). Microorganisms have the potential to degrade different pollutants to less-toxic forms, as described in Table 16.2.

TABLE 16.2
Microorganisms Used for the Degradation of Toxic Compounds

S. No	Microorganism Involved	Toxic Compound	References
1.	*Klebsiella pneumonia* (RS-13)	N, N-dimethyl-pphenylenediamine	Malik (2021)
2.	*Burkholderia* sp., *Myceliophthorathermophila*	Polycyclic aromatic hydrocarbons	Mohapatra and Phale (2021)
3.	*Phanerochaetechrysosporium*	4,4 dibromodiphenyl ether	Sen et al. (2019)
4.	*Acinetobacter* sp., *Microbacterium* sp., *Pseudomonas* sp. and *Ralstonia* sp.	Aromatic hydrocarbons	Basu et al. (2018)
5.	*Alcaligenes odorans, Corynebacterium propinquum, B. subtilis*, and *P. aeruginosa*	Phenol	Gaur et al. (2018)
6.	*Bacillus, Staphylococcus*	Endosulfan	Liu et al. (2016)
7.	*B. cereus, Exiguobacterium indicum, A. baumanii and E. aurantiacums*	Azo dyes effluents	Raquel et al. (2013)
8.	*B. firmus, Staphylococcus aureus, B. macerans, K. oxytoca*	Vat dyes	Sangkharak et al. (2020)
9.	*B. subtilis* strain NAP1, NAP2, NAP4	Oil-based based paints	Phulpoto et al. (2016)
10.	*P. cepacia, B. cereus, B. coagulans, B. cereus A, Serratia ficaria*	Diesel oil	Miri et al. (2022)

(Continued)

TABLE 16.2 (*Continued*)
Microorganisms Used for the Degradation of Toxic Compounds

S. No	Microorganism Involved	Toxic Compound	References
11.	*Alcaligenes odorans*, *Corynebacterium propinquum*, *P. aeruginosa*, *Fusarium* sp.	Oils	Pande et al. (2020)
12.	*Saccharomyces cerevisiae*, *Cunninghamella elegans*	Mercury, nickel and lead	Duc et al. (2021)
13.	*Aerococcus* sp., *Rhodopseudomonas palustris*	Lead, chromium, and cadmium	Sravya and Sangeetha (2022)
14.	*Chlorella* spp. *Spirulina* sp. (HD-104)	$Cr_2O_7^{22}$	Geetha et al. (2021)
15.	*Spirulina* sp., *Ulothrix tenuissima* and *C. reinhardtii*	Cr^{51}	Aregbesola et al. (2020)
16.	*Oscillatoria laete-virens*, *Arthrospira platensis*, *Pseudochlorococcumtypicum* and *Spirogyra insignis*	Pb^{21}	Song et al. (2019)

16.2.2.2 Plants

Different organisms may detect metal pollution in water, air, and soil, including micro- and macroalgae, lichens, mosses, tree bark, fungi, and the leaves of higher plants. Common bioindicators for primary producers include species of algae such as *Macrocystis pyrifera*, *Lobophora variegata*, *Antithamnion cruciatum*, and *Catenella nipae*. For instance, *Klebsormidium*-dominated algal mats have been discovered to be robust markers of higher iron concentrations in water. At the same time, the presence of *Fucus vesiculosus* has been described to demonstrate heavy metal contamination in marine environments. Mosses like *Pleurozium schreberi*, *Bryum argenteum*, and *Bryum capillare* may be used to monitor the appearance of metals in soil. Due to their deciduous nature and high cation exchange capacity, they can store high levels of heavy metals without suffering severe harm (Petrovic and Krivokapic, 2020). Heavy metals get entry into the bark of trees, roots, leaves, fruits, and so on. Nickel accumulates in hydrophytic plants in their roots and shoots; therefore, these plants act as bioindicators of heavy metal pollution.

16.2.2.3 Aquatic Animals

In the aquatic environment, fishes such as piranha, mullets, perches, and aquatic animals, such as water snakes, crabs, snails, starfishes, limpets, shellfishes, sea urchins, bivalves, amphipods, oysters, mussels, are used for bioindication because metals like Hg, Zn, Cd, Fe, and Co accumulate in the liver and gills. They can be analyzed to determine their concentrations and exhibit shell alterations, larval development, growth, and other physiologic indicators (Chiarelli and Roccheri, 2014).

16.2.2.4 Terrestrial Animals

Various animal groups such as insects, earthworms, birds, and even higher animals are some of the most prevalent bioindicators of several metals, including Cu, Hg, Cd, As, Zn, Pb, Mn, and Fe. The bioaccumulation of heavy metals in their bodies alters their receptivity and development; laboratory techniques may examine these changes. Roundworms and earthworms, for example, are used because of the changes in their digestive system and sensitivity to environmental shifts that they reveal (Xiao, et al., 2022). Flying foxes (*Pteropus poliocephalus* and *P. alecto*) and other chordates are also used for Cd bioindication (Pulscher et al., 2020).

16.3 REMEDIATION OF TOXIC POLLUTANTS USING DIFFERENT STRATEGIES

In heavy metal remediation, their concentration/levels need to be minimized to protect human health and the environment. A remediation method should be selected-based on the sample/contamination site characteristics, the degree of heavy metal toxicity, and the permissible limits prescribed. There are three broad categories of remediation methods: physical, chemical, and biological (Khalid et al., 2017).

16.3.1 Physical Methods

It is well-known that physical methods opting for degradation are costly and time-consuming. The heat treatment of most metal-contaminated substrates is an effective physical treatment method. Vitrification is when soil is heated to high temperatures (1,400°C–2,100°C) so that organic matter can decompose or volatilize to form vitreous materials, including oxide solids (Trifunović, 2021). This particular technology has been applied both in *ex situ* (soils are excavated and subjected to pretreatment) and *in situ* methods (electrodes are intercalated into contaminated areas vertically to heat the soil). The major disadvantage of this approach is that it invests a lot of energy. According to the study, there are many advantages to treating sewage sludge at 300°C–400°C over other approaches because many inorganic elements may be mobilized and available after treatment (Shi et al., 2013). It has been shown that the higher the applied temperature, the greater the metals' leaching. Nevertheless, due to higher temperatures, organic matter and nutrients are lost more quickly, preventing their subsequent use as fertilizer. Over the past decade, electro-remediation has been extensively considered as an alternative to chemical and physical methods. Electrodialysis has proven effective for eliminating a few metals, particularly Ni (56%) and Zn (85%), as well as for recovering phosphorus from sludge, but it has proven super effective in acidic conditions (Ebbers et al., 2015; Camargo et al., 2016). Remediation of toxicants by physical means includes soil replacement, washing, and vitrification. Heavy metal contamination can be minimized by washing contaminated soil. *Ex situ* washing is achieved via physically excavating soil and soaking it in a reactor, but this strategy can also be done *in situ* by forcing a washing solution through the soil matrix (Pavel and Gavrilescu, 2008).

16.3.2 Chemical Methods

Different chemical agents are used to immobilize or extract heavy metals via chemical remediation. Immobilization reduces the mobility and solubility of the metals in soil because they become less soluble. Precipitation, sorption, and complexations reduce metal toxicity using organic and inorganic immobilization methods. Minerals, phosphates, zeolites, clays, and organic compost are commonly used as immobilizers (Palansooriya et al., 2020). Immobilization processes often combine soil capping methods to encapsulate metals in a specific area. During extraction, chemical agents are added to increase the metal's solubility, bringing it into the aqueous phase for extraction/removal. The method typically utilizes distinct surfactants, chelating agents, and organic eluents. Research has also been conducted on the role of potassium phosphate in soil arsenic extraction. Phytoremediation is often combined with chemically induced metal extraction processes and is more efficient than mechanical metal extraction (Priya et al., 2023).

16.3.3 Biological Treatment Methods

With the aid of diverse intestinal flora, earthworms consume and digest waste throughout the vermicomposting process, excreting a humidified material that is uniform and low in chemicals (Suthar et al., 2014). Vermicomposting can also be used to stabilize sewage sludge, reducing organic carbon levels and increasing phosphorus levels so that waste can be used more efficiently as fertilizer. Another one is the use of biosurfactant, which is a biological agent used to remove metal from various substrates. By interaction with insoluble toxicants, these substances are transmitted to the liquid phase, where they can be removed. Heteroatoms often dominate a biosurfactant's structure, and their functional groups (carbonyl, hydroxyl, amine, etc.) interact with metal ions, so that the metal is removed (Lawniczak et al., 2013). Another method is bioleaching which is used to remove metals present in river sediments, sewage sludge, and soils. This method's simplicity, efficiency, and economic viability have made it promising for the removal of metals. *Acidithiobacillus* genus, *A. thiooxidans* and *A. ferrooxidans*, is the most used microorganisms in the bioleaching process (Sarkodie et al., 2022). Another evolving approach is phytoremediation which uses rhizospheric microorganisms associated with plant roots in toxic pollutants bioremediation (Kumar et al., 2023). There is a direct relationship between phytoremediation success and the type of toxicity involved.

16.4 ROLE OF BIOTECHNOLOGY IN BIOREMEDIATION OF HEAVY METALS

For the effective bioremediation of activated sludge, water, and soil, biotechnology-based methods are employed, and among those, genetically engineered microorganisms (GEM) draws attention to the field of bioremediation. GEM is modified through recombinant-DNA technology to produce a strain with a particular character for the degradation of diverse types of environmental pollutants (Azad et al., 2014). Some other approaches are also explained in the following sections.

16.4.1 Bioaugmentation

Bioaugmentation is a process that involves the inclusion of microbial biomass, including bacteria, fungi, and enzymes into toxic material containment zones, which may be adapted to the green environment. This process can significantly enhance the removal efficiency of a pollutant in a particular area, as well as minimize the amount of time and money required to remove contaminants (Ahmad et al., 2022). Further research on how various functional bacteria interact with one another concerning their resilience to diverse stress aspects, enzyme activities, and system robustness is required to pick the most effective bioaugmentation technique (Chen et al., 2022).

The impact of bioaugmentation of active sludge with feasible brewing waste yeast biomass on the organization of microbial communities was investigated in the presence of benzalkonium chloride. During the treatment that lasted ten days, adding up yeast biomass nurtured its feasibility. It helped lessen BAC's inhibitory impact on Bacilli that were present in activated sludge. Nevertheless, the bioaugmentation prompted an increase in the rate of bacterial reproduction as well as microbial metabolism. At the phyla extent, two main taxa, namely Proteobacteria and Firmicutes, were detected in the active sludge, with the abundance in non-incubated controls and all incubated samples ranging between 22%–36% and 27%–35%, respectively. This indicates that Firmicutes and Proteobacteria are the predominant types of bacteria found in active sludge (Zorža et al., 2022).

16.4.2 Metagenomics in Bioremediation Process

Metagenomic techniques are often divided into sequence-based metagenomics and function-based techniques. These methods have increased our understanding of many noncultivable microorganisms' environments, enhancing our knowledge of the microbial world. The function-focused metagenomic approach promotes the discovery of novel genes and provides the option for genetic research, which is an enthralling prospect for discovering novel ecosystems (Tripathi et al., 2018). Environmental materials may be analyzed for directly extracted genetic material using metagenomics. Thus, in a niche environment, metagenomic research based on sequence and function gives information about microbial communities of noncultivable species. Possible uses include encoding gene clusters for pollutant-degrading enzymes and discovering new microorganisms (Datta et al., 2020). Metagenomics is a method that allows researchers to analyze DNA from ambient materials without performing isolation and cultivation of the species being studied.

The metagenomic analysis illustrated in Figure 16.1 investigates bacteria that produce new genes and contributes to discovering various processes and techniques that may be used to improve the remediation approach (Navgire et al., 2022).

16.4.3 Microbial Consortia and Community Interactions

Microbes having similar properties tend to congregate and organize themselves into communities that serve various purposes in their natural environments. According to the findings of several studies, the steadiness of microbial consortia is positively correlated with their level of genetic diversity (Hempson et al., 2019). This is because

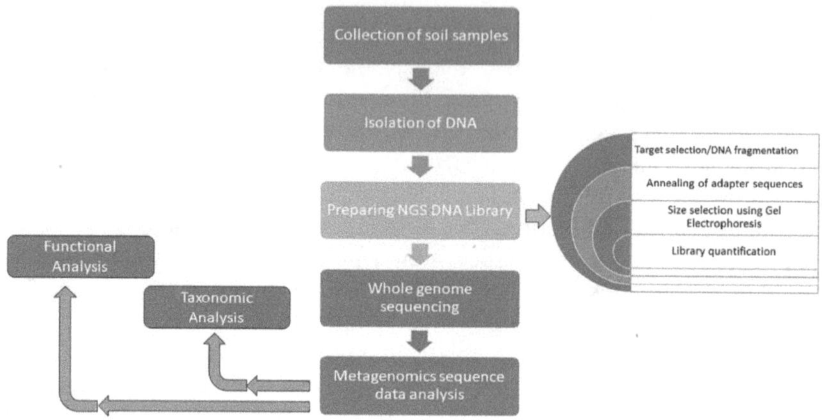

FIGURE 16.1 Workflow of steps involved in DNA isolation from contaminated sites via metagenomics.

microbial communities that perform various activities are resilient due to the greater availability of nutrients and immense support from crux partner species. This enables these communities to remain ingenious by environmental stressors and to perform their duties more effectively.

In recent years, the engineered microbial consortium has been utilized to investigate the degradation of organic contaminants and the removal of metal ions (Li et al., 2021). The assemblage of several microbial genera and species is a necessary step in engineering microbial consortia, which is a challenging and involved process. Studies on some engineered microbial consortia have been described in Table 16.3 for bioremediation.

TABLE 16.3
Examples of Engineered Microbial Consortia Used for Bioremediation

Microorganism	Toxicants	Bioremediation Efficacy	References
Streptomyces sp. M7, A5, A11, MC1	Cr(VI) Lindane	Removal of 86% Cr(VI) initially by 50 mg/kg concentration in soil Removal of 46% of lindane initially by 25 mg/kg concentration in soil	Polti et al. (2014)
Bacillus sp. FW1, *Novosphin-gobium Ochrobactrum* sp. PW1, *Mycobacterium spp.* PO1 and PO2, and *pentaromativorans* PY1	Pyrene	degradation rate is three-fold greater compared to individual degrader	Wanapaisan et al. (2018)
Arthrobacter sp. DNS10, *Bacillus subtilis* DNS4, *Arthrobacter* sp. DNS9, *Variovorax* sp. DNS12	Atrazine	Removal of 100% of atrazine initially at 100 mg/L	Zhang et al. (2012)

16.5 GENETIC AND METABOLIC ENGINEERING

According to Butt et al. (2018), molecular scissors have enormous potential applications in a diverse array of study fields relating to plants, animals, and microorganisms. Similarly, Bier et al. (2018) described a method for editing that entails targeting by a self-prepared guide sequence that is complementary to the sequence of the gene of interest, creating a gap at a spot, repairing the break using homologous recombination, and then performing modifications like insertion and deletion of a desired fragment sequence. Genome engineering via genetic editing tools led to next-level applications of microbes in different fields, including agriculture, food production, medicine, and animal feed (Yadav et al., 2018). The gene-editing techniques can significantly enhance bioremediation approaches like transforming harmful substances into less-toxic substances (Hussain et al., 2018). There are three main types of gene-editing tools that are zinc-finger nucleases (ZFNs), transcription activator-like effector nucleases (TALENs), and CRISPR/Cas system (Gaj et al., 2013).

16.5.1 Zinc-Finger Nucleases

The first step toward developing technology that could alter genomes was the discovery of ZFNs. These are noted as restriction enzymes that have been synthetically created to perform site-specific genome editing (Khalil et al., 2020). The ZFNs were key targeted nuclease that attained widespread usage (Urnov et al., 2010). These are fusions of a specifically created Cys2-His2 zinc-finger protein with the FokI restriction endonuclease's cleavage domain. They act as dimers, with each monomer recognizing a particular "half-site" sequence on the DNA through the zinc-finger DNA-binding domain. This "half-site" sequence is generally between 9 and 18 base pairs (bps). The FokI cleavage domain may be responsible for the dimerization of the ZFN proteins. This domain breaks DNA inside a spacer sequence of 5–7 bps and is located between the two-flanking zinc-finger binding sites. Each ZFN region comprises three or four zinc-finger domains, each consisting of around 30 amino acid residues arranged in a pattern (Carroll, 2011). The residues responsible for facilitating DNA recognition are found inside the alpha-helical domain. These residues usually interact with three bps of DNA, but occasional overlap may occur from a neighboring field. Many zinc-finger domains that recognize individual DNA triplets are found using techniques such as phage display and other similar approaches. It is feasible to build polydactyl zinc-finger proteins by fusing in tandem with these domains by employing a conventional linker peptide. These proteins can target a large variety of different DNA sequences.

16.5.2 Transcription Activator-Like Effector Nucleases

Bacterial effectors are referred to as TALE proteins. TALENs, much like ZFNs, have a commutable structure both in terms of their shape and their function. They are made up of an amino-terminal TALE DNA-binding domain that is joined with a carboxy-terminal FokI cleavage domain. This FokI cleavage domain cuts inside a 12- to 19-bp spacer region, which separates each TALE-binding site and is responsible for mediating the formation of dimers of TALEN proteins, much as it does for ZFNs. TALEs are commonly constructed to distinguish between 12 and 20bps of DNA, in which most bases often contribute to better gene-editing specificity. The TALE-binding domain is made up of a succession of repeated parts, each of which is about 34 residues long. Repeats make contact with DNA via the amino acid residues located at positions 12 and 13 and are referred to as repeat variable direct residues (RVDs) (Jankele and Svoboda, 2014).

Several approaches might accomplish the construction of TALEN DNA-binding domains; however, Golden Gate assembly is the most fundamental technique. High-throughput TALE assembly techniques, such as the FLASH assembly, iterative capped assembly, and ligation-independent cloning, have also been developed.

16.5.3 CRISPR–Cas System

The most advanced tool to be added to the arsenal for editing genomes is called CRISPR–Cas9, and it is used by bacteria to help them develop their kind of adaptive immunity (Marraffini and Sontheimer, 2010). Through RNA-guided DNA cleavage carried out by the protein Cas, the CRISPR type II system offers bacterial cells defense against the introduction of foreign DNA in the form of viruses and plasmids. When foreign DNA is integrated into the CRISPR locus, it is transcribed to (crRNA) CRISPR RNA that annealed to trans-activating (tracrRNA) CRISPR RNA to drive the sequence-specific destruction of pathogenic DNA by Cas9 protein. Foreign DNA can be integrated into the CRISPR locus using the Cas9 protein. In 2012, Jinek and coworkers showed that for the Cas9 protein to recognize a target, all that is required is a seed sequence to be present inside crRNA and a protospacer-adjacent motif (PAM) that is conserved to be located upstream of the crRNA binding site (Jinek et al., 2012). This system has now been reduced so that it may be used for genetic engineering. It currently comprises only Cas9 nuclease and a single guide RNA (gRNA) enclosing crucial crRNA and tracrRNA components. CRISPR-Cas9 has evolved as the most versatile and user-friendly, convenient platform for genome editing because the gRNA solely controls target site identification. This removes the need to build new proteins to recognize each new target site, making it one of the primary reasons for its rise to prominence. The PAM motif, which is needed for DNA cleavage and is recognized by the Cas9 nuclease, must be placed directly downstream of the gRNA target site for Cas9 to recognize the target site (Gleditzsch et al., 2019). This is the single significant constraint for Cas9 target site recognition. For example, the PAM sequence for the *Streptococcus pyogenes* Cas9 is 5′-NGG-3′ (although it is possible to accept 5′-NAG-3′ in certain circumstances).

16.5.4 Genetic Engineering in Bioremediation Processes

Using genetically engineered organisms (GEOs) for bioremediation is a low-cost, eco-friendly, and socially acceptable approach for permanently eradicating toxic compounds. While GEO for bioremediation shows promise, it is not yet commercialized and has very limited approval for environmental discharge (Macek et al., 2008). Using technological protections and appropriate regulatory processes, such as

TABLE 16.4
Genetically Modified Bacteria for Bioremediation of Heavy Metals

Heavy Metal	Expressed Gene	Genetically Engineered Bacteria	References
As	Metalloregulatory protein ArsR	*E. coli* strain	Kostal et al. (2004)
Cd^{2+}	SpPCS	*E. coli*	Kang et al. (2007)
Cr6+	CrR	*Methylococcuscapsulatus*	Al Hasin et al. (2010)
Cr	Chromate reductase ChrR	*P. putida* strain	Ackerley et al. (2004)
Hg	Organomurcurial lyase	*E. coli* strain	Murtaza et al. (2002)
	Hg^{2+} transporter	*E. coli* JM109	Zhao et al. (2005)
	Organomercurial lyase	*Pseudomonas* K-62	Kiyono et al. (2006)
	mer	*Achromobactersp* AO22	Ng et al. (2009)
Ni	Phytochelatin synthase (PCS)	*P. fluorescens* 4F39	Lopez et al. (2002) and Sriprang et al. (2003)

conducting thorough risk evaluations and regular monitoring, scientists might guarantee the safe use of GEO for bioremediation (Azad et al., 2014). Bioremediation has been accomplished using genetically engineered bacteria and recombinant-DNA/RNA technologies. It is the best way to take full advantage of microbial cells for the biotransformation of hazardous environmental pollutants (Kumar et al., 2023). To improve bioremediation techniques, scientists have modified microbial genes to generate novel metabolic pathways. Bacterial cells may reduce their exposure to potentially harmful heavy metals by using their metal regulatory genes to alter their gene expression. Genetically engineered bacteria that produce metallothioneins (MT) may hasten the buildup of heavy metals. Some of the most promising emerging strategies for cleaning up metal-contaminated locations include genetically modifying endophytes and rhizospheric bacteria to degrade contaminants in the soil as plants grow (Mosa et al., 2016). When phytochelatin 20 is expressed upon the cell surface, bacteria like *Moreaxella sp.* and *Escherichia coli* collect 25-fold Cd or Hg compared with wild-type strains. Table 16.4 shows some examples of genetically modified bacteria for the bioremediation of heavy metals.

16.6 CONCLUSION

The anthropogenic activities are solely responsible for heavy metal contamination, which harms human and ecological health. In the last few decades, technologies and industrial manufacturing have developed to such an extent that some harmful side effects have been observed, like soil degradation and pollution. This problem has been addressed through the development of several remediation technologies over the years. Using microorganisms, bioremediation can be carried out more quickly due to some of the intricacies involved in conventional remediation methods. The book chapter examines the environmental effects of heavy metal ions due to human activities, the health hazards associated with such contamination, and how plants and microbes can effectively remove polluted environments through enzymatic reactions and mechanisms. Based on the literature surveyed, it has been concluded that bioremediation is a more efficient and cost-effective alternative to physico-chemical methods for removing heavy metals from polluted sites. Transgenic microbes play a significant role in detoxifying and degrading heavy metal contaminants and xenobiotics. Still, more studies need to be conducted to improve their endurance when liberated into the environment for bioremediation. It is possible to choose the most appropriate treatment strategy for a specific site based on three basic principles: whether the pollutant can be transformed into less-toxic products biologically, whether it is bioavailable to microorganisms, and whether it can be optimized for bioprocesses. Biotechnology has a bright future when it comes to developing bioprocesses for degradation of toxicants from environment as a result of advances in bioinformatics.

REFERENCES

Ackerley, D. F. (2004). Mechanism of chromate reduction by the *Escherchia coli* protein NfsA, the role of different chromate reductases in minimizing oxidative stress during chromate reduction. *Environmental Microbiology*, *6*, 851–860.

Ahmad, I., Abdullah, N., Koji, I., Yuzir, A., Mohamad, S. E., Show, P. L., ... & Khoo, K. S. (2022). The role of restaurant wastewater for producing bioenergy towards a circular bioeconomy: A review on composition, environmental impacts, and sustainable integrated management. *Environmental Research*, *214*, 113854.

Al Hasin, A., Gurman, S. J., Murphy, L. M., Perry, A., Smith, T. J., & Gardiner, P. H. (2010). Remediation of chromium (VI) by a methane-oxidizing bacterium. *Environmental Science & Technology*, *44*(1), 400–405.

Aregbesola, O. A., Kumar, A., Mokoena, M. P., & Olaniran, A. O. (2020). Cloning, overexpression, purification, characterization and structural modelling of a metabolically active Fe^{2+} dependent 2, 6-dichloro-p-hydroquinone 1, 2-dioxygenase (CpsA) from *Bacillus cereus* strain AOA-CPS_1. *International Journal of Biological Macromolecules*, *161*, 247–257.

Azad, M. A. K., Amin, L., &Sidik, N. M. (2014). Genetically engineered organisms for bioremediation of pollutants in contaminated sites. *Chinese Science Bulletin*, *59*, 703–714.

Bae, J., Lim, J. W., & Kim, T. (2018). Reusable and storable whole-cell microbial biosensors with a microchemostat platform for in situ on-demand heavy metal detection. *Sensors and Actuators B: Chemical*, *264*, 372–381.

Basu, S., Rabara, R. C., Negi, S., & Shukla, P. (2018). Engineering PGPMOs through gene editing and systems biology: A solution for phytoremediation. *Trends in Biotechnology*, *36*(5), 499–510.

Bier, E., Harrison, M. M., O'Connor-Giles, K. M., & Wildonger, J. (2018). Advances in engineering the fly genome with the CRISPR-Cas system. *Genetics*, *208*(1), 1–18.

Butt, H., Jamil, M., Wang, J. Y., Al-Babili, S., & Mahfouz, M. (2018). Engineering plant architecture via CRISPR/Cas9-mediated alteration of strigolactone biosynthesis. *BMC Plant Biology*, *18*, 1–9.

Camargo, F. P., Sérgio Tonello, P., dos Santos, A. C. A., & Duarte, I. C. S. (2016). Removal of toxic metals from sewage sludge through chemical, physical, and biological treatments—a review. *Water, Air, & Soil Pollution*, *227*, 1–11.

Carroll, D. (2011). Genome engineering with zinc-finger nucleases. *Genetics*, *188*(4), 773–782.

Chen, Y., Wang, S., Geng, N., Wu, Z., Xiong, W., & Su, H. (2022). Artificially constructing mixed bacteria system for bioaugmentation of nitrogen removal from saline wastewater at low temperature. *Journal of Environmental Management*, *324*, 116351.

Chiarelli, R., & Roccheri, M. C. (2014). Marine invertebrates as bioindicators of heavy metal pollution. *Open Journal of Metal, 4*, 93–106.

Chopra, S., & Kumar, D. (2022a). Characteristics and growth kinetics of biomass of Citrobacter freundii strains PYI-2 and *Citrobacter portucalensis* strain YPI-2 during the biodegradation of Ibuprofen. *International Microbiology*, *25*(3), 615–628.

Chopra, S., & Kumar, D. (2022b). Characterization and biodegradation of ibuprofen by *Bacillus siamensis* strain DSI-1 isolated from wastewater. *RendicontiLincei. ScienzeFisiche e Naturali*, *33*(3), 643–652.

Chopra, S., & Kumar, D. (2023). Characterization and biodegradation of paracetamol by biomass of *Bacillus licheniformis* strain PPY-2 isolated from wastewater. *RendicontiLincei. ScienzeFisiche e Naturali*, *34*(2), 491–501.

Duc, H. D., Hung, N. V., & Oanh, N. T. (2021). Anaerobic Degradation of Endosulfans by a Mixed Culture of Pseudomonas sp. and Staphylococcus sp. *Applied Biochemistry and Microbiology*, *57*(3), 327–334.

Ebbers, B., Ottosen, L. M., & Jensen, P. E. (2015). Electrodialytic treatment of municipal wastewater and sludge for the removal of heavy metals and recovery of phosphorus. *Electrochimica Acta*, *181*, 90–99.

Gaj, T., Gersbach, C. A., & Barbas, C. F. (2013). ZFN, TALEN, and CRISPR/Cas-based methods for genome engineering. *Trends in Biotechnology*, *31*(7), 397–405.

Gao, G., Fang, D., Yu, Y., Wu, L., Wang, Y., & Zhi, J. (2017). A double-mediator based whole cell electrochemical biosensor for acute biotoxicity assessment of wastewater. *Talanta*, *167*, 208–216.

Gaur, N., Narasimhulu, K., &PydiSetty, Y. (2018). Recent advances in the bio-remediation of persistent organic pollutants and its effect on environment. *Journal of Cleaner Production*, *198*, 1602–1631.

Gautam, M., Mishra, S., & Agrawal, M. (2022). Bioindicators of soil contaminated with organic and inorganic pollutants. In: S. Tiwari, S. Agrawal (Eds.), *New Paradigms in Environmental Biomonitoring Using Plants* (pp. 271–298). Amsterdam: Elsevier. https://doi.org/10.1016/B978-0-12-824351-0.00001-8.

Gavrilaş, S., Ursachi, C. Ş., Perţa-Crişan, S., & Munteanu, F. D. (2022). Recent rends in biosensors for environmental quality monitoring. *Sensors*, *22*(4), 1513.

Geetha, N., Bhavya, G., Abhijith, P., Shekhar, R., Dayananda, K., & Jogaiah, S. (2021). Insights into nanomycoremediation: Secretomics and mycogenic biopolymer nanocomposites for heavy metal detoxification. *Journal of Hazardous Materials*, *409*, 124541.

Gleditzsch, D., Pausch, P., Müller-Esparza, H., Özcan, A., Guo, X., Bange, G., & Randau, L. (2019). PAM identification by CRISPR-Cas effector complexes: Diversified mechanisms and structures. *RNA Biology*, *16*(4), 504–517.

Hempson, G. P., Archibald, S., Donaldson, J. E., & Lehmann, C. E. (2019). Alternate grassy ecosystem states are determined by palatability–flammability trade-offs. *Trends in Ecology & Evolution*, *34*(4), 286–290.

Holt, E. A., & Miller, S. W. (2010). Bioindicators: Using organisms to measure environmental impacts. *Nature Education Knowledge*, 3(10), 8.

Hussain, I., Aleti, G., Naidu, R., Puschenreiter, M., Mahmood, Q., Rahman, M. M., … & Reichenauer, T. G. (2018). Microbe and plant assisted-remediation of organic xenobiotics and its enhancement by genetically modified organisms and recombinant technology: A review. *Science of the Total Environment*, *628*, 1582–1599.

Jankele, R., & Svoboda, P. (2014). TAL effectors: Tools for DNA targeting. *Briefings in Functional Genomics*, *13*(5), 409–419.

Jinek, M., Chylinski, K., Fonfara, I., Hauer, M., Doudna, J. A., & Charpentier, E. (2012). A programmable dual-RNA–guided DNA endonuclease in adaptive bacterial immunity. *Science, 337*(6096), 816–821.

Kang, S. H., Singh, S., Kim, J. Y., Lee, W., Mulchandani, A., & Chen, W. (2007). Bacteria metabolically engineered for enhanced phytochelatin production and cadmium accumulation. *Applied and Environmental Microbiology, 73*(19), 6317–6320.

Khalid, S., Shahid, M., Niazi, N. K., Murtaza, B., Bibi, I., & Dumat, C. (2017). A comparison of technologies for remediation of heavy metal contaminated soils. *Journal of Geochemical Exploration, 182*, 247–268.

Khalil, A. M. (2020). The genome editing revolution. *Journal of Genetic Engineering and Biotechnology, 18*(1), 68.

Kiyono. Masako, & Hidemitsu, P. H. (2006). Remediation of hazardous waste-contaminated soils. *Journal of Health Sciences, 52*(3), 199–204.

Kostal, J., Yang, R., Wu, C. H., Mulchandani, A., & Chen, W. (2004). Enhanced arsenic accumulation in engineered bacterial cells expressing ArsR. *Applied and Environmental Microbiology, 70*(8), 4582–4587.

Kumar, D., Malik, S., Rani, R., Kumar, R., & Duhan, J. S. (2023). Behavior, risk, and bioremediation potential of heavy metals/metalloids in the soil system. *RendicontiLincei. ScienzeFisiche e Naturali, 34*(3), 809–831.

Lawniczak, Ł., Marecik, R., & Chrzanowski, Ł. (2013). Contributions of biosurfactants to natural or induced bioremediation. *Applied Microbiology and Biotechnology, 97*, 2327–2339.

Li, X., Wu, S., Dong, Y., Fan, H., Bai, Z., & Zhuang, X. (2021). Engineering microbial consortia towards bioremediation. *Water, 13*(20), 2928.

Liu, Z., Shao, B., Zeng, G., Chen, M., Li, Z., Liu, Y., … & Yan, M. (2016). Effects of rhamnolipids on the removal of 2, 4, 2, 4-tetrabrominated biphenyl ether (BDE-47) by *Phanerochaetechrysosporium*analyzed with a combined approach of experiments and molecular docking. *Chemosphere, 210*, 922–930.

López, A., Lázaro, N., Morales, S., & Marqués, A. M. (2002). Nickel biosorption by free and immobilized cells of Pseudomonas fluorescens 4F39: A comparative study. *Water, Air, and Soil Pollution, 135*, 157–172.

Lovley, D. R. (2003). Cleaning up with genomics: Applying molecular biology to bioremediation. *Nature Reviews Microbiology, 1*(1), 35–44.

Macek, T., Kotrba, P., Svatos, A., Novakova, M., Demnerova, K., &Mackova, M. (2008). Novel roles for genetically modified plants in environmental protection. *Trends in Biotechnology, 26*(3), 146–152.

Malik, J. A. (Ed.). (2021). *Handbook of Research on Microbial Remediation and Microbial Biotechnology for Sustainable Soil*. IGI Global.

Malik, S., & Kumar, D. (2023). Perspectives of nanomaterials in microbial remediation of heavy metals and their environmental consequences: A review. *Biotechnology and Genetic Engineering Reviews, 39*(1), 1–48.

Marraffini, L. A., & Sontheimer, E. J. (2010). CRISPR interference: RNA-directed adaptive immunity in bacteria and archaea. *Nature Reviews Genetics, 11*(3), 181–190.

Miri, S., Rasooli, A., Brar, S. K., Rouissi, T., & Martel, R. (2022). Biodegradation of p-xylene—A comparison of three psychrophilic Pseudomonas strains through the lens of gene expression. *Environmental Science and Pollution Research, 29*, 21465–21479.

Mohapatra, B., & Phale, P. S. (2021). Microbial degradation of naphthalene and substituted naphthalenes: Metabolic diversity and genomic insight for bioremediation. *Frontiers in Bioengineering and Biotechnology, 9*, 602445.

Mosa, K. A., Saadoun, I., Kumar, K., Helmy, M., & Dhankher, O. P. (2016). Potential biotechnological strategies for the cleanup of heavy metals and metalloids. *Frontiers in Plant Science, 7*, 183531.

Murtaza, I., Dutt, A., & Ali, A. (2002). Biomolecular engineering of Escherichia coli organo-mercurial lyase gene and its expression. *Indian Journal of Biotechnology, 1*, 117–120.

Navgire, G. S., Goel, N., Sawhney, G., Sharma, M., Kaushik, P., Mohanta, Y. K., ... & Al-Harrasi, A. (2022). Analysis and interpretation of metagenomics data: An approach. *Biological Procedures Online*, *24*(1), 18.

Ng, S. P., Davis, B., Palombo, E. A., & Bhave, M. (2009). A Tn 5051-like mer-containing transposon identified in a heavy metal tolerant strain Achromobacter sp. AO22. *BMC Research Notes*, *2*, 1–7.

Palansooriya, K. N., Shaheen, S. M., Chen, S. S., Tsang, D. C., Hashimoto, Y., Hou, D., ... & Ok, Y. S. (2020). Soil amendments for immobilization of potentially toxic elements in contaminated soils: A critical review. *Environment International*, *134*, 105046.

Pande, V., Pandey, S. C., Sati, D., Pande, V., & Samant, M. (2020). Bioremediation: An emerging effective approach towards environment restoration. *Environmental Sustainability*, *3*, 91–103.

Pannu, R., & Kumar, D. (2023a). Biodegradation of lindane (γ-Hexachlorocyclohexane) and other isomers by *Bacillus subtilis* strain Mz-13i. *Biocatalysis and Agricultural Biotechnology*, *48*, 102630.

Pannu, R., & Kumar, D. (2023b). Characterization of bacteria capable of degrading lindane (γ-hexachlorocyclohexane) and study the dynamics of biodegradation. *Microbiology*, *92*(3), 434–447.

Pavel, L. V., & Gavrilescu, M. (2008). Overview of ex situ decontamination techniques for soil cleanup. *Environmental Engineering & Management Journal (EEMJ)*, *7*(6).

Petrovic, D., & Krivokapic, S. (2020). The effect of Cu, Zn, Cd, and Pb accumulation on biochemical parameters (proline, chlorophyll) in the water caltrop (*Trapa natans* L.), Lake Skadar, Montenegro. *Plants*, *9*(10), 1287.

Phulpoto, A. H., Qazi, M. A., Mangi, S., Ahmed, S., & Kanhar, N. A. (2016). Biodegradation of oil-based paint by Bacillus species monocultures isolated from the paint warehouses. *International Journal of Environmental Science and Technology*, *13*, 125–134.

Polti, M. A., Aparicio, J. D., Benimeli, C. S., & Amoroso, M. J. (2014). Simultaneous bioremediation of Cr (VI) and lindane in soil by actinobacteria. *International Biodeterioration & Biodegradation*, *88*, 48–55.

Priya, A. K., Muruganandam, M., Ali, S. S., & Kornaros, M. (2023). Clean-up of heavy metals from contaminated soil by phytoremediation: A multidisciplinary and eco-friendly approach. *Toxics*, *11*(5), 422.

Pulscher, L. A., Gray, R., McQuilty, R., Rose, K., Welbergen, J., & Phalen, D. N. (2020). Investigation into the utility of flying foxes as bioindicators for environmental metal pollution reveals evidence of diminished lead but significant cadmium exposure. *Chemosphere*, *254*, 126839.

Rafeeq, H., Afsheen, N., Rafique, S., Arshad, A., Intisar, M., Hussain, A., ... & Iqbal, H. M. (2023). Genetically engineered microorganisms for environmental remediation. *Chemosphere*, *310*, 136751.

Rani, R., Malik, S., Kumar, D., Kumar, R., Mukherjee, S., Saharan, P. S., & Duhan, J. S. (2024). Advances in the role of microorganisms, waste management strategies and policies on microplastic abatement in the era of bio-circular economy. Sustainable Chemistry and Pharmacy, *39*, 101595.

Raquel, S., Natalia, G., Luis Fernando, B., & Maria Carmen, M. (2013). Biodegradation of high-molecular-weight polycyclic aromatic hydrocarbons by a wood-degrading consortium at low temperatures. *FEMS Microbiology Ecology*, *83*(2), 438–449.

Rathnayake, I. V. N., Megharaj, M., & Naidu, R. (2021). Green fluorescent protein based whole cell bacterial biosensor for the detection of bioavailable heavy metals in soil environment. *Environmental Technology & Innovation*, *23*, 101785.

Renu, Agarwal, M., & Singh, K. (2017). Methodologies for removal of heavy metal ions from wastewater: An overview. *Interdisciplinary Environmental Review*, *18*(2), 124–142.

Saharan, P., Kumar, V., Kaushal, I., Mittal, A., Shukla, S. K., Kumar, D., … & Om, H. (2023). A comprehensive review on the metal-based green valorized nanocomposite for the remediation of emerging colored organic waste. *Environmental Science and Pollution Research*, *30*(16), 45677–45700.

Sangkharak, K., Choonut, A., Rakkan, T., & Prasertsan, P. (2020). The degradation of phenanthrene, pyrene, and fluoranthene and its conversion into medium-chain-length polyhydroxyalkanoate by novel polycyclic aromatic hydrocarbon-degrading bacteria. *Current Microbiology*, *77*, 897–909.

Sarkodie, E. K., Jiang, L., Li, K., Yang, J., & Liu, H. (2022). A review on the bioleaching of toxic metal (loid) s from contaminated soil: Insight into the mechanism of action and the role of influencing factors. *Frontiers in Microbiology*, *13*, 1049277.

Sen, S. K., Patra, P., Das, C. R., Raut, S., & Raut, S. (2019). Pilot-scale evaluation of bio-decolorization and biodegradation of reactive textile wastewater: An impact on its use in irrigation of wheat crop. *Water Resources and Industry*, *21*, 100106.

Shi, W., Liu, C., Ding, D., Lei, Z., Yang, Y., Feng, C., & Zhang, Z. (2013). Immobilization of heavy metals in sewage sludge by using subcritical water technology. *Bioresource Technology*, *137*, 18–24.

Song, J., Zhang, S., Xie, Y., & Li, Q. (2019). Purification and characteristics of an aflatoxin B1 degradation enzyme isolated from *Pseudomonas aeruginosa*. *FEMS Microbiology letters*, *366*(5), fnz034.

Sravya, K., & Sangeetha, S. (2022). Feasibility study on bioremediation techniques to contaminated soils. *Materials Today: Proceedings*, *51*, 2556–2560.

Sriprang, R., Hayashi, M., Ono, H., Takagi, M., Hirata, K., & Murooka, Y. (2003). Enhanced accumulation of Cd2+ by a *Mesorhizobium* sp. transformed with a gene from *Arabidopsis thaliana* coding for phytochelatin synthase. *Applied and Environmental Microbiology*, *69*(3), 1791–1796.

Suthar, S., Sajwan, P., & Kumar, K. (2014). Vermiremediation of heavy metals in wastewater sludge from paper and pulp industry using earthworm *Eisenia fetida*. *Ecotoxicology and Environmental Safety*, *109*, 177–184.

Thakur, M., Medintz, I. L., & Walper, S. A. (2019). Enzymatic bioremediation of organophosphate compounds: Progress and remaining challenges. *Frontiers in Bioengineering and Biotechnology*, *7*, 289.

Thakur, Y., Grover, A., & Sinha, R. (2023). Differential distribution of macroinvertebrate associated with water quality. *World Water Policy*, *9*(1), 84–112.

Thriene, B., Weege, K. H., & Schulz, S. (2012). Heavy metal resistance of bacteria as biological indicator for environmental pollution. In: *Environmental Hygiene II* (pp. 83–86). Berlin, Heidelberg: Springer.

Trifunović, V. (2021). Vitrification as a method of soil remediation. *Zaštitamaterijala*, *62*(3), 166–179.

Tripathi, M., Singh, D. N., Vikram, S., Singh, V. S., & Kumar, S. (2018). Metagenomic approach towards bioprospection of novel biomolecule (s) and environmental bioremediation. *Annual Research & Review in Biology*, *22*(2), 1–12.

Tsopela, A., Laborde, A., Salvagnac, L., Ventalon, V., Bedel-Pereira, E., Séguy, I., … & Launay, J. (2016). Development of a lab-on-chip electrochemical biosensor for water quality analysis based on microalgal photosynthesis. *Biosensors and Bioelectronics*, *79*, 568–573.

Tucci, M., Grattieri, M., Schievano, A., Cristiani, P., & Minteer, S. D. (2019). Microbial amperometric biosensor for online herbicide detection: Photocurrent inhibition of Anabaena variabilis. *Electrochimica Acta*, *302*, 102–108.

Urnov, F. D., Rebar, E. J., Holmes, M. C., Zhang, H. S., & Gregory, P. D. (2010). Genome editing with engineered zinc finger nucleases. *Nature Reviews Genetics*, *11*(9), 636–646.

Wanapaisan, P., Laothamteep, N., Vejarano, F., Chakraborty, J., Shintani, M., Muangchinda, C., … & Pinyakong, O. (2018). Synergistic degradation of pyrene by five culturable bacteria in a mangrove sediment-derived bacterial consortium. *Journal of Hazardous Materials*, *342*, 561–570.

Xiao, R., Ali, A., Xu, Y., Abdelrahman, H., Li, R., Lin, Y., … & Zhang, Z. (2022). Earthworms as candidates for remediation of potentially toxic elements contaminated soils and mitigating the environmental and human health risks: A review. *Environment International*, *158*, 106924.

Yadav, D., & Kumar, P. (2019). Phytoremediation of hazardous radioactive wastes. In: Hosam El-Din Saleh (Ed.), *Assessment and Management of Radioactive and Electronic Wastes* (pp. 29–43). IntechOpen.

Yadav, R., Kumar, V., Baweja, M., & Shukla, P. (2018). Gene editing and genetic engineering approaches for advanced probiotics: A review. *Critical Reviews in Food Science and Nutrition*, *58*(10), 1735–1746.

Zhang, Y., Cao, B., Jiang, Z., Dong, X., Hu, M., & Wang, Z. (2012). Metabolic ability and individual characteristics of an atrazine-degrading consortium DNC5. *Journal of Hazardous Materials*, *237*, 376–381.

Zhao, X. W., Zhou, M. H., Li, Q. B., Lu, Y. H., He, N., Sun, D. H., & Deng, X. (2005). Simultaneous mercury bioaccumulation and cell propagation by genetically engineered *Escherichia coli*. *Process Biochemistry*, *40*(5), 1611–1616.

Zorža, L., Kalniņš, M., Gudrā, D., Megnis, K., Fridmanis, D., Rapoport, A., & Muter, O. (2022, February). Changes in bacterial community structure in wastewaters in the presence of *Saccharomyces cerevisiae* and benzalkonium chloride. *IOP Conference Series: Earth and Environmental Science*, *987*(1), 012010.

17 Effective Microorganisms as a Key to Biotechnologically Emerging Eco-Remedial Technology for Restoration of Deteriorated Soil

Rani Devi, Alok Kumar Saran, and Pierre Navarro

17.1 INTRODUCTION

Ever-expanding human population, unsustainable agricultural practices, haphazard urbanization and industrialization and moreover globalization and modernization are all human activities that have a considerable negative impact on the world environment, as shown in Figure 17.1. These are the root causes of deforestation, soil degradation, depletion of natural resources, and pollution of water and air. These problems with adverse impacts on the environment will inevitably harm people's health, wellbeing, and quality of life (Wang et al., 2021; Akoijam and Joshi, 2022). As a result, it is vital to regulate, monitor, and adopt procedures that are environmentally sustainable to control and reverse these worst and mostly irreversible negative impacts (VoPham et al., 2017; Vannier, Agler and Hacquard, 2019; Wang et al., 2020).

Out of the entire life forms, although microorganisms are the tiniest residents of our globe but are still responsible for a significant portion of the processes that go into forming and sustaining the ecosystems of our planet (Van et al., 2017; Van Deynze et al., 2018; Sokol et al., 2022). Microbes are often thought of as disease-causing pathogens; nevertheless, there is an entire universe of benign microbes that contribute to the ecological balance and environmental health (Wolfe et al., 2008; Yin et al., 2013). Beneficial bacteria, such as those that may be found in effective microorganisms (EMs), make a positive contribution to the ecological systems in which they are located (Turner et al., 1991; Valero-Rello et al., 2017; Wang et al., 2021; Ware et al., 2021).

DOI: 10.1201/9781032642567-20

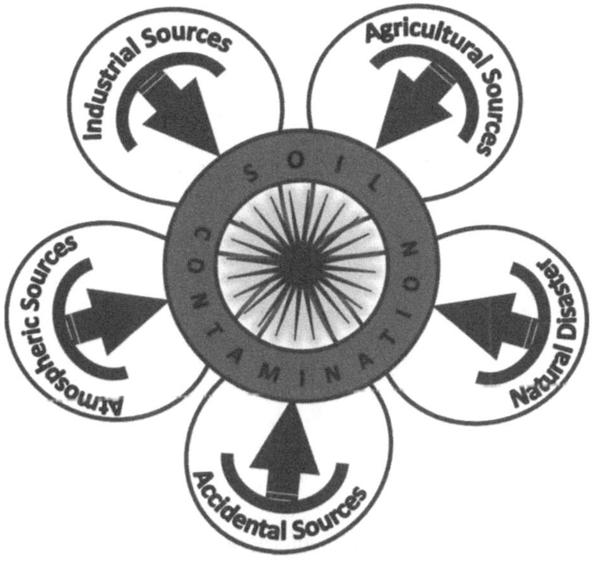

FIGURE 17.1 Soil contamination and major sources of pollution for the same.

17.2 EFFECTIVE MICROORGANISMS

The microorganisms are not a single entity but rather a diversity of microorganisms that collaborate to play their role in the soil environment and work together with other life forms as shown in Figure 17.2. Dr. Teruo Higa, a Japanese scientist, developed the idea of the use of effective and efficient microorganisms in the latter half of

FIGURE 17.2 The way microorganisms work together naturally in soil and interact with other life forms.

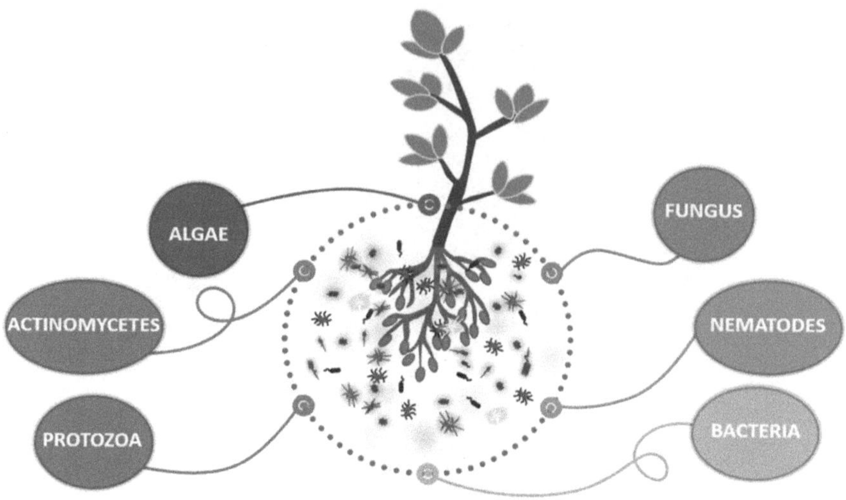

FIGURE 17.3 Way of working of effective microorganisms to achieve the desired goal for soil sustainability and desired productivity.

the 20th century. This is how we interpret the concept of EMs today (Tamang et al., 1988; Sørensen et al., 2005; Smith et al., 2016; Smith and Fratamico, 2018; Sparks et al., 2021).

This consortium's members were chosen after thorough consideration of their potential to increase the fertility of the soil, raise crop yields, and revitalize ecosystems (Stauder et al., 2019; Norris et al., 2020; Sportcs et al., 2021). These microorganisms carry out their functions through a variety of processes, including fermentation, the fixation of nitrogen, and the synthesis of enzymes and organic acids that help maintain soil fertility, soil detoxification, bioremediation, and eco-restoration, as shown in Figure 17.3 (Buntgen et al., 2022; Gould et al., 2022; Sayers et al., 2022; Rafeeq et al., 2023).

17.2.1 Challenges due to EM

Although the widespread use of EMs presents several hurdles despite that these organisms provide a few viable answers to the problems of soil and environmental degradation (Shang et al., 2012; Reganold and Wachter, 2016; Sheinman et al., 2021; Sharma et al., 2022). These difficulties can be faced at three broad platforms the technological, the economic level, and the scientific levels.

The development of technologies for effective manufacture and deployment of EM is one of the technological challenges now being faced. Important factors to consider too to from a technological perspective include achieving consistency in EM production and maintaining the material's sustainability over the long term (Richards et al., 2011; Pineda et al., 2017; Saez-Sandino et al., 2023).

To find solutions to these problems, there needs to be continual research, as well as collaboration between scientists and researchers to make this consortium more viable technologically and economically.

17.2.2 Mechanism of Operation of EM

EMs are members of different taxonomic groups, with a particular emphasis on lactic acid bacteria, yeast, and photosynthetic bacteria. These microorganisms could fulfill a unique function inside the EM consortium, which contributes to the organization's overall effectivity and efficiency (Arny et al., 1976; Lindemann, 1987; Baertlein et al., 1992; Clough and Bent, 1998; Burden et al., 2000; Ambrosio et al., 2017). Let us go a little deeper into the categorization of these essential components in the following sections.

17.2.3 Different Functions of EM

Nutrient Cycling in the Environment: Decomposition of organic debris, release of humic chemicals, and increased availability of nutrients to plants and other microbes are three of the most important aspects of the process of nutrient cycling that takes place within EM, as shown in Figure 17.4 (Cavalieri, 1991; Antonenka et al., 2005; Altieri et al., 2015; Breakfield et al., 2021; Rebello et al., 2021).

Decomposition of Organic Matter: Decomposers particularly yeast and bacteria that produce lactic acid are effective in breaking down more complicated organic molecules into simpler, straightforward ones. As a result of this process of decomposition, key minerals such as nitrogen, phosphorus, and micronutrients are released into the environment. These nutrients are necessary for the growth of plants.

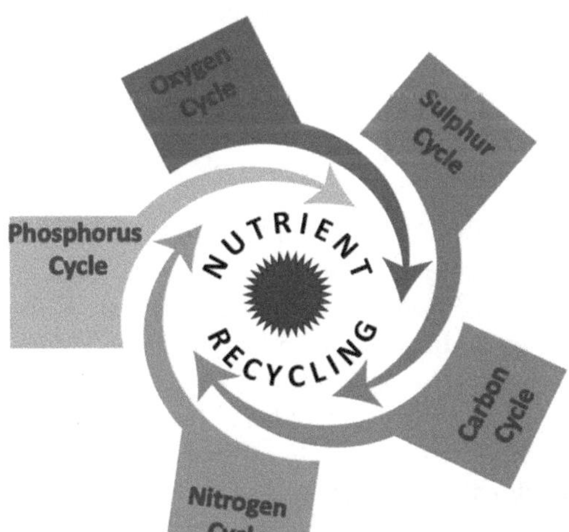

FIGURE 17.4 Different functions of effective microorganisms and nutrient cycling in the environment.

This accessibility of nutrients is vital for agricultural output as well as for the health of the environment, as it enables the growth of vegetation and contributes to the maintenance of ecological equilibrium. Additionally, EMs are effective for the transformation of nonbiodegradable (or refractory) organic dissolved carbon (RODC) into biodegradable organic dissolved carbon (BODC).

Production of Humic Substances: Complex chemical molecules that are abundant in nutrients are known as humic substances, that is, humic and fulvic acids. These compounds are well-known for their capacity to improve the structure of the soil as well as the fertility of the soil (Lindow, 1987; Lampel et al., 1994; Mascarin and Jaronski, 2016; Huang et al., 2017; Newlands, 2018; Reardon, 2019).

These benefits help to contribute to the stability of the soil and minimize erosion, which ultimately makes the soil more resistant to the effects of environmental stresses (Marx, 1987; Ramos et al., 2014; Mus et al., 2016; Hart et al., 2018; Gornish et al., 2020).

Nutrient Availability: Nutrients are liberated from the organic or mineral forms in which they normally exist, making them more readily available for uptake by plants (Marrone, 2002; Puławska et al., 2017; Montaño López et al., 2022).

The enhanced availability of nutrients to plants, in conjunction with the increased nutrient content in humic substances, leads to the development of vegetation that is both healthier and more robust (López et al., 2017; Mohanram and Kumar, 2019; Priyadarshini et al., 2021). This, in turn, contributes to increased crop yields as well as sustainable ecosystems.

Disease Suppression: The capacity of EM to outcompete and inhibit the proliferation of potentially harmful pathogens is an essential component of its activity, particularly in the context of agricultural and ecological settings (Lemaux, 2008; Marcet-Houben and Gabaldón, 2010; Peralta-Yahya et al., 2012; Mitani et al., 2014; Arruda et al., 2021).

Soil Health and Fertility: EM can be applied to soil in several different ways, including by spraying it on, irrigating it with EM solution, or mixing it with compost. The application of EM can assist in the restoration of the microbial variety of soil or enhance soil microorganism's diversity, which, in turn, can improve the health and fertility of the soil (Lu et al., 2008; Meyer et al., 2017; Pei and Schmidt, 2018; Li et al., 2020).

Soil Restoration through EMs: The process of recovering healthy soil by increasing the number of EMs already present in the soil is referred to as "soil restoration through effective microbes." Reduced tilling, crop rotations that include perennial legumes, cover cropping, adding helpful microbes, suppressing harmful microbes, and feeding the soil with microorganisms are all tactics that can be utilized in the process of restoring bacteria in the soil. To bring back the functionality of the soil, it is necessary to maintain a constant supply of fresh root exudates or organic wastes (Peng et al., 2003; Kirchmann et al., 2008; Lupatini et al., 2017; Meyer et al., 2021). For this reason, it is vital to employ ecologically friendly soil management practices like cover crops, and crop rotation in conjunction with the application of EM to ensure long-term soil health and fertility (Mehrabi et al., 2011; Lori et al., 2017; Hebert et al., 2018).

Pest and Disease: The use of an EMs-based approach is a substitute for the use of chemical pesticides and is widely regarded as the most successful and economical option available so far (Bextine et al., 2004; Grace and Ewart, 2008; Chandel and Singh, 2011; Fu et al., 2017; Ducange et al., 2021).

EM is a mixture of naturally existing beneficial microorganisms such as lactic acid bacteria, yeast, and photosynthetic bacteria that can assist enhance soil structure, increase the availability of nutrients, and inhibit dangerous diseases (McCary et al., 2019; Kuzma and Grieger, 2020; Pankievicz et al., 2021). EM can also assist to reduce and minimize the use of synthetic pesticides and fertilizers, both of which have the potential to harm the ecosystem and the health of the soil. However, the effectiveness of EM is determined by several factors, including the kind of soil, the environment, and the management practices that are put into place.

EM in Sustainable Agriculture: These microorganisms have the potential to enhance soil composition, augment the accessibility of nutrients, enhance crop productivity, ameliorate soil health and mitigate reliance on synthetic fertilizers and pesticides and inhibit detrimental diseases (Lawrence and Retchless, 2009; Koch et al., 2011; Pankievicz et al., 2019; Mawarda et al., 2020). *Rhizobia*, *Mycorrhizae*, *Azospirillum*, *Bacillus*, *Pseudomonas*, *Trichoderma*, and *Streptomyces* species are among the often advocated and employed advantageous microorganisms in global agriculture, as shown in Figure 17.5 (Gilbertson et al., 2003; Herridge et al., 2008; Lau and Suwa, 2016; Panke-Buisse et al., 2017; Kong et al., 2018; George et al., 2022; Sáez-Sandino et al., 2023).

17.3 CASE STUDIES OF USE OF EMs

EMs have been employed in several studies that have been carried out in various regions of the world to enhance agricultural productivity while improving the general

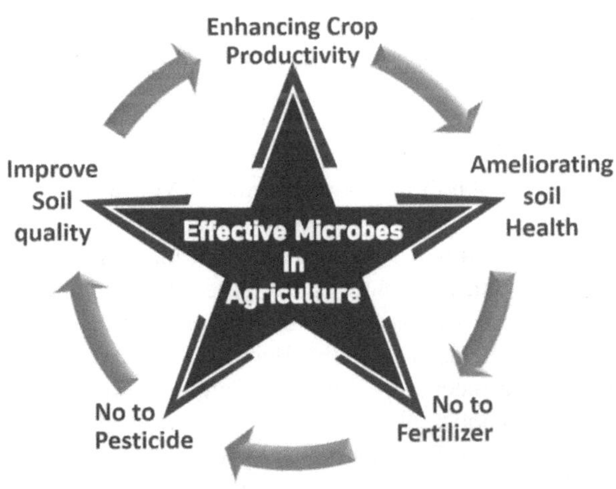

FIGURE 17.5 Effective microorganisms and their role for maintaining soil fertility and sustainable production

health of the soil. These studies have been carried out to improve the overall health of the soil (VoPham et al., 2017; Sharma et al., 2021, 2022; Sheinman et al., 2021). The following is a collection of examples that illustrate how EM can be employed in a variety of agricultural settings.

17.3.1 Regional Case Studies

A research investigation carried out in the region of Haryana, India, involved the identification and isolation of bacteria that exhibit tolerance to high levels of salt. These bacteria were specifically obtained from saline soils located in the districts of Jhajjar and Panipat, Haryana, India. The bacteria with the potential of salt tolerance were assessed for their ability to promote plant growth through activities such as phosphate solubilization, hydrogen cyanide (HCN) generation, indole acetic acid (IAA) synthesis, and ammonia production. These isolates exhibited notable salt tolerance and showed the potential to promote plant growth. The use of rhizobacteria that exhibit tolerance to salt and promote plant development, often known as plant growth-promoting rhizobacteria (PGPR), has been identified as a potential strategy to mitigate the adverse impacts of salt stress on plants. These PGPR strains, isolated from saline soil have demonstrated the ability to enhance many physiological functions of plants, including growth and yield, while also providing resistance against diseases (Sharma et al., 2021).

Tomato plants' development and production were improved with the help of EM in another experiment. The research was carried out at ICAR, Central Soil Salinity Research Institute, Karnal, Haryana, India. In this study, the use of EM resulted in much greater plant growth and yield than the control group. The findings of this research show that EM can increase harvest yields in sodic soil (Damodaran et al., 2023).

Hence, the utilization of microbial inoculants to mitigate stresses and augment plant yield presents a viable and eco-friendly approach for effectively managing saline soil and improving agricultural output at a reduced cost.

17.3.2 National Case Studies

In a study conducted in Andhra Pradesh, India, EM was utilized to boost the development and harvest of maize plants. The investigation was carried out at S.V. Agricultural College in Tirupati, Andhra Pradesh, India, in the agronomy department. When compared with the control group, the growth and yield of the maize crops treated with EM rose dramatically. The findings of this research showed that EM can increase harvest yields potentially (Babu et al., 2023).

As part of a research project at the Indian Agricultural Research Institute, Delhi, India, EM was utilized to boost the development and productivity of rice plants. The research was carried out at the Division of Microbiology of IARI. The results showed that the application of EM considerably boosted the growth and yield of rice crops compared with the control group. The findings of this research show that EM can increase harvest yields (Prasanna et al., 2012).

17.3.3 International Case Studies

Researchers in the Department of Plant Health, Rural Engineering, and Soils, São Paulo State University, Brazil, discovered that some plant growth-promoting bacteria (PGPBs) may have the potential to enhance phosphate availability to plants and produce phytohormones which contribute to crop development, quality, and yield. Hence utilizing the EM consortium may improve both the development of sugarcane crops and the yield of those crops. When compared with the control group, the findings showed that the utilization of EM resulted in a significant increase in the growth and production of sugarcane crops (Rosa et al., 2022).

Researchers in Italy discovered that using EM helped their tomato plants grow larger and produce more fruit when applied to those plants. When compared with the control group 10, the findings demonstrated that the use of EM had a significant impact on increasing both the growth and yield of tomato crops (Sellitto et al., 2019).

17.4 FUTURE PROSPECTS OF EMs

There is promising evidence that EMs can increase crop productivity, improve soil fertility, and decrease reliance on synthetic fertilizers and pesticides. Thus, the application of EM has excellent future potential. The use of EM is projected to rise in the future years due to the rise in the demand and necessity of sustainable farming practices. Restoring degraded ecosystems and cleaning up polluted soils are two further uses for EM as shown in Figure 17.6.

Greenhouse gas emissions from agricultural soils can be reduced with the use of EM, which is a useful tool in the fight against climate change (Ligon et al., 2000; Oliver and Solomon, 2008; Lankau, 2011; Kikuchi et al., 2012; Herzig et al., 2014; Massart et al., 2015; Lovett and St. Leger, 2018; Schelkle and Galland, 2020; George et al., 2022).

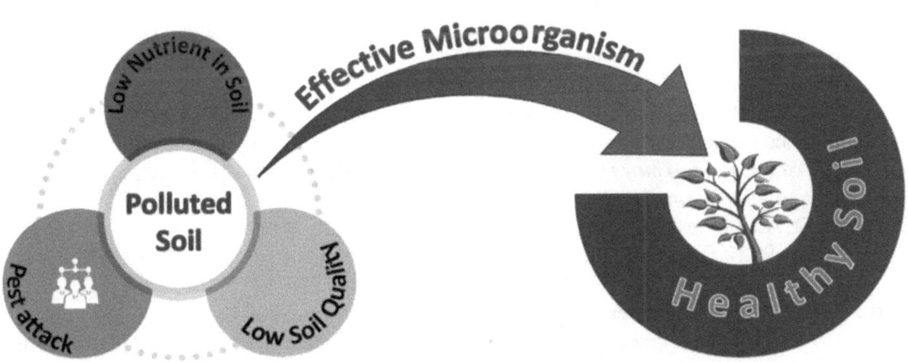

FIGURE 17.6 Effective microorganisms and their efficient role in restoring degraded ecosystems and cleaning up polluted soils and better crop productivity.

17.5 CONCLUSION

With the aid of EM, more sustainable agricultural practices that can withstand the effects of climate change can be established. These practices can be developed by enhancing the health and fertility of the soil and reducing reliance on synthetic fertilizers and pesticides. The introduction of EM has the potential to transform the microbiology of soil. The use of EM makes it possible to extract nutrients necessary for plant growth from organic matter that has decomposed. Two more applications for EM are the restoration of degraded ecosystems and the clean-up of dirty soils. As a result, the widespread application of EM in agricultural settings is required to facilitate the establishment of resilient food systems.

REFERENCES

Abu-Qaoud, H., Al-Fares, H., Shtaya, M. J., & Shawarb, N. (2021). Effect of effective microorganisms on wheat growth under salt stress condition. *Chilean Journal of Agricultural Research*, *81*(3), 351–356.

Akoijam, N., & Joshi, S. R. (2022). Conservation metagenomics: Understanding microbiomes for biodiversity sustenance and conservation. In: *Molecular Genetics and Genomics Tools in Biodiversity Conservation*, Singapore: Springer Nature (pp. 31–61).

Altieri, M. A., Nicholls, C. I., Henao, A., & Lana, M. A. (2015). Agroecology and the design of climate change-resilient farming systems. *Agronomy for Sustainable Development*, *35*(3), 869–890.

Ambrosio, R., Ortiz-Marquez, J. C. F., & Curatti, L. (2017). Metabolic engineering of a diazotrophic bacterium improves ammonium release and biofertilization of plants and microalgae. *Metabolic Engineering*, *40*, 59–68.

Antonenka, U., Nölting, C., Heesemann, J., & Rakin, A. (2005). Horizontal transfer of *Yersinia* high-pathogenicity island by the conjugative RP4 *attB* target-presenting shuttle plasmid. *Molecular Microbiology*, *57*(3), 727–734. https://doi.org/10.1111/j.1365-2958.2005.04722.x

Arny, D. C., Lindow, S. E., & Upper, C. D. (1976). Frost sensitivity of *Zea mays* increased by application of *Pseudomonas syringae*. *Nature*, *262*(5566), 282–284.

Arruda, B., George, P. B. L., Robin, A., De L. C. Mescolotti, D., Herrera, W. F. B., Jones, D. L., & Andreote, F. D. (2021). Manipulation of the soil microbiome regulates the colonization of plants by arbuscular mycorrhizal fungi. *Mycorrhiza*, *31*(5), 545–558. https://doi.org/10.1007/s00572-021-01044-3

Babu, P. H., Lalitha, D. K. J., Sreekanth, D. B., Kumar, D. M. S., & Umar, D. N. (2023). Influence of plant growth promoting rhizo microorganisms on yield attributes and grain yield of maize (*Zea mays* L.) under water stress in Krishna Zone of Andhra Pradesh. *The Pharma Innovation*, *12*(6), 4481–4484.

Baertlein, D. A., Lindow, S. E., Panopoulos, N. J., Lee, S. P., Mindrinos, M. N., & Chen, T. H. H. (1992). Expression of a bacterial ice nucleation gene in plants. *Plant Physiology*, *100*(4), 1730–1736. https://doi.org/10.1104/pp.100.4.1730

Bextine, B., Lauzon, C., Potter, S., Lampe, D., & Miller, T. A. (2004). Delivery of a genetically marked *Alcaligenes* sp. to the glassy-winged sharpshooter for use in a paratransgenic control strategy. *Current Microbiology*, *48*(5), 327–331. https://doi.org/10.1007/s00284-003-4178-2

Breakfield, N. W., Collett, D., & Frodyma, M. E. (2021). Plant growth-promoting microbes: An industry view. *Emerging Topics in Life Sciences*, *5*(2), 317–324. https://doi.org/10.1042/ETLS20200313

Buntgen, U., Piermattei, A., Krusic, P. J., Esper, J., Sparks, T., & Crivellaro, A. (2022). Plants in the UK flower a month earlier under recent warming. *Proceedings of the Royal Society B: Biological Sciences*, *289*(1968), 20212456.

Burden, J. P., Hails, R. S., Windass, J. D., Suner, M.-M., & Cory, J. S. (2000). Infectivity, Speed of kill, and productivity of a baculovirus expressing the itch mite toxin Txp-1 in second and fourth instar larvae of *Trichoplusia ni*. *Journal of Invertebrate Pathology*, *75*(3), 226–236. https://doi.org/10.1006/jipa.1999.4921

Cavalieri, L. F. (1991). Scaling-up field testing of modified microorganisms. *BioScience*, *41*(8), 568–574. https://doi.org/10.2307/1311610

Chandel, A. K., & Singh, O. V. (2011). Weedy lignocellulosic feedstock and microbial metabolic engineering: Advancing the generation of 'Biofuel.' *Applied Microbiology and Biotechnology*, *89*(5), 1289–1303. https://doi.org/10.1007/s00253-010-3057-6

Clough, S. J., & Bent, A. F. (1998). Floral dip: A simplified method for *Agrobacterium* -mediated transformation of *Arabidopsis thaliana*. *The Plant Journal*, *16*(6), 735–743. https://doi.org/10.1046/j.1365-313x.1998.00343.x

Damodaran, T., Jha, S. K., Kumari, S., Gupta, G., Mishra, V. K., Sharma, P. C., Gopal, R., Singh, A., & Jat, H. S. (2023). Development of halotolerant microbial consortia for salt stress mitigation and sustainable tomato production in sodic soils: An enzyme mechanism approach. *Sustainability*, *15*(6), 5186. https://doi.org/10.3390/su15065186

Ducange, P., Verdina, T., Stiro, F., Grottola, A., Orlando, G., Delvecchio, G., & Mastropasqua, R. (2021). *Beauveria bassiana* keratitis: Management of an atypical clinical presentation. *Medical Mycology Case Reports*, *33*, 1–4. https://doi.org/10.1016/j.mmcr.2021.05.001

Florea, S., Jaromczyk, J., & Schardl, C. L. (2021). Non-transgenic CRISPR-mediated knockout of entire ergot alkaloid gene clusters in slow-growing asexual polyploid fungi. *Toxins*, *13*(2), 153. https://doi.org/10.3390/toxins13020153

Fu, L., Penton, C. R., Ruan, Y., Shen, Z., Xue, C., Li, R., & Shen, Q. (2017). Inducing the rhizosphere microbiome by biofertilizer application to suppress banana Fusarium wilt disease. *Soil Biology and Biochemistry*, *104*, 39–48. https://doi.org/10.1016/j.soilbio.2016.10.008

George, D. R., Hornstein, E. D., Clower, C. A., Coomber, A. L., Dillard, D., Mugwanya, N., Pezzini, D. T., & Rozowski, C. (2022). Lessons for a SECURE future: Evaluating diversity in crop biotechnology across regulatory regimes. *Frontiers in Bioengineering and Biotechnology*, *10*, 886765.

Gilbertson, M., Haffner, G. D., Drouillard, K. G., Albert, A., & Dixon, B. (2003). Immunosuppression in the northern leopard frog (*Rana pipiens*) induced by pesticide exposure. *Environmental Toxicology and Chemistry*, *22*(1), 101–110. https://doi.org/10.1002/etc.5620220113

Gornish, E. S., Franklin, K., Rowe, J., & Barberán, A. (2020). Buffelgrass invasion and glyphosate effects on desert soil microbiome communities. *Biological Invasions*, *22*(8), 2587–2597. https://doi.org/10.1007/s10530-020-02268-8

Gould, F., Amasino, R. M., Brossard, D., Buell, C. R., Dixon, R. A., Falck-Zepeda, J. B., Gallo, M. A., Giller, K. E., Glenna, L. L., Griffin, T., Magraw, D., Mallory-Smith, C., Pixley, K. V., Ransom, E. P., Stelly, D. M., & Stewart, C. N. (2022). Toward product-based regulation of crops. *Science*, *377*(6610), 1051–1053.

Grace, J. K., & Ewart, D. M. (2008). Recombinant cells of *Pseudomonas fluorescens*: A highly palatable encapsulation for delivery of genetically engineered toxins to subterranean termites (Isoptera: Rhinotermitidae). *Letters in Applied Microbiology*, *23*(3), 183–186. https://doi.org/10.1111/j.1472-765X.1996.tb00060.x

Hart, M. M., Antunes, P. M., Chaudhary, V. B., & Abbott, L. K. (2018). Fungal inoculants in the field: Is the reward greater than the risk? *Functional Ecology*, *32*(1), 126–135. https://doi.org/10.1111/1365-2435.12976

Hebert, P. D. N., Braukmann, T. W. A., Prosser, S. W. J., Ratnasingham, S., deWaard, J. R., Ivanova, N. V., Janzen, D. H., Hallwachs, W., Naik, S., Sones, J. E., & Zakharov, E. V. (2018). A sequel to Sanger: Amplicon sequencing that scales. *BMC Genomics*, *19*(1), 219. https://doi.org/10.1186/s12864-018-4611-3

Herridge, D. F., Peoples, M. B., & Boddey, R. M. (2008). Global inputs of biological nitrogen fixation in agricultural systems. *Plant and Soil*, *311*(1–2), 1–18. https://doi.org/10.1007/s11104-008-9668-3

Herzig, V., Bende, N. S., Alam, Md . S., Tedford, H. W., Kennedy, R. M., & King, G. F. (2014). Methods for deployment of spider venom peptides as bioinsecticides. In: *Advances in Insect Physiology*, Amsterdam: Elsevier (Vol. 47, pp. 389–411). https://doi.org/10.1016/B978-0-12-800197-4.00008-7

Huang, W., Tsai, L., Li, Y., Hua, N., Sun, C., & Wei, C. (2017). Widespread of horizontal gene transfer in the human genome. *BMC Genomics*, *18*(1), 274. https://doi.org/10.1186/s12864-017-3649-y

Kikuchi, Y., Hayatsu, M., Hosokawa, T., Nagayama, A., Tago, K., & Fukatsu, T. (2012). Symbiont-mediated insecticide resistance. *Proceedings of the National Academy of Sciences*, *109*(22), 8618–8622. https://doi.org/10.1073/pnas.1200231109

Kirchmann, H., Thorvaldsson, G., Bergström, L., Gerzabek, M., Andrén, O., Eriksson, L.-O., & Winninge, M. (2008). Fundamentals of organic agriculture: Past and present. In: H. Kirchmann & L. Bergström (Eds.), *Organic Crop Production – Ambitions and Limitations*, Berlin: Springer Netherlands (pp. 13–37). https://doi.org/10.1007/978-1-4020-9316-6_2

Koch, A. M., Antunes, P. M., Kathryn Barto, E., Cipollini, D., Mummey, D. L., & Klironomos, J. N. (2011). The effects of arbuscular mycorrhizal (AM) fungal and garlic mustard introductions on native AM fungal diversity. *Biological Invasions*, *13*(7), 1627–1639. https://doi.org/10.1007/s10530-010-9920-7

Kong, Z., Hart, M., & Liu, H. (2018). Paving the way from the lab to the field: Using synthetic microbial consortia to produce high-quality crops. *Frontiers in Plant Science*, 9, 1467. https://doi.org/10.3389/fpls.2018.01467

Kuzma, J., & Grieger, K. (2020). Community-led governance for gene-edited crops. *Science*, *370*(6519), 916–918. https://doi.org/10.1126/science.abd1512

Lampel, J. S., Canter, G. L., Dimock, M. B., Kelly, J. L., Anderson, J. J., Uratani, B. B., Foulke, J. S., & Turner, J. T. (1994). Integrative cloning, expression, and stability of the cryIA(c) gene from Bacillus thuringiensis subsp. Kurstaki in a recombinant strain of Clavibacter xyli subsp. Cynodontis. *Applied and Environmental Microbiology*, *60*(2), 501–508. https://doi.org/10.1128/aem.60.2.501-508.1994

Lankau, R. A. (2011). Resistance and recovery of soil microbial communities in the face of *Alliaria petiolata* invasions. *New Phytologist*, *189*(2), 536–548. https://doi.org/10.1111/j.1469-8137.2010.03481.x

Lau, J. A., & Suwa, T. (2016). The changing nature of plant–microbe interactions during a biological invasion. *Biological Invasions*, *18*(12), 3527–3534. https://doi.org/10.1007/s10530-016-1245-8

Lawrence, J. G., & Retchless, A. C. (2009). The interplay of homologous recombination and horizontal gene transfer in bacterial speciation. In: M. B. Gogarten, J. P. Gogarten, & L. C. Olendzenski (Eds.), *Horizontal Gene Transfer*, London: Humana Press (Vol. 532, pp. 29–53). https://doi.org/10.1007/978-1-60327-853-9_3

Lemaux, P. G. (2008). Genetically engineered plants and foods: A scientist's analysis of the issues (Part I). *Annual Review of Plant Biology*, *59*(1), 771–812. https://doi.org/10.1146/annurev.arplant.58.032806.103840

Li, L.-G., Huang, Q., Yin, X., & Zhang, T. (2020). Source tracking of antibiotic resistance genes in the environment—Challenges, progress, and prospects. *Water Research*, *185*, 116127. https://doi.org/10.1016/j.watres.2020.116127

Ligon, J. M., Hill, D. S., Hammer, P. E., Torkewitz, N. R., Hofmann, D., Kempf, H.-J., & Pée, K.-H. V. (2000). Natural products with antifungal activity from Pseudomonas biocontrol bacteria. *Pest Management Science*, *56*(8), 688–695.

Lindemann, J. (1987). Competition between ice nucleation-active wild type and ice nucleation-deficient deletion mutant strains of *Pseudomonas syringae* and *P. fluorescens* biovar i and biological control of frost injury on strawberry blossoms. *Phytopathology*, *77*(6), 882. https://doi.org/10.1094/Phyto-77-882

Lindow, S. E. (1987). Competitive exclusion of epiphytic bacteria by ice – *Pseudomonas syringae* mutants. *Applied and Environmental Microbiology*, *53*(10), 2520–2527. https://doi.org/10.1128/aem.53.10.2520-2527.1987

López, J. H., Krainer, S., Engert, A., Schuehly, W., Riessberger-Gallé, U., & Crailsheim, K. (2017). Sublethal pesticide doses negatively affect survival and the cellular responses in American foulbrood-infected honeybee larvae. *Scientific Reports*, *7*(1), 40853. https://doi.org/10.1038/srep40853

Lori, M., Symnaczik, S., Mäder, P., De Deyn, G., & Gattinger, A. (2017). Organic farming enhances soil microbial abundance and activity—A meta-analysis and meta-regression. Plos One, 12(7), e0180442. https://doi.org/10.1371/journal.pone.0180442

Lovett, B., & St. Leger, R. J. (2018). Genetically engineering better fungal biopesticides. *Pest Management Science*, *74*(4), 781–789. https://doi.org/10.1002/ps.4734

Lu, D., Pava-Ripoll, M., Li, Z., & Wang, C. (2008). Insecticidal evaluation of *Beauveria bassiana* engineered to express a scorpion neurotoxin and a cuticle degrading protease. *Applied Microbiology and Biotechnology*, *81*(3), 515–522. https://doi.org/10.1007/s00253-008-1695-8

Lupatini, M., Korthals, G. W., De Hollander, M., Janssens, T. K. S., & Kuramae, E. E. (2017). Soil microbiome is more heterogeneous in organic than in conventional farming system. *Frontiers in Microbiology*, 7. https://doi.org/10.3389/fmicb.2016.02064

Marcet-Houben, M., & Gabaldón, T. (2010). Acquisition of prokaryotic genes by fungal genomes. *Trends in Genetics*, *26*(1), 5–8. https://doi.org/10.1016/j.tig.2009.11.007

Marrone, P. G. (2002). An effective biofungicide with novel modes of action. *Pesticide Outlook*, *13*(5), 193–194. https://doi.org/10.1039/b209431m

Marx, J. L. (1987). Assessing the risks of microbial release: As genetically engineered microbes move into the field, risk assessment becomes a fact of life for biotechnology researchers. *Science*, *237*(4821), 1413–1417. https://doi.org/10.1126/science.3114879

Mascarin, G. M., & Jaronski, S. T. (2016). The production and uses of *Beauveria bassiana* as a microbial insecticide. *World Journal of Microbiology and Biotechnology*, *32*(11), 177. https://doi.org/10.1007/s11274-016-2131-3

Massart, S., Martinez-Medina, M., & Jijakli, M. H. (2015). Biological control in the microbiome era: Challenges and opportunities. *Biological Control*, *89*, 98–108. https://doi.org/10.1016/j.biocontrol.2015.06.003

Mawarda, P. C., Le Roux, X., Dirk Van Elsas, J., & Salles, J. F. (2020). Deliberate introduction of invisible invaders: A critical appraisal of the impact of microbial inoculants on soil microbial communities. *Soil Biology and Biochemistry*, *148*, 107874. https://doi.org/10.1016/j.soilbio.2020.107874

McCary, M. A., Zellner, M., & Wise, D. H. (2019). The role of plant–mycorrhizal mutualisms in deterring plant invasions: Insights from an individual-based model. *Ecology and Evolution*, *9*(4), 2018–2030. https://doi.org/10.1002/ece3.4892

Mehrabi, R., Bahkali, A. H., Abd-Elsalam, K. A., Moslem, M., Ben M'Barek, S., Gohari, A. M., Jashni, M. K., Stergiopoulos, I., Kema, G. H. J., & De Wit, P. J. G. M. (2011). Horizontal gene and chromosome transfer in plant pathogenic fungi affecting host range. *FEMS Microbiology Reviews*, *35*(3), 542–554. https://doi.org/10.1111/j.1574-6976.2010.00263.x

Meyer, M., Cox, J. A., Hitchings, M. D. T., Burgin, L., Hort, M. C., Hodson, D. P., & Gilligan, C. A. (2017). Quantifying airborne dispersal routes of pathogens over continents to safeguard global wheat supply. *Nature Plants*, *3*(10), 780–786. https://doi.org/10.1038/s41477-017-0017-5

Meyer, R. S., Ramos, M. M., Lin, M., Schweizer, T. M., Gold, Z., Ramos, D. R., Shirazi, S., Kandlikar, G., Kwan, W.-Y., Curd, E. E., Freise, A., Parker, J. M., Sexton, J. P., Wetzer, R., Pentcheff, N. D., Wall, A. R., Pipes, L., Garcia-Vedrenne, A., Mejia, M. P., & … Wayne, R. K. (2021). The CALeDNA program: Citizen scientists and researchers inventory California's biodiversity. *California Agriculture*, *75*(1), 20–32.

Mitani, A., Shiraishi, A., Miyamoto, H., Sunada, A., Ueda, A., Asari, S., Zheng, X., Yamamoto, Y., Hara, Y., & Ohashi, Y. (2014). Fungal keratitis caused by *Beauveria bassiana*: Drug and temperature sensitivity profiles: A case report. *BMC Research Notes*, *7*(1), 677. https://doi.org/10.1186/1756-0500-7-677

Mohanram, S., & Kumar, P. (2019). Rhizosphere microbiome: Revisiting the synergy of plant-microbe interactions. *Annals of Microbiology*, *69*(4), 307–320. https://doi.org/10.1007/s13213-019-01448-9

Montaño López, J., Duran, L., & Avalos, J. L. (2022). Physiological limitations and opportunities in microbial metabolic engineering. *Nature Reviews Microbiology*, *20*(1), 35–48. https://doi.org/10.1038/s41579-021-00600-0

Mus, F., Crook, M. B., Garcia, K., Garcia Costas, A., Geddes, B. A., Kouri, E. D., Paramasivan, P., Ryu, M.-H., Oldroyd, G. E. D., Poole, P. S., Udvardi, M. K., Voigt, C. A., Ané, J.-M., & Peters, J. W. (2016). Symbiotic nitrogen fixation and the challenges to its extension to nonlegumes. *Applied and Environmental Microbiology*, *82*(13), 3698–3710.

Newlands, N. K. (2018). Model-based forecasting of agricultural crop disease risk at the regional scale, integrating airborne inoculum, environmental, and satellite-based monitoring data. *Frontiers in Environmental Science*, *6*, 63.

Norris, A. L., Lee, S. S., Greenlees, K. J., Tadesse, D. A., Miller, M. F., & Lombardi, H. A. (2020). Template plasmid integration in germline genome-edited cattle. *Nature Biotechnology*, *38*(2), 163–164. https://doi.org/10.1038/s41587-019-0394-6

Oliver, R. P., & Solomon, P. S. (2008). Recent fungal diseases of crop plants: Is lateral gene transfer a common theme. *Molecular Plant-Microbe Interactions*, *21*(3), 287–293. https://doi.org/10.1094/MPMI-21-3-0287

Panke-Buisse, K., Lee, S., & Kao-Kniffin, J. (2017). Cultivated sub-populations of soil microbiomes retain early flowering plant trait. *Microbial Ecology*, *73*(2), 394–403. https://doi.org/10.1007/s00248-016-0846-1

Pankievicz, V. C. S., Do Amaral, F. P., Ané, J.-M., & Stacey, G. (2021). Diazotrophic bacteria and their mechanisms to interact and benefit cereals. *Molecular Plant-Microbe Interactions*, *34*(5), 491–498. https://doi.org/10.1094/MPMI-11-20-0316-FI

Pankievicz, V. C. S., Irving, T. B., Maia, L. G. S., & Ané, J.-M. (2019). Are we there yet? The long walk towards the development of efficient symbiotic associations between nitrogen-fixing bacteria and non-leguminous crops. BMC Biology, *17*(1), 99. https://doi.org/10.1186/s12915-019-0710-0

Pei, L., & Schmidt, M. (2018). Fast-growing engineered microbes: New concerns for gain-of-function research? *Frontiers in Genetics*, *9*, 207. https://doi.org/10.3389/fgene.2018.00207

Peng, R., Xiong, A., Li, X., Fuan, H., & Yao, Q. (2003). A -endotoxin encoded in Pseudomonas fluorescens displays a high degree of insecticidal activity. *Applied Microbiology and Biotechnology*, *63*(3), 300–306. https://doi.org/10.1007/s00253-003-1343-2

Peralta-Yahya, P. P., Zhang, F., Del Cardayre, S. B., & Keasling, J. D. (2012). Microbial engineering for the production of advanced biofuels. *Nature*, *488*(7411), 320–328. https://doi.org/10.1038/nature11478

Pineda, A., Kaplan, I., & Bezemer, T. M. (2017). Steering soil microbiomes to suppress aboveground insect pests. *Trends in Plant Science*, *22*(9), 770–778. https://doi.org/10.1016/j.tplants.2017.07.002

Prasanna, R., Joshi, M., Rana, A., Shivay, Y. S., & Nain, L. (2012). Influence of co-inoculation of bacteria-cyanobacteria on crop yield and C–N sequestration in soil under rice crop. *World Journal of Microbiology and Biotechnology*, *28*(3), 1223–1235. https://doi.org/10.1007/s11274-011-0926-9

Priyadarshini, P., Choudhury, S., Tilgam, J., Bharati, A., & Sreeshma, N. (2021). Nitrogen fixing cereal: A rising hero towards meeting food security. *Plant Physiology and Biochemistry*, *167*, 912–920. https://doi.org/10.1016/j.plaphy.2021.09.012

Puławska, J., Kałużna, M., Warabieda, W., & Mikiciński, A. (2017). Comparative transcriptome analysis of a lowly virulent strain of Erwinia amylovora in shoots of two apple cultivars – susceptible and resistant to fire blight. *BMC Genomics*, *18*(1), 868. https://doi.org/10.1186/s12864-017-4251-z

Rafeeq, H., Afsheen, N., Rafique, S., Arshad, A., Intisar, M., Hussain, A., Bilal, M., & Iqbal, H. M. N. (2023). Genetically engineered microorganisms for environmental remediation. *Chemosphere*, *310*, 136751. https://doi.org/10.1016/j.chemosphere.2022.136751

Ram, H., Malik, S. S., Dhaliwal, S. S., Kumar, B., & Singh, Y. (2015). Growth and productivity of wheat affected by phosphorus-solubilizing fungi and phosphorus levels. *Plant, Soil and Environment*, *61*(3), 122–126. https://doi.org/10.17221/982/2014-PSE

Ramos, L. S., Lehman, B. L., Peter, K. A., & McNellis, T. W. (2014). Mutation of the *Erwinia amylovora argD* gene causes arginine auxotrophy, nonpathogenicity in apples, and reduced virulence in pears. *Applied and Environmental Microbiology*, *80*(21), 6739–6749. https://doi.org/10.1128/AEM.02404-14

Reardon, S. (2019). Gene edits to 'CRISPR babies' might have shortened their life expectancy. *Nature*, *570*(7759), 16–17. https://doi.org/10.1038/d41586-019-01739-w

Rebello, S., Nathan, V. K., Sindhu, R., Binod, P., Awasthi, M. K., & Pandey, A. (2021). Bioengineered microbes for soil health restoration: Present status and future. *Bioengineered*, *12*(2), 12839–12853. https://doi.org/10.1080/21655979.2021.2004645

Reganold, J. P., & Wachter, J. M. (2016). Organic agriculture in the twenty-first century. *Nature Plants*, *2*(2), 15221. https://doi.org/10.1038/nplants.2015.221

Richards, T. A., Dacks, J. B., Jenkinson, J. M., Thornton, C. R., & Talbot, N. J. (2006). Evolution of filamentous plant pathogens: Gene exchange across eukaryotic kingdoms. *Current Biology*, *16*(18), 1857–1864.

Richards, T. A., Leonard, G., Soanes, D. M., & Talbot, N. J. (2011). Gene transfer into the fungi. *Fungal Biology Reviews*, *25*(2), 98–110.

Rodgers, V. L., Wolfe, B. E., Werden, L. K., & Finzi, A. C. (2008). The invasive species *Alliaria petiolata* (garlic mustard) increases soil nutrient availability in northern hardwood-conifer forests. *Oecologia*, *157*(3), 459–471. https://doi.org/10.1007/s00442-008-1089-8

Rosa, P. A. L., Galindo, F. S., Oliveira, C. E. D. S., Jalal, A., Mortinho, E. S., Fernandes, G. C., Marega, E. M. R., Buzetti, S., & Teixeira Filho, M. C. M. (2022). Inoculation with Plant growth-promoting bacteria to reduce phosphate fertilization requirement and enhance technological quality and yield of sugarcane. *Microorganisms*, 10(1), 192. https://doi.org/10.3390/microorganisms10010192

Sandino, T., Delgado-Baquerizo, M., Egidi, E., & Singh, B. K. (2023). New microbial tools to boost restoration and soil organic matter. *Microbial Biotechnology*, *16*(11), 2019–2025. https://doi.org/10.1111/1751-7915.14325

Sayers, E. W., Bolton, E. E., Brister, J. R., Canese, K., Chan, J., Comeau, D. C., Connor, R., Funk, K., Kelly, C., Kim, S., Madej, T., Marchler-Bauer, A., Lanczycki, C., Lathrop, S., Lu, Z., Thibaud-Nissen, F., Murphy, T., Phan, L., Skripchenko, Y., Tse, T., Wang, J., Williams, R., Trawick, B. W., Pruitt, K. D., & Sherry, S. T. (2022). Database resources of the national center for biotechnology information. *Nucleic Acids Research*, 50(D1), D20–D26. https://doi.org/10.1093/nar/gkab1112

Schelkle, B., & Galland, Q. (2020). Microbiome research: Open communication today, microbiome applications in the future. *Microorganisms*, 8(12), 1960. https://doi.org/10.3390/microorganisms8121960

Sellitto, V. M., Golubkina, N. A., Pietrantonio, L., Cozzolino, E., Cuciniello, A., Cenvinzo, V., Florin, I., & Caruso, G. (2019). Tomato yield, quality, mineral composition and antioxidants as affected by beneficial microorganisms under soil salinity induced by balanced nutrient solutions. *Agriculture*, *9*(5), 110.

Shang, Y., Duan, Z., Huang, W., Gao, Q., & Wang, C. (2012). Improving UV resistance and virulence of *Beauveria bassiana* by genetic engineering with an exogenous tyrosinase gene. *Journal of Invertebrate Pathology*, *109*(1), 105–109.

Sharma, A., Dev, K., Sourirajan, A., & Choudhary, M. (2021). Isolation and characterization of salt-tolerant bacteria with plant growth-promoting activities from saline agricultural fields of Haryana, India. *Journal of Genetic Engineering and Biotechnology*, *19*(1), 99. https://doi.org/10.1186/s43141-021-00186-3

Sharma, P., Bano, A., Singh, S. P., Sharma, S., Xia, C., Nadda, A. K., Lam, S. S., & Tong, Y. W. (2022). Engineered microbes as effective tools for the remediation of polyaromatic aromatic hydrocarbons and heavy metals. *Chemosphere*, *306*, 135538. https://doi.org/10.1016/j.chemosphere.2022.135538

Sheinman, M., Arkhipova, K., Arndt, P. F., Dutilh, B. E., Hermsen, R., & Massip, F. (2021). Identical sequences found in distant genomes reveal frequent horizontal transfer across the bacterial domain. *Elife*, 10, e62719. https://doi.org/10.7554/eLife.62719

Smith, J. L., & Fratamico, P. M. (2018). Emerging and re-emerging foodborne pathogens. *Foodborne Pathogens and Disease*, *15*(12), 737–757. https://doi.org/10.1089/fpd.2018.2493

Smith, R. J., Paterson, J. S., Launer, E., Tobe, S. S., Morello, E., Leijs, R., Marri, S., & Mitchell, J. G. (2016). Stygofauna enhance prokaryotic transport in groundwater ecosystems. Scientific Reports, *6*(1), 32738. https://doi.org/10.1038/srep32738

Sokol, N. W., Slessarev, E., Marschmann, G. L., Nicolas, A., Blazewicz, S. J., Brodie, E. L., Firestone, M. K., Foley, M. M., Hestrin, R., Hungate, B. A., Koch, B. J., Stone, B. W., Sullivan, M. B., Zablocki, O., LLNL Soil Microbiome Consortium, Trubl, G., McFarlane, K., Stuart, R., Nuccio, E., … & Pett-Ridge, J. (2022). Life and death in the soil microbiome: How ecological processes influence biogeochemistry. *Nature Reviews Microbiology*, *20*(7), 415–430. https://doi.org/10.1038/s41579-022-00695-z

Sørensen, S. J., Bailey, M., Hansen, L. H., Kroer, N., & Wuertz, S. (2005). Studying plasmid horizontal transfer in situ: A critical review. *Nature Reviews Microbiology*, *3*(9), 700–710. https://doi.org/10.1038/nrmicro1232

Sparks, T. C., Storer, N., Porter, A., Slater, R., & Nauen, R. (2021). Insecticide resistance management and industry: The origins and evolution of the Insecticide Resistance Action Committee (IRAC) and the mode of action classification scheme. *Pest Management Science*, *77*(6), 2609–2619. https://doi.org/10.1002/ps.6254

Sportes, A., Hériché, M., Boussageon, R., Noceto, P.-A., Van Tuinen, D., Wipf, D., & Courty, P. E. (2021). A historical perspective on mycorrhizal mutualism emphasizing arbuscular mycorrhizas and their emerging challenges. *Mycorrhiza*, *31*(6), 637–653. https://doi.org/10.1007/s00572-021-01053-2

Stauder, C. M., Nuss, D. L., Zhang, D.-X., Double, M. L., MacDonald, W. L., Metheny, A. M., & Kasson, M. T. (2019). Enhanced hypovirus transmission by engineered super donor strains of the chestnut blight fungus, *Cryphonectria parasitica*, into a natural population of strains exhibiting diverse vegetative compatibility genotypes. *Virology*, *528*, 1–6. https://doi.org/10.1016/j.virol.2018.12.007

Tamang, R. K., Jha, G. J., Gupta, M. K., Chauhan, H. V. S., & Tiwary, B. K. (1988). In vivo immunosuppression by synthetic pyrethroid (cypermethrin) pesticide in mice and goats. *Veterinary Immunology and Immunopathology*, *19*(3–4), 299–305.

Valero-Rello, A., López-Sanz, M., Quevedo-Olmos, A., Sorokin, A., & Ayora, S. (2017). Molecular mechanisms that contribute to horizontal transfer of plasmids by the bacteriophage SPP1. *Frontiers in Microbiology*, 8, 1816. https://doi.org/10.3389/fmicb.2017.01816

Van Der Putten, W. H., Klironomos, J. N., & Wardle, D. A. (2007). Microbial ecology of biological invasions. *The International Society for Microbial Ecology Journal*, *1*(1), 28–37. https://doi.org/10.1038/ismej.2007.9

Van Deynze, A., Zamora, P., Delaux, P.-M., Heitmann, C., Jayaraman, D., Rajasekar, S., Graham, D., Maeda, J., Gibson, D., Schwartz, K. D., Berry, A. M., Bhatnagar, S., Jospin, G., Darling, A., Jeannotte, R., Lopez, J., Weimer, B. C., Eisen, J. A., Shapiro, H.-Y., Ané, J.-M., & Bennett, A. B. (2018). Nitrogen fixation in a landrace of maize is supported by a mucilage-associated diazotrophic microbiota. *PLOS Biology*, 16(8), e2006352. https://doi.org/10.1371/journal.pbio.2006352

Vannier, N., Agler, M., & Hacquard, S. (2019). Microbiota-mediated disease resistance in plants. *PLOS Pathogens*, *15*(6), e1007740.

VoPham, T., Bertrand, K. A., Hart, J. E., Laden, F., Brooks, M. M., Yuan, J.-M., Talbott, E. O., Ruddell, D., Chang, C.-C. H., & Weissfeld, J. L. (2017). Pesticide exposure and liver cancer: A review. *Cancer Causes & Control*, *28*(3), 177–190. https://doi.org/10.1007/s10552-017-0854-6

Wang, M., Bian, Z., Shi, J., Wu, Y., Yu, X., Yang, Y., Ni, H., Chen, H., Bian, X., Li, T., Zhang, Y., Jiang, L., & Tu, Q. (2020). Effect of the nitrogen-fixing bacterium *Pseudomonas protegens* CHA0-ΔretS-nif on garlic growth under different field conditions. *Industrial Crops and Products*, *145*, 111982. https://doi.org/10.1016/j.indcrop.2019.111982

Wang, W.-H., Thitithanyanont, A., Urbina, A. N., & Wang, S.-F. (2021). Emerging and re-emerging diseases. *Pathogens*, *10*(7), 827.

Ware, I. M., Van Nuland, M. E., Yang, Z. K., Schadt, C. W., Schweitzer, J. A., & Bailey, J. K. (2021). Climate-driven divergence in plant-microbiome interactions generates range-wide variation in bud break phenology. *Communications Biology*, *4*(1), 748. https://doi.org/10.1038/s42003-021-02244-5

Yin, D., Wang, N., Xia, F., Li, Q., & Wang, W. (2013). Impact of biocontrol agents Pseudomonas fluorescens 2P24 and CPF10 on the bacterial community in the cucumber rhizosphere. *European Journal of Soil Biology*, *59*, 36–42. https://doi.org/10.1016/j.ejsobi.2013.09.001

Yuan, K., Reckling, M., Ramirez, M. D. A., Djedidi, S., Fukuhara, I., Ohyama, T., Yokoyama, T., Bellingrath-Kimura, S. D., Halwani, M., Egamberdieva, D., & Ohkama-Ohtsu, N. (2020). Characterization of rhizobia for the improvement of soybean cultivation at cold conditions in central Europe. *Microbes and Environments*, 35(1). https://doi.org/10.1264/jsme2.ME19124

Zhang, J., Liang, Z., Wang, C., & Li, S. (2021). Compound effective microorganisms treatment increases growth and yield of vegetables. *Mdrsjrns*, *23*(4), 943–954.

Part IV

Industrial Biotechnology

18 Various Nano Supports for Immobilizing Cellulases

The Plethora of Opportunities for Diverse Industrial Applications

Shashi, Parul, and Vinita Hooda

18.1 INTRODUCTION

Cellulases, a group of saccharifying enzymes sourced from fungi, bacteria, insects, and plants, consist of endoglucanases, exoglucanases, and β-glucosidases, collectively hydrolyzing cellulose into fermentable sugars (Karmakar et al., 2011). Despite their wide-ranging applications across industries like food, paper, detergent, and biorefinery (Sulyman et al., 2020), challenges such as cost and vulnerability to inactivation hinder their industrial use. Immobilization, overcoming these obstacles, reduces costs and enhances performance under mild conditions (Mohamad et al., 2015). Various techniques, including adsorption, covalent bonding, entrapment, and cross-linking, have been developed (Bilal et al., 2018). Physical adsorption has wide commercial potential due to its low cost and simplicity. However, enzyme activity is reduced due to leakage of the immobilized enzyme from support due to weak interactions such as hydrophobic and van der Waals after a few cycles (Sassolas et al., 2012). Covalent immobilization comprises a strong multipoint attachment between the support and the enzyme's surface, making the overall process irreversible. However, a significant decrease in the enzyme's activity upon its multipoint covalent attachment might occur due to the restriction on the enzyme's mobility, conformational changes on the carrier surface, or mass transfer effects after several cycles (Marrazza, 2014).

Over the years, cellulase has been immobilized on several supports such as carbon (Ahmad & Khare, 2018), chitosan microspheres (Miao et al., 2016), clay-poly composites (Bayramoglu et al., 2013), nonporous silica particles (Kheirolomoom & Kazemi, 2007), mesoporous silica (Califano & Costantini, 2020), polyvinyl alcohol beads (Liao et al., 2010), and metals: titanium, zinc, and zirconium (Ahmad & Sardar, 2014; Xu et al., 2018; Ahmed et al., 2018) via adsorption, cross-linking, and covalent bonding. As cellulose is an insoluble substrate, mass transfer is the primary barrier for

DOI: 10.1201/9781032642567-22

hydrolysis, and enzyme and substrate have limited accessibility to each other, particularly when the enzyme is immobilized on a carrier (Varalakshmi & Prasad, 2013). As a result, much interest has been developed in the search for a better-supporting matrix to increase the accessibility of the enzyme to cellulose (Saeed et al., 2022).

Recently, a new and rapidly maturing field of nanotechnology has given better choices for immobilizing cellulase on nanocarriers. The dimensions of these nanocarriers are between 1 and 100 nm and possess high surface area, better enzyme loading capacity, and mass transfer resistance that enhance enzyme operational stability, shelf life, and efficiency of the overall process. Also, the immobilization of enzymes on these nanocarriers can reduce protein unfolding via multipoint attachment with nano support and boost enzyme stability and activity. All these potential features make nanocarriers the preferred support for application in various fields, such as biomedicine, agriculture, polluted water treatment, cosmetics, electronics, and construction (Husain, 2017). To date, several nanomaterials and their nanocomposites have been used as a suitable support for cellulase immobilization due to their miscellaneous features. The type, nature, and properties of nano matrix, such as large surface area, biocompatibility, biodegradability, nontoxicity, and easy accessibility selected for immobilization, are crucial for the performance of an immobilized enzyme system (Rajnish et al., 2021).

This review explores cellulase immobilization on nano supports and their potential industrial applications, addressing the pressing need for effective enzyme immobilization techniques in diverse fields.

18.2 CELLULASE

18.2.1 Hydrolytic Mechanism of Cellulase

Cellulase is a member of the glycoside hydrolases family that breaks the β-1,4-glucosidic linkage in cellulose to produce glucose by the action of tri-enzymatic complex: endoglucanase, exoglucanase, and β-glucosidase.

$$\text{Cellulose} \xrightarrow{\text{Endoglucanase}} \beta\text{- Cellobiose} \xrightarrow{\text{Exoglucanase}} \text{Glucan units} \xrightarrow{\beta\text{-glucosidase}} \text{Glucose}$$

The endoglucanase enzyme cleaves the amorphous areas of cellulose and produces new chain ends by randomly disrupting the internal connections. Exoglucanase is effective against crystalline cellulose and targets the freshly made chain ends, which have glucose or cellobiose units. Ultimately, β-glucosidase hydrolyzes cellobiose to glucose from the non-reducing ends; it exhibits no activity against crystalline or amorphous cellulose (Karmakar et al., 2011).

18.3 IMMOBILIZATION

Immobilization anchors enzymes to a solid support, enabling their reuse over multiple cycles while retaining catalytic activity. Various techniques like entrapment, ionic and covalent bonding, adsorption, and cross-linking shown in Figure 18.1 have

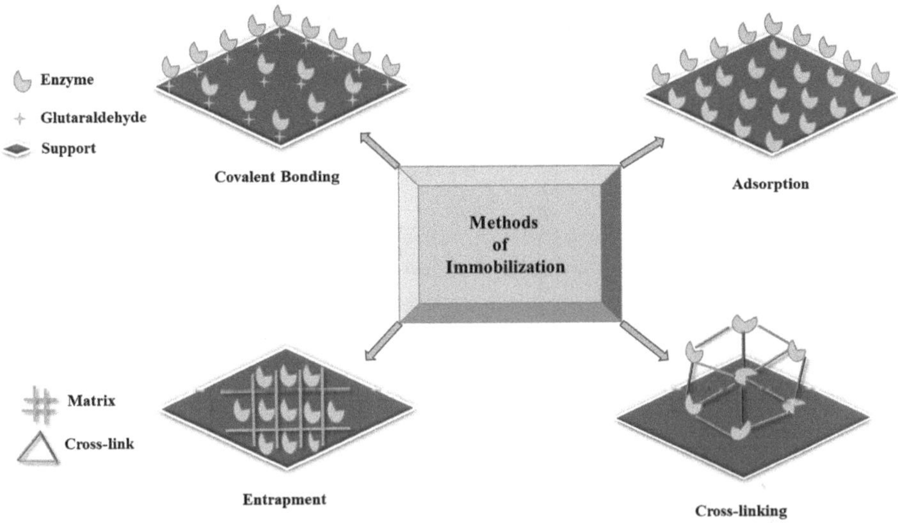

FIGURE 18.1 Various methods of immobilization including covalent bonding, adsorption, entrapment, and cross-linking.

been developed to achieve this (Junior et al., 2016). For cellulase, immobilization ensures reusability during cellulose hydrolysis (Junior et al., 2016). The choice of support material is crucial, with factors like accessibility, cost, surface area, and enzyme affinity influencing functionality (Sirisha et al., 2016). Support materials stabilize enzyme structure, resulting in improved stability and kinetic properties of the immobilized enzyme (Ahmad & Khare, 2018). While organic–inorganic and metal-based supports have been traditionally used, nanomaterials are gaining popularity due to their high surface area, microbial resistance, enhanced loading capacity, and stability (Rajnish et al., 2021). As a result, nanomaterials are increasingly preferred for enzyme immobilization.

18.4 NANOMATERIALS FOR CELLULASE IMMOBILIZATION

In pursuit of cellulose hydrolysis for various industrial applications, a range of nanomaterials such as nanoparticles, nanotubes, nanofibers, and nanocomposites have been harnessed for immobilizing cellulases. This section provides an overview of the diverse types of nanomaterials used in cellulase immobilization.

18.4.1 Magnetic Nanomaterials

The decline in enzyme activity after each assay is a significant concern in enzyme immobilization. Utilizing magnetic nanoparticles (MNPs) as carriers for enzyme immobilization presents a strategic solution to this challenge. MNPs offer a large surface area conducive to effective immobilization and can be easily recovered using an external magnetic field (Khoshnevisan et al., 2016). Enzymes can adhere to the surface of MNPs through noncovalent bonding mechanisms, including electrostatic interactions, hydrophobicity, van der Waals forces, and π–π stacking. However, a key issue with noncovalent binding is enzyme leakage from the surface of MNPs. To address this, glutaraldehyde (GA), a crosslinker soluble in aqueous solvents, is employed. GA facilitates strong inter- and intracovalent interactions, mitigating the problem of enzyme leakage (Han et al., 2018). While MNPs exhibit reduced activity after each cycle, they remain functional for several cycles before a significant decline in activity is observed.

Zhang et al. (2014) successfully immobilized cellulase onto Fe_3O_4–chitosan nanoparticles using glutaraldehyde as a crosslinker. The remarkable magnetic stability of MNPs facilitated the straightforward separation of the immobilized cellulase from the reaction mixture. The immobilized cellulase achieved an impressive loading of over 112.3 mg/g on the Fe_3O_4–chitosan nanoparticles, exhibiting an activity of 5.23 IU/mg. Moreover, it demonstrated a wider range of pH and temperature stability than the free enzyme, retaining 50% of its initial activity even after ten reuse cycles.

Similarly, Sanchez et al. (2017) conducted a comparable study wherein cellulase was immobilized on chitosan-coated MNPs. They observed a notable shift in the optimum temperature of the immobilized cellulase from 50°C to 60°C, attributed to the formation of hydrogen bonds and weak intermolecular forces with the carrier material. This structural reinforcement resisted the temperature-induced changes. Additionally, the pH profile of the immobilized enzyme shifted from 4.5–5.0 to 5.5, influenced by the altered hydronium ion concentration in the reaction environment, which in turn affected the ionization degree of enzyme active site amino acid residues. Storage stability assessments revealed that while the free enzyme lost 70% of its activity after 28 days, the immobilized enzyme retained an impressive 65% of its initial activity. Furthermore, even after 15 cycles and exposure to 4 hours of incubation at elevated temperatures of 60°C and 70°C, the immobilized enzyme retained 80% and 50% of its activity, respectively, outperforming the free enzyme, which retained only 20%.

In their subsequent investigation, Ladole et al. (2017) explored the use of hyperactivated cellulase, subjected to low-frequency ultrasound waves, for immobilization onto MNPs (cellulase@MNPs). They observed a remarkable enhancement in catalytic activity, nearly 3.6-fold higher than the control, under the specified conditions of 24 kHz ultrasonic frequency, 6 W power, and 6 minutes of incubation. Optimal immobilization of cellulase was achieved at a ratio of 1:6 (MNPs:enzyme), while the highest activity recovery, at 59%, was attained at a ratio of 1:3. However, at higher immobilization ratios, the enzyme's activity decreased due to steric hindrance between enzyme molecules, obstructing access to active sites on cellulase@MNPs. Glutaraldehyde concentration played a crucial role, with an optimal value of 10 mM resulting in a maximum activity recovery of 65%. Beyond this concentration, glutaraldehyde induced conformational changes in the active site, leading to enzyme denaturation and rigidification due to excessive cross-linking. Thermal stability, as indicated by the deactivation energy (E_d), was significantly higher for cellulase@MNPs (48.17 kJ/mol) compared to free cellulase (28.06 kJ/mol). The kinetics parameters K_m and V_{max} for free cellulase, cellulase@MNPs, and sonicated cellulase@MNPs were measured, showing variations attributed to changes in enzyme conformation and active site exposure upon immobilization. Upon repeated use, the residual activity of cellulase@MNPs decreased by 46% after the seventh cycle. The highest hydrolysis yield was achieved at 24 hours, with cellulase@MNPs exhibiting hydrolysis yields of 42% and 35% before and after sonication, respectively.

Abbaszadeh and Hejazi (2019) embarked on immobilizing cellulase from *Aspergillus niger* onto amine-functionalized Fe_3O_4 nanoparticles coordinated with copper. They observed remarkable stability in the immobilized enzyme, retaining 73% of its initial activity after five reuses and preserving 84% of its activity even after 8 days of storage, compared with 70% for the free enzyme. Notably, immobilized and free enzymes exhibited maximum activity at pH 3.0, characteristic of the natural habitat for cellulase-producing microorganisms. Although the optimal pH for free and immobilized enzymes was 5.0, the immobilized enzyme displayed a 15% increase in relative activity compared with the free form. However, magnetite nanoparticles are prone to rapid oxidation, diminishing their magnetic responsiveness. To counteract this challenge, Lima et al. (2017) encapsulated cellulase within polymer nanospheres (PMMA) containing magnetite nanoparticles. This straightforward immobilization technique conferred remarkable stability, enabling storage of the enzyme for up to 75 days without activity loss. Moreover, the immobilized enzyme exhibited enhanced stability and activity at elevated temperatures, retaining 31% of its initial activity even after approximately eight cycles of reuse.

Porous silica nanoparticles (PSNs) are favored for enzyme immobilization due to their large surface area, pore volume and pore size. When combined with MNPs, they offer even greater utility. In a study by Poorakbar et al. (2018), AuNPs were integrated with PSNs to immobilize cellulase, enhancing stability by 57% with a binding efficacy of 76% using glutamic acid and polyethylene glycol as crosslinkers. Although activity declined to 28% after five cycles, the immobilized cellulase remained significantly more stable. Liao et al. (2010) employed a microemulsion system coupled with a ball mill reactor to immobilize cellulase onto polyvinyl alcohol/Fe_2O_3 nanoparticles. This preparation demonstrated remarkable efficacy against microcrystalline cellulose (MCC) and retained 40% of its initial activity after four reuse cycles. In another approach, Zhang et al. (2015) modified magnetic nanospheres by introducing silane groups such as 3-(2-aminoethyl aminopropyl)-triethoxysilane, 3-(2-aminoethyl aminopropyl)-trimethoxysilane and 3-aminopropyltriethoxysilane (AEAPTMES) for cellulase immobilization. Functionalized magnetic nanospheres exhibited superior immobilization due to their higher surface charge, recovering 87% of the enzyme activity with the AEAPTMES group. Remarkably, the enzyme retained stable activity even at higher temperatures.

Sillu and Agnihotri (2019) successfully immobilized cellulase onto halloysite nanotubes (HNTs) and iron oxide nanoparticles, imparting magnetic properties to the system. Remarkably, after seven cycles of reuse, the immobilized enzyme retained 68.2% of its activity, and with the aid of an external magnet, 100% of the enzyme was recovered from the reaction mixture. Even after 20 days of storage, the immobilized cellulase maintained 81.2% activity, whereas free cellulase retained only 29.6%.

MNPs have gained widespread use due to their ease of retrieval. Most notably, cellulases immobilized on these magnetic materials retained over 30% of their initial activity even after five or more cycles of use, while the free enzyme exhibited significant activity loss (Rajnish et al., 2021). Moreover, the immobilized enzymes demonstrated enhanced stability, prolonging their shelf life. Covalent connections facilitated by crosslinkers like glutaraldehyde further improved thermal stability. These attributes render MNPs highly reliable supports for enzyme immobilization, with applications spanning various industries.

18.4.2 Carbon-based Nanomaterials

Carbon nanotubes (CNTs) stand out as the most commonly used carbon-based nanostructures, prized for their superior electrical properties. They are particularly well-suited for enzyme immobilization, notably in biofuel production. Noncovalent approaches are preferred for enzyme immobilization on CNTs as they mitigate enzyme denaturation. Among CNT variants, multiwall carbon nanotubes (MWCNTs) are often chosen for cellulase immobilization due to their cost-effectiveness, ease of dispersibility, and multiple graphite layers (Feng & Ji, 2011). MWCNTs were employed for cellulase immobilization, achieving a remarkable 98% immobilization efficiency under optimal conditions of pH 4.5, temperature 30°C, and 1 mL of 10 mg/mL *N*-ethyl-*N*-(3-dimethyl aminopropyl). Similarly, Ahmad and Khare (2018) utilized MWCNTs for cellulase immobilization using 1-ethyl-3-(3-dimethylamino-propyl) carbodiimide (EDC) and *N*-hydroxysuccinimide (NHS), maintaining 75% of the enzyme's initial activity over six reuse cycles.

In a study by Li et al. (2019), cellulase was immobilized on sodium alginate–carbon nanotube composites, demonstrating superior temperature tolerance compared to the free enzyme. At 40°C, the free cellulase retained 84.1% activity, while the immobilized enzyme maintained 92.3% activity at 50°C. Furthermore, after seven reuse cycles, the immobilized enzyme retained 71.5% of its activity. Gokhale et al. (2013) explored graphene coated with maghemite for cellulase immobilization. Graphene, akin to carbon nanotubes but in a two-dimensional form, boasts immense surface area, cost-effectiveness, and heat conduction properties. After four cycles of use, the immobilized enzyme retained up to 55% of its initial activity, highlighting the efficacy of graphene-based immobilization strategies.

Daoud et al. (2010) utilized commercial activated carbon for immobilizing *Aspergillus niger* cellulase through adsorption. Activated carbon was chosen for its large surface area and suitable pore volume conducive to enzyme immobilization. At cellulase concentrations of 2 and 16 g/L, the activated carbon could adsorb 90.2% and 43.2% of the enzyme, respectively. Remarkably, 70% of the enzyme's original activity was retained after five reuse cycles. In another study by Jabasingh and Nachiyar (2012), *Aspergillus nidulans* SU04 cellulase was immobilized onto modified activated carbon (MAC). At a MAC concentration of 16 mg/mL, the immobilized enzyme maintained 70% activity even after ten reuse cycles. Immobilization of cellulases on carbon nanomaterials has shown improvements in reusability, with most studies demonstrating that the immobilized enzymes could retain up to 70% of their initial activity even after five cycles of reuse. Additionally, these nanomaterials exhibit remarkable loading capacity and enhanced stability against fluctuations in pH and reaction temperature, making them promising candidates as enzyme supports (Samuel et al., 2020).

18.4.3 Carbon-Composite Nanomaterials

Composite nanomaterials are made up of various compounds, each contributing to specific features that aid in enzyme immobilization or enhance its activity. In these materials, enzyme immobilization typically occurs at the outer layer, while the inner layers serve additional functions such as providing hybrid functional groups to improve stability and reusability. MNPs are commonly included in composite nanomaterials to facilitate enzyme recycling.

Han et al. (2018) developed magnetic double-shell hybrid microspheres (Fe_3O_4@SiO_2@p(NIPAM-*co*-GMA)) for cellulase immobilization. The microspheres contained NIPAM (*N*-isopropylacrylamide) for temperature-sensitive capability and glycidyl methacrylate (GMA) on the surface for covalent binding with cellulase. The immobilized enzyme retained 65.6% of its initial activity after six usage cycles and 59.6% activity after 40 days of storage at pH 5.0 and 60°C. Asar et al. (2020) synthesized magnetic graphene by incorporating Fe_3O_4 and coating it with chitosan nanocomposites (Fe_3O_4/GO/CS/NC) to immobilize cellulase. After immobilization, the enzyme retained 78% of its initial activity, and even after eight cycles, 49% of the original activity remained. Moreover, after 45 days of storage at pH 5.0 and 50°C, the enzyme maintained 64% activity.

In a study similar to Li et al. (2015), cellulase immobilization was achieved using the Fe_3O_4@SiO_2@GO nanocomposite. This nanocomposite was prepared by linking NH_2-functionalized Fe_3O_4@SiO_2 amino groups with the graphene oxide (GO) carboxyl groups. The immobilized enzyme retained 80% of its initial activity after seven reuse cycles. Its high specific surface area and magnetic solid response make it an excellent support for enzyme immobilization.

Zhang et al. (2018) employed chitosan and magnetic coal fly ash (MCFA) to prepare two supports for cellulase immobilization. Support I comprised nonporous MCFA@chitosan gel microcomposites, while Support II consisted of porous MCFA/chitosan gel beads. Cellulase was immobilized using glutaraldehyde. Supports I and II yielded 219.8 and 246.4 mg of glucose generation, respectively. After ten usage cycles, 69.9% and 75.5% of the activity remained in Supports I and II.

Han et al. (2018) used magnetic graphene molded with 4arm PEG NH_2 [molecular weight (MW) 5 or 10 K] to enhance loading capacity and catalytic activity of cellulase immobilization. Immobilized cellulase exhibited a relative activity of 76% (5 K) and 80% (10 K) at pH 3.0, compared with only 20% for the free enzyme. Storage stability after 14 days was 65% and 70% for immobilized cellulase on GO with 4arm PEG NH_2 (MW 5 or 10 K), whereas free cellulase retained only 50%. After seven reuses, cellulase bound to GO with 4arm PEG NH_2 (MW 5 or 10 K) retained 40% and 50% of its activity, respectively.

Bayramoglu et al. (2013) employed a composite of clay-poly(glycidylmethacrylate) to immobilize cellulase via adsorption and covalent bonding. The covalently attached enzyme retained 79% activity after 12 use cycles, whereas the adsorbed enzyme was only 37% active. Jia et al. (2017) combined cross-linked enzyme aggregates (CLEA) with magnetic iron oxide nanoparticles for cellulase immobilization. After six reuse cycles, the resulting support retained 74% of its initial activity.

These carbon-based nanocomposite materials boast a remarkable loading capacity, effectively boosting enzyme activity. Even after undergoing ten usage cycles, clay-poly(glycidylmethacrylate) support, retain a substantial portion of their initial activity. Moreover, their magnetic properties enable easy recovery with minimal loss, making them a promising choice as support for cellulase applications across various industries.

18.4.4 Metal-Organic Frameworks

The porosity of the support material plays a crucial role in enzyme immobilization by influencing the enzyme loading capacity. Metal-organic frameworks (MOFs) offer an excellent solution in this regard. MOFs are composed of organic linkers and metal ions, forming a hybrid porous coordination polymer. These frameworks are constructed using a variety of metal ions and organic ligands, resulting in highly porous, three-dimensional networks with remarkable thermal stability and tunable pores.

Enzyme immobilization on MOFs provides an ideal environment for preserving the enzyme's physiochemical properties and stability. Ahmed et al. (2018) demonstrated this by immobilizing cellulase onto NH_2-functionalized zirconium MOF (UIO-66), achieving an enzyme loading capacity of 350 mg/g. The immobilized enzyme exhibited improved pH and thermal stability, retaining 85% activity at pH 5.0 and 55°C. After ten cycles, 72% of the activity was still preserved, and even after 30 days of storage at 4°C, the enzyme retained 65% of its activity.

In a similar study, Qi and colleagues (2018) immobilized cellulase onto magnetic UiO-66-NH_2, leading to an enhanced hydrolysis yield by 18.7%. Notably, the immobilized enzyme showed increased tolerance to vanillin and formic acid, two common cellulase enzyme inhibitors found in lignocellulosic materials, at a concentration of 5 g/L.

Wang et al. (2019) successfully immobilized β-glucosidase and cellulase onto Cu-MOF, resulting in the β-G@Cu(PABA) biocomposite. Remarkably, even after ten cycles, this biocomposite retained 90% of its activity, demonstrating outstanding reusability. When coimmobilized with cellulase, the β-G@Cu(PABA) biocomposite achieved a glucose yield of 98% from the hydrolysis of carboxymethylcellulose (CMC), doubling the yield compared with coimmobilized cellulase alone. After eight cycles of CMC hydrolysis, both the β-G@Cu(PABA) biocomposite and coimmobilized cellulase retained 70% of their activity.

Zhou et al. (2019) employed physical adsorption to immobilize cellulase on UIO-66-NH_2, ZIF-8, MIL-100-Fe, and PCN-250. These MOFs were prepared using a solvent replacement procedure followed by vacuum drying. To evaluate immobilized cellulase activity, CMC and filter paper were used as substrates in the presence of 1-ethyl-3-methylimidazolium diethylphosphate. Among the MOFs studied, ZIF-8 exhibited the highest loading capacity at 176.16 mg/g, followed by UIO-66-NH_2 with 146.8 mg/g. MIL-100-Fe and PCN-250 showed enzyme loadings of 119.88 and 136.45 mg/g, respectively. Notably, cellulase immobilized on ZIF-8 demonstrated solid resistance to ionic liquids and a remarkable 112.59% increase in activity. Moreover, the hydrolyzing power of lignocellulosic waste increased by 92% when cellulase was immobilized onto ZIF-8.

In another study, Han et al. (2019) immobilized cellulase onto phosphotungstic acid-supported MIL-101(Cr) functionalized with electronegative groups such as -NH_2, -NO_2, and -Cl. Despite unaffected stability, the addition of electronegative groups significantly enhanced the hydrolyzing capability of the metal-organic framework. For instance, using PTA@MIL-101–NO_2 as the cellulase support led to a 16.2% increase in glucose yield.

MOFs emerge as the top choice for cellulase immobilization due to their cost-effectiveness, high enzyme loading capacity, enhanced stability, reusability, and wide applicability in hydrolyzing lignocellulosic waste and producing biofuel in biorefineries. Table 18.1 provides an overview of cellulase immobilized on different nanomaterials and their diverse applications across various industries.

18.5 INDUSTRIAL APPLICATION OF CELLULASES

Cellulases have become indispensable in various industries such as textile, detergent, pharmaceutical, medical, brewery, wine, agricultural, food, and feed. They are used in treating fruit juices and pulps to eliminate unwanted colors and in the detergent industry for bio-stoning jeans, as color brighteners, and softeners. Additionally, they play a crucial role in industrial waste treatment and biomass pretreatment to enhance food nutritional content. Despite their significance, the use of cellulases in these industries is limited due to their low stability, short shelf life, restricted reusability, and susceptibility to harsh chemical conditions. To overcome these limitations, cellulases are immobilized on nano supports, which significantly improve their stability, reusability, and applicability in various industrial sectors.

Cellulase immobilization on various nano supports and their varied uses in different industries is given in Figure 18.2.

18.6 CONCLUSION

Cellulases have become indispensable in various industries, spanning from textile to agriculture, owing to their versatile applications. Immobilizing cellulases on nano supports significantly enhances their stability, reusability, and effectiveness across diverse industrial applications. In the food industry, cellulase, often combined with pectinase and hemicellulase, aids in olive oil extraction, improves food nutritional quality and extends shelf life. Similarly, in baking, cellulase enzymes derived from *Aspergillus niger* and *Trichoderma reesei* contribute to dough refinement by breaking down cellulose polymers into glucose. In animal feed formulations, cellulases enhances feed conversion rates by breaking down cellulose, hemicellulose and lignin, thereby boosting weight gain and milk production in livestock. Additionally, in the textile industry, cellulases replace harsh chemicals in dyeing and printing processes, significantly reducing environmental pollution. Moreover, in detergent formulations, cellulase enzymes maintain fabric cleanliness and improve aesthetics by breaking down amorphous cellulose during stonewashing processes. The paper and pulp industry benefits from cellulases by enabling a more sustainable approach to pulp production, with cellulase-assisted pulping yielding superior-quality pulp compared to mechanical methods. Cellulases also facilitate efficient deinking of waste paper, offering recycled paper with properties akin to freshly prepared paper. Furthermore, in the brewery and wine industries, cellulases enhance malt extraction, beer filterability and wine clarity by breaking down β-glucans and aiding in maceration processes.

TABLE 18.1
An overview of Cellulase Immobilized on Different Nanomaterials and Their Diverse Applications across Various Industries

S. No	Source of Cellulase	Support	Conjugation	Optimum pH	Temperature	Storage Stability (Activity Retained)	Reusability (Activity Retained)	Industrial Applications	References
1.	–	Fe_3O_4-chitosan NPs	Covalent	5.0	50°C	–	50% activity after 10 cycles	Glucose production	Zhang et al. (2014)
2.	*Trichoderma reesei*	Chitosan-coated MNPs	Covalent	5.0	60°C	65% activity after 28 days	50% activity after 17 cycles	Glucose production	Sanchez et al. (2017)
3.	–	Hyperactivated cellulase@MNPs	Covalent	5.0	60°C	–	46% activity after 7 cycles	Hydrolysis of *Napier grass*	Ladole et al., (2017)
4.	*Aspergillus niger*	amine-functionalized Fe_3O_4 nanoparticles with copper as a ligand	Adsorption	3.0	50°C	84% activity after 8 days	73% activity after 5 cycles	CMC hydrolysis	Abbaszadeh and Hajezi (2019)
5.	–	MNPs-PMMA	Encapsulation	6.0	55°C	100% activity after 75 days			Lima et al. (2017)
6.	*Penicillium funiculosum*	Magnetic gold mesoporous silica nanoparticles (mAu@PSNs)	Adsorption	6.0	50°C	–	28% activity after 5 cycles	CMC hydrolysis	Poorakbar et al. (2018)
7.	–	Polyvinyl alcohol/Fe_2O_3 nanoparticles	Adsorption	4.5	50°C	–	40% activity after 4 cycles	MCC hydrolysis	Liao et al. (2010)
8.	–	Amino-silane modified magnetic nanosphere	Covalent	5.0	50°C	–	–	CMC hydrolysis	Zhang et al. (2015)
9.	–	Magnetic Halloysite Nanotubes (MHNTs)	Covalent	6.0	50°C	81.2% activity after 20 days	68.2% activity after 7 cycles	Cellulose hydrolysis	Sillu and Agnihotri (2019)
10.	*Trichoderma reesei*	Functionalised Multiwalled Carbon Nanotubes (MWCNT)	Covalent	4.5	30°C	–	–	CMC hydrolysis	Liao et al. (2010)

(Continued)

TABLE 18.1 (*Continued*)
An overview of Cellulase Immobilized on Different Nanomaterials and Their Diverse Applications across Various Industries

S. No	Source of Cellulase	Support	Conjugation	Optimum pH	Temperature	Storage Stability (Activity Retained)	Reusability (Activity Retained)	Industrial Applications	References
11.	*Trichoderma reesei*	Multiwalled carbon nanotubes and sodium alginate	Adsorption	3.0	40°C	71.2% activity after 30 days	71.5% activity after 7 cycles	Straw hydrolysis	Li et al. (2019)
12.	–	Magneto responsive graphene nano support (MNP–GNPs)	Adsorption	5.1	50°C	–	55% activity after 4 cycles	CMC hydrolysis	Gokhale et al. (2013)
13.	*Aspergillus niger*	Activated carbon	Adsorption	4.8	30°C	–	70% activity after 5 cycles	MCC hydrolysis	Daoud et al. (2010)
14.	*Aspergillus nidulans*	Modified activated carbon (MAC)	Adsorption	5.0	50°C	–	70% activity after 10 cycles	CMC hydrolysis	Jabasingh and Nachiyar, (2012)
15.	*Trichoderma viride*	Fe_3O_4@SiO_2@p (NIPAM-*co*-GMA)	Covalent	5.0	60°C	59.6 activity after 40 days	65.6% activity after 6 cycles	CMC hydrolysis	Han et al. (2018)
16.	*Trichoderma reesei*	Fe_3O_4/GO/CS/NC	Covalent	5.0	50°C	64% activity after 45 days	49% activity after 8 cycles	MCC hydrolysis	Asar et al. (2020)
17.	–	Fe_3O_4@SiO_2@GO	Covalent	4.0	50°C	–	80% activity after 7 cycles	CMC hydrolysis	Li et al. (2015)
18.	–	MCFA@chitosan gel microcomposites	Covalent	4.0	70°C	–	69.9% activity after 10 cycles	CMC hydrolysis	Zhang et al. (2018)
19.	–	MCFA/chitosan gel beads	Covalent	3.0	70°C	–	75.5% activity after 10 cycles	CMC hydrolysis	Zhang et al. (2018)
20.	*Trichoderma viride*	Graphene oxide (GO) with 4arm PEG NH_2	Covalent	4.0	40°C	46% activity after 30 days	60% activity after 6 cycles	CMC hydrolysis	Han et al. (2018)
21.	*Trichoderma reesei*	Clay-poly (glycidylmethacrylate) composite	Covalent	6.0	55°C	86% activity after 60 days	79% activity after 12 cycles	CMC hydrolysis	Bayramoglu et al. (2013)

(*Continued*)

TABLE 18.1 (*Continued*)

An overview of Cellulase Immobilized on Different Nanomaterials and Their Diverse Applications across Various Industries

S. No	Source of Cellulase	Support	Conjugation	Optimum pH	Temperature	Storage Stability (Activity Retained)	Reusability (Activity Retained)	Industrial Applications	References
22.	*Trichoderma reesei*	Clay-poly (glycidylmethacrylate) composite	Adsorption	5.5	55°C	64% activity after 60 days	37% activity after 12 cycles	CMC hydrolysis	Bayramoglu et al. (2013)
23.	*Trichoderma reesei*	Magnetic CLEA	Covalent	5.0	60°C	–	74% activity after 6 cycles	Lignocellulosic biomass hydrolysis	Jia et al., (2017)
24.	*Trichoderma reesei*	NH_2-UIO-66	Adsorption	5.0	55°C	65% activity after 30 days	72% activity after 10 cycles	CMC hydrolysis	Ahmed et al. (2018)
25.	–	Magnetic UiO-66-NH_2	Covalent	5.0	50°C	–	70% activity after 5 cycles	MCC hydrolysis	Qi and Wan (2018)
26.	*Trichoderma viride*	Cu-MOF	Encapsulation	5.0	50°C	–	70% activity after 8 cycles	CMC hydrolysis	Wang et al., (2019)
27.	–	PTA@MIL-101–NO_2	–	–	–	–	–	Lignocellulose hydrolysis	Han et al., 2019)

NPs,Nanoparticles; MNPs,Magnetic nanoparticles; PMMA,Poly(methyl methacrylate); mAu@PSNs,Magnetic gold mesoporous silica nanoparticles; MHNTs,Magnetic halloysite nanotubes; MWCNT,Multiwalled carbon nanotubes; MNP–GNPs,Magneto responsive graphene nano support; MAC,Modified activated carbon; NIPAM and GMA,*N*-isopropylacrylamide and glycidyl methacrylate; Fe_3O_4/GO/CS/NC,magnetic grapheme coated with chitosan nanocomposites; MCFA,magnetic coal fly ash; GO,Graphene oxide; PEG,Polyethylene glycol; CLEA,cross-linked enzyme aggregates; NH_2-UIO-66,Amine-functionalized zirconium metal-organic framework; Cu-MOF,copper metal-organic framework; PTA@MIL-101–NO_2,phosphotungstic acid-supported chromium metal-organic framework.

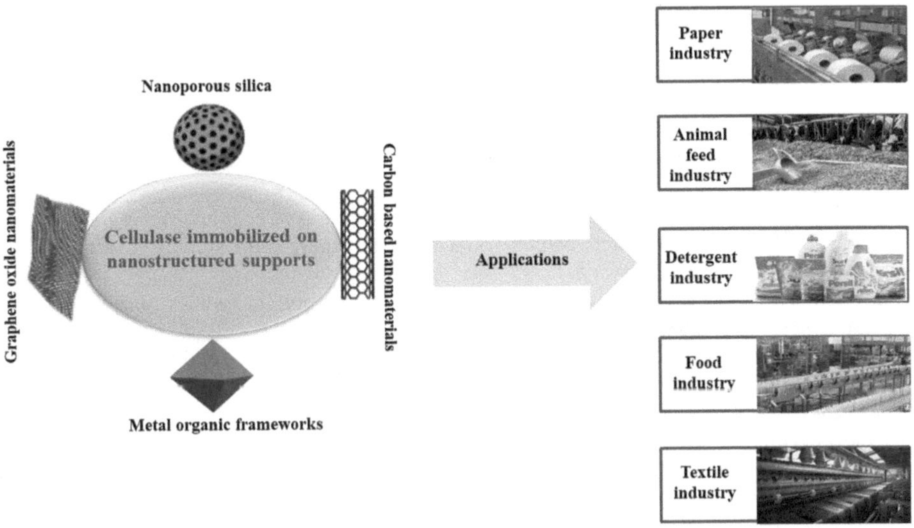

FIGURE 18.2 Cellulase immobilization on various nano supports and their varied uses in different industries.

In agriculture, fungal cellulases offer sustainable alternatives to chemical pesticides by inhibiting pathogen growth through the degradation of pathogen cell walls.

To conclude, the immobilization of cellulases on nano supports unlocks their full potential, enabling a wide range of applications across various industries. This approach not only addresses the challenges associated with free cellulase enzymes but also contributes to sustainable and environmentally friendly industrial practices, thereby paving the way for a more efficient and eco-conscious future.

REFERENCES

Abbaszadeh, M., & Hejazi, P. (2019). Metal affinity immobilization of cellulase on Fe3O4 nanoparticles with copper as ligand for biocatalytic applications. *Food Chemistry*, *290*, 47–55.

Ahmad, R., & Khare, S. K. (2018). Immobilization of *Aspergillus niger* cellulase on multiwall carbon nanotubes for cellulose hydrolysis. *Bioresource Technology*, *252*, 72–75.

Ahmad, R., & Sardar, M. (2014). Immobilization of cellulase on TiO_2 nanoparticles by physical and covalent methods: A comparative study. *Indian Journal of Biochemistry and Biophysics*, *51*, 314–320.

Ahmed, I. N., Yang, X. L., Dubale, A. A., Li, R. F., Ma, Y. M., Wang, L. M., & Xie, M. H. (2018). Hydrolysis of cellulose using cellulase physically immobilized on highly stable zirconium based metal-organic frameworks. *Bioresource Technology*, *270*, 377–382.

Anuradha Jabasingh, S., & Valli Nachiyar, C. (2012). Immobilization of *Aspergillus nidulans* SU04 cellulase on modified activated carbon: Sorption and kinetic studies. *Journal of Thermal Analysis and Calorimetry*, *109*(1), 193–202.

Asar, M. F., Ahmad, N., & Husain, Q. (2020). Chitosan modified Fe3O4/graphene oxide nanocomposite as a support for high yield and stable immobilization of cellulase: Its application in the saccharification of microcrystalline cellulose. *Preparative Biochemistry & Biotechnology*, *50*(5), 460–467.

Bayramoglu, G., Senkal, B. F., & Arica, M. Y. (2013). Preparation of clay–poly (glycidyl methacrylate) composite support for immobilization of cellulase. *Applied Clay Science*, *85*, 88–95.

Bilal, M., Rasheed, T., Zhao, Y., Iqbal, H. M., & Cui, J. (2018). "Smart" chemistry and its application in peroxidase immobilization using different support materials. *International Journal of Biological Macromolecules*, *119*, 278–290.

Califano, V., & Costantini, A. (2020). Immobilization of cellulolytic enzymes in mesostructured silica materials. *Catalysts*, *10*(6), 706.

Daoud, F. B. O., Kaddour, S., & Sadoun, T. (2010). Adsorption of cellulase *Aspergillus niger* on a commercial activated carbon: Kinetics and equilibrium studies. *Colloids and Surfaces B: Biointerfaces*, *75*(1), 93–99.

Feng, W., & Ji, P. (2011). Enzymes immobilized on carbon nanotubes. *Biotechnology Advances*, *29*(6), 889–895.

Gokhale, A. A., Lu, J., & Lee, I. (2013). Immobilization of cellulase on magnetoresponsive graphene nano-supports. *Journal of Molecular Catalysis B: Enzymatic*, *90*, 76–86.

Han, J., Rong, J., Wang, Y., Liu, Q., Tang, X., Li, C., & Ni, L. (2018). Immobilization of cellulase on thermo-sensitive magnetic microspheres: Improved stability and reproducibility. *Bioprocess and Biosystems Engineering*, *41*, 1051–1060.

Han, J., Wang, L., Wang, Y., Dong, J., Tang, X., Ni, L., & Wang, L. (2018). Preparation and characterization of Fe3O4-NH2@ 4-arm-PEG-NH2, a novel magnetic four-arm polymer-nanoparticle composite for cellulase immobilization. *Biochemical Engineering Journal*, *130*, 90–98.

Han, J., Wang, Y., Wan, J., & Ma, Y. (2019). Catalytic hydrolysis of cellulose by phosphotungstic acid–supported functionalized metal-organic frameworks with different electronegative groups. *Environmental Science and Pollution Research*, *26*, 15345–15353.

Husain, Q. (2017). Nanomaterials immobilized cellulolytic enzymes and their industrial applications. *JSM Biochemistry & Molecular Biology*, *4*, 1029.

Jia, J., Zhang, W., Yang, Z., Yang, X., Wang, N., & Yu, X. (2017). Novel magnetic cross-linked cellulase aggregates with a potential application in lignocellulosic biomass bioconversion. *Molecules*, *22*(2), 269.

Karmakar, M., & Ray, R. R. (2011). Current trends in research and application of microbial cellulases. *Research Journal of Microbiology*, *6*(1), 41.

Kheirolomoom, A., & Kazemi, A. (2007). Immobilization of cellulase on non-porous ultrafine silica particles. *Scientia Iranica*, *14*(4), 379–383.

Khoshnevisan, K., Barkhi, M., Ghasemzadeh, A., Tahami, H. V., & Pourmand, S. (2016). Fabrication of coated/uncoated magnetic nanoparticles to determine their surface properties. *Materials and Manufacturing Processes*, *31*(9), 1206–1215.

Ladole, M. R., Mevada, J. S., & Pandit, A. B. (2017). Ultrasonic hyperactivation of cellulase immobilized on magnetic nanoparticles. *Bioresource Technology*, *239*, 117–126.

Li, L. J., Xia, W. J., Ma, G. P., Chen, Y. L., & Ma, Y. Y. (2019). A study on the enzymatic properties and reuse of cellulase immobilized with carbon nanotubes and sodium alginate. *Amb Express*, *9*(1), 112.

Li, Y., Wang, X. Y., Jiang, X. P., Ye, J. J., Zhang, Y. W., & Zhang, X. Y. (2015). Fabrication of graphene oxide decorated with Fe_3O_4@ SiO_2 for immobilization of cellulase. *Journal of Nanoparticle Research*, *17*, 1–12.

Liao, H., Chen, D., Yuan, L., Zheng, M., Zhu, Y., & Liu, X. (2010). Immobilized cellulase by polyvinyl alcohol/Fe_2O_3 magnetic nanoparticle to degrade microcrystalline cellulose. *Carbohydrate Polymers*, *82*(3), 600–604.

Lima, J. S., Araújo, P. H., Sayer, C., Souza, A. A., Viegas, A. C., & de Oliveira, D. (2017). Cellulase immobilization on magnetic nanoparticles encapsulated in polymer nanospheres. *Bioprocess and Biosystems Engineering*, *40*, 511–518.

Marrazza, G. (2014). Piezoelectric biosensors for organophosphate and carbamate pesticides: A review. *Biosensors*, *4*(3), 301–317.

Miao, X., Pi, L., Fang, L., Wu, R., & Xiong, C. (2016). Application and characterization of magnetic chitosan microspheres for enhanced immobilization of cellulase. *Biocatalysis and Biotransformation*, *34*(6), 272–282.

Mohamad, N. R., Marzuki, N. H. C., Buang, N. A., Huyop, F., & Wahab, R. A. (2015). An overview of technologies for immobilization of enzymes and surface analysis techniques for immobilized enzymes. *Biotechnology & Biotechnological Equipment*, *29*(2), 205–220.

Poorakbar, E., Shafiee, A., Saboury, A. A., Rad, B. L., Khoshnevisan, K., Ma'mani, L., & Hosseini, M. (2018). Synthesis of magnetic gold mesoporous silica nanoparticles core shell for cellulase enzyme immobilization: Improvement of enzymatic activity and thermal stability. *Process Biochemistry*, *71*, 92–100.

Qi, B., Luo, J., & Wan, Y. (2018). Immobilization of cellulase on a core-shell structured metal-organic framework composites: Better inhibitors tolerance and easier recycling. *Bioresource Technology*, *268*, 577–582.

Quilles Junior, J. C., Ferrarezi, A. L., Borges, J. P., Brito, R. R., Gomes, E., Da Silva, R., & Boscolo, M. (2016). Hydrophobic adsorption in ionic medium improves the catalytic properties of lipases applied in the triacylglycerol hydrolysis by synergism. *Bioprocess and Biosystems Engineering*, *39*, 1933–1943.

Rajnish, K. N., Samuel, M. S., Datta, S., Chandrasekar, N., Balaji, R., Jose, S., & Selvarajan, E. (2021). Immobilization of cellulase enzymes on nano and micro-materials for breakdown of cellulose for biofuel production: A narrative review. *International Journal of Biological Macromolecules*, *182*, 1793–1802.

Saeed, T., Naeem, A., Din, I. U., Farooq, M., Khan, I. W., Hamayun, M., & Malik, T. (2022). Synthesis of chitosan composite of metal-organic framework for the adsorption of dyes; kinetic and thermodynamic approach. *Journal of Hazardous Materials*, *427*, 127902.

Samuel, M. S., Suman, S., Selvarajan, E., Mathimani, T., & Pugazhendhi, A. (2020). Immobilization of Cu_3 (btc) 2 on graphene oxide-chitosan hybrid composite for the adsorption and photocatalytic degradation of methylene blue. *Journal of Photochemistry and Photobiology B: Biology*, *204*, 111809.

Sanchez-Ramirez, J., Martinez-Hernandez, J. L., Segura-Ceniceros, P., Lopez, G., Saade, H., Medina-Morales, M. A., ... & Ilyina, A. (2017). Cellulases immobilization on chitosan-coated magnetic nanoparticles: Application for Agave Atrovirens lignocellulosic biomass hydrolysis. *Bioprocess and Biosystems Engineering*, *40*, 9–22.

Sassolas, A., Blum, L. J., & Leca-Bouvier, B. D. (2012). Immobilization strategies to develop enzymatic biosensors. *Biotechnology Advances*, *30*(3), 489–511.

Sillu, D., & Agnihotri, S. (2019). Cellulase immobilization onto magnetic halloysite nanotubes: Enhanced enzyme activity and stability with high cellulose saccharification. *ACS Sustainable Chemistry & Engineering*, *8*(2), 900–913.

Sirisha, V. L., Jain, A., & Jain, A. (2016). Enzyme immobilization: An overview on methods, support material, and applications of immobilized enzymes. *Advances in Food and Nutrition Research*, *79*, 179–211.

Sulyman, A. O., Igunnu, A., & Malomo, S. O. (2020). Isolation, purification and characterization of cellulase produced by *Aspergillus niger* cultured on *Arachis hypogaea* shells. *Heliyon*, *6*(12).

Varalakshmi, V., & Prasad, D. S. R. (2013). Enzyme immobilization: The fascinating challenge in biotechnology. *Journal of Environmental Science, Computer Science and Engineering & Technology*, *2*(3), 840–853.

Wang, L., Zhi, W., Wan, J., Han, J., Li, C., & Wang, Y. (2019). Recyclable β-glucosidase by one-pot encapsulation with Cu-MOFs for enhanced hydrolysis of cellulose to glucose. *ACS Sustainable Chemistry & Engineering*, *7*(3), 3339–3348.

Xu, W., Sun, Z., Meng, H., Han, Y., Wu, J., Xu, J., & Zhang, X. (2018). Immobilization of cellulase proteins on zeolitic imidazolate framework (ZIF-8)/polyvinylidene fluoride hybrid membranes. *New Journal of Chemistry*, *42*(21), 17429–17438.

Zhang, L., Qiao, X., Hu, L., Yang, C., Liu, Q., Wei, C., & Liu, C. (2018). Preparation and evaluation of coal fly ash/chitosan composites as magnetic supports for highly efficient cellulase immobilization and cellulose bioconversion. *Polymers*, *10*(5), 523.

Zhang, L., Qiu, J., Wu, X., Zhang, W., Sakai, E., & Wei, Y. (2014). Preparation of magnetic chitosan nanoparticles as support for cellulase immobilization. *Industrial & Engineering Chemistry Research*, *53*(9), 3448–3454.

Zhang, W., Qiu, J., Feng, H., Zang, L., & Sakai, E. (2015). Increase in stability of cellulase immobilized on functionalized magnetic nanospheres. *Journal of Magnetism and Magnetic Materials*, *375*, 117–123.

Zhou, M., Ju, X., Zhou, Z., Yan, L., Chen, J., Yao, X., & Li, L. Z. (2019). Development of an immobilized cellulase system based on metal–organic frameworks for improving ionic liquid tolerance and in situ saccharification of bagasse. *ACS Sustainable Chemistry & Engineering*, *7*(23), 19185–19193.

19 Oleaginous Algae—A Promising Industrial Source for Essential Fatty Acids
Developments and Scope

Manju Nehra, Amanjyoti, Vikash Nain, Kawaljeet Sandhu, and Rahul Thory

19.1 INTRODUCTION

A group of photosynthetic organisms known as algae originated before higher plants. Algae are typically defined as organisms with chlorophyll in their cells but without distinct roots, stalks, leaves, or other tissues and organs (Lee, 2018). For thousands of years, people have used algae like Aphanizomenon, Spirulina, and Nostoc as food (Jensen et al., 2001). According to Chen et al. (2021), oleaginous algae are the main source of nutrients for many aquatic organisms and are the primary source of water productivity on Earth. Certain algae are especially well-suited for animal food, and preliminary processing methods have been developed (Becker, 2003a; Amorim et al., 2020).

Numerous useful compounds, including proteins, lipids, carbohydrates, mannose, vitamin, terpenoids, and chitin, can be produced by algae (Capek et al., 2023), the number of research has been published as shown in Figure 19.1. Some other critical fatty acids are also produced by certain microalgae, including gamma-linolenic acid (GLA), docosahexaenoic acid (DHA), eicosapentaenoic acid (EPA), and alpha-linolenic acid (ALA) (Xia et al., 2020) (Figure 19.1).

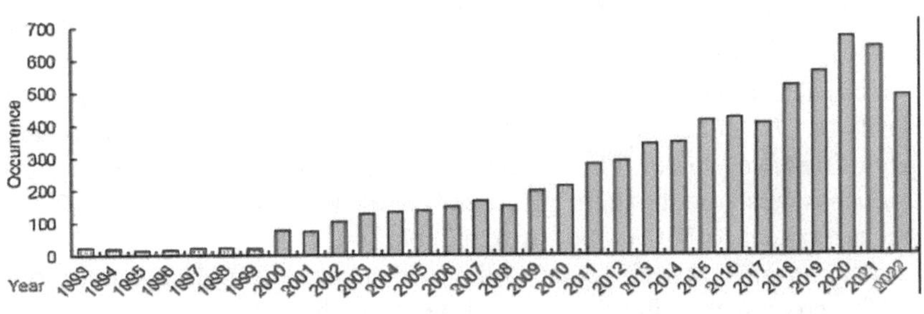

FIGURE 19.1 Several research articles on algae-PUFAs published in the Web of Science in the last several years.

DOI: 10.1201/9781032642567-23

According to the data, the main topics of study in algae research are lipids and fatty acids. Lipids make up between 1% and 40% of the dry weight of algae on average, but in certain species, the amount can reach 80%–85% (Chen and Wang, 2021; Becker, 2003b). Free fatty acids, glycolipids, phospholipids, and glycerides occur in microalgae cells (Becker, 2003b). Some microalgae creatures, such as diatom *Tisochrysis lutea* and *Nitzschia laevis* (Lu et al., 2019), are capable of producing both PUFA and pigment fucoxanthin at the same time. Algae have great potential not only for food and health applications but also for the production of biodiesel, hydrogen, nitrogen removal, flame-retardant substances, and biological material (Chen et al., 2023; Wang et al., 2019). However, due to the high cost of cultivation, removal, and transformation, the commercial benefits of algal biodiesel are limited when compared to renewable sources of energy (Chen et al., 2020). As a result, since 2014, high-value biological products and ecological uses have become the focus of microalgae studies (Rashid et al., 2018).

Chen et al. (2021) reported that polyunsaturated fatty acids (PUFAs), such as ARA, EPA, docosapentaenoic acid (DPA), linolenic acid (LA), and decosahexonic acid (DHA), are currently being created from algae as novel sources (Cui et al., 2021). Even though fish are the source of the underwater fish oil PUFAs that are already on the market, fish consume the oil that algae accumulate in their stomachs (Ebm et al., 2021). Based on how frequently they were found, the microorganisms linked to polyunsaturated fats (PUFAs) were ranked, with diatoms and green algae appearing more frequently. *Chlamydomonas*, *Chlorella*, *Haematococcus*, *Nannochloropsis*, *Dunaliella*, and *Scenedesmus* were the green algae that were reported with greater quantitative data. *Thalassiosira rotula*, *Skeletononema costatum*, and *Phaeodactylum tricornutum* were the most frequently reported diatoms. It implies that most study on the generation of PUFAs in these algae is focused on these algae. Diatoms can typically accumulate significant amounts of long-chain PUFA in their cells, regardless of whether they are supplied with organic carbon sources, although their pace of biomass generation is often quite slow (Cheah et al., 2022; Marella et al., 2021). Astaxanthin can be collected by certain green algae, including *Chromochloris zoffingiensis*, but the majority of the PUFAs accumulated are C_{18} PUFAs (Liu et al., 2016). Therefore, several research teams are also trying to contribute to astaxanthin and PUFA in oleaginous algal cells to make it more inexpensive for industrial manufacturing (Pawar et al., 2021).

19.2 PUFA SOURCES FOUND IN FOOD

According to Simopoulos et al. (2000), the Center for Genetics, Nutrition, and Health at the National Institutes of Health in the United States suggests consuming 2.22 g of ALA daily. Low concentrations of ALA can be found in everyday foods, such as a variety of fresh vegetables. Plants typically contain 100 mg/100 g (fresh weight) of ALA, which is mostly found in the glycolipids of the leaves (Ayerza, 1995). Common sources and contents of ALA are shown in Table 19.1.

The majority of ALA-rich goods on the market come from deep-sea fish oil and selected few crops such as flaxseed. DHA is not something that marine fish can make on their own. According to Brett and MÜller-Navarra (1997), the DHA that is present in their bodies instead comes from marine microorganisms like phytoplankton and

TABLE 19.1
Content of Unsaturated Fatty Acid (%) in Total Lipid of Common Foods

Source	OA	LA	GLA	ALA	EPA	DHA	References
Flaxseed oil	21	13	Less than 1	43–61	<1	<1	Aguillon-Paez et al. (2010)
Refined Canola seed oil	61	25	<1	7	<1	<1	Özcan et al. (2021)
Refined linoleic sunflower oil	34	54	–	–	–	–	Aşkın and Kaya (2020)
Oyster nut oil	9	49	<1	<1	<1	<1	Mwakasege et al. (2011)
Extra-virgin olive oil	55–83	3–21	–	<1	<1	<1	Gunstone (2011)
Soybean oil	22	56	–	7		–	Rodriguez-Miranda et al. (2019)
Walnuts oil	8	61	–	12	–	–	Copolovici et al. (2017)
Butter	15–18	1–2	<1	<1	–	–	Faraji Sarabmirza et al. (2023)
Milk	39–45	9–13	<1	<1	–	–	Anzhany et al. (2021)
Beef	40–42	2–3	<1	<1	<1	<1	Najar-Villarreal et al. (2019)

DW. Dry weight

TABLE 19.2
Essential Fatty Acid Content of Various Algae

Organisms	GLA	ALA	EPA	DHA	References
Acutodesmus obliquus	–	38	–	–	Othman et al. (2019)
Aurantiochytrium sp.	–	–	1	42	Chauhan et al. (2023)
Chlamydomonas reinhardtii	–	44	–	–	Zheng et al. (2022)
Chlorella vulgaris	–	35.4	–	–	Othman et al. (2019)
Chlorococcum amblystomatis	4	31	–	–	Correia et al. (2020)
Codium fragile hariot	–	14.2	3.0 4.4	–	Schmid et al. (2014)
Codium tomentosum	3	12	2	12	Ferreira et al. (2022)
Crypthecodinium cohnii	–	–	–	37–40	Moniz et al. (2021)
Isochrysis galbana	–	6	3	10	Señoráns et al. (2020)
Isochrysis zhangjiangensis	–	11	2	14	Li Y et al. (2019)
Isochrysis galbana	–	11	7	14	Ge et al. (2023)
Nannochloropsis oceanica	–	–	27	–	Matsui et al. (2022)
Oocystis pusilla	–	23	–	–	Vidyashankar et al. (2015)
Phaeodactylum tricornutum	0.8	–	27–30	0.7	Kadalag et al. (2022)
Scenedesmus sp.	–	42	–	–	Meng et al. (2019)
Schizochytrium limacinum	–	1	1	45	Rodriguez-Espana et al. (2022)
Arthrospira platensis	34.5	–	–	–	Yang X et al. (2019)
Spirulina platensis	15.8	–	–	–	Yang X et al. (2019)

builds up in fish, mollusks, and reptiles as a result of the marine food chain (Nichols, 2003). Studies have revealed that some marine algae, fungi, and bacteria may synthesize PUFAs (Ratledge, 2001). As seen by Table 19.2, some algae, in particular, can accumulate ALA at levels between 10% and 40% of the total lipid fraction.

19.3 OUTLOOK

Several health and medicinal advantages of essential PUFAs. People become more conscious of health protection as their quality of life rises, and they focus more on diet and nutrition. By giving people access to high-quality PUFAs, we can improve their health and well-being while also indirectly lowering the cost of healthcare for the general public. To industrialize algae edible oil, it is imperative to investigate the technical model of large-scale production of these high-yield PUFA algae, acquire practical industrial technology skills, and advance the commercialization of algal resources.

19.5 SOURCE OF PUFAS (SYNTHESIS)

PUFAs are straight-chain fatty acids that have two or more double bonds. PUFAs are categorized into ω-x fatty acids, which include omega-9, omega-6, and omega-3 (also known as n-9, n-6, and n-3) depending on the location of the first double bond from the methyl end of the fatty acid (Djuricic and Calder, 2021). As shown in Figure 19.2, the biosynthesis of ω-3, ω-6, and ω-9 PUFAs starts with acetyl-CoA and necessitates a sequence of carbon chain-elongation and desaturation events. The production of ω-3 and ω-6 PUFAs occurs via two different processes, and it is not possible to directly convert intermediates from one pathway into intermediates from the other (Napier, 2002). While ω-6 PUFAs like GLA and ARA are formed from LA (9,12-linoleic acid), ω-3 PUFAs like DHA and EPA are derived from ALA. Higher creatures and microbes both synthesize PUFAs. Fatty acid synthase complex catalyzes the formation of LA from acetyl-CoA and malonyl-CoA. LA is subsequently further metabolized via the ω-6 and ω-3 pathways to produce PUFAs (Lehninger, 1982).

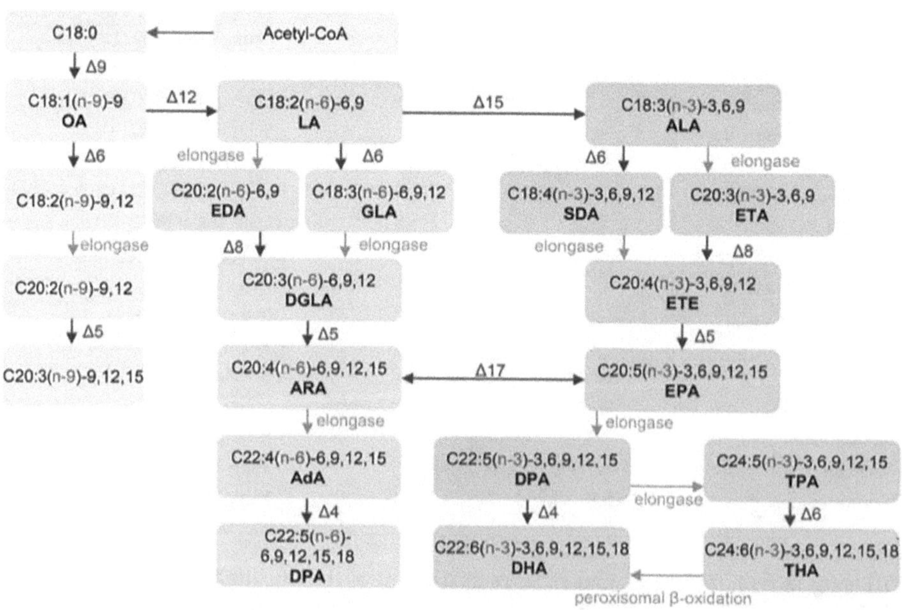

FIGURE 19.2 The general pathways of PUFAs synthesis.

As individuals lack Δ12 and Δ15 dehydrogenases, humans and other mammals are unable to synthesize ω-3 and ω-6 PUFAs from scratch (Napier, 2002). As a result, food is the only source of LA and ALA (Djuricic and Calder, 2021). There is a metabolic competition process between the synthesis routes of ω-3 and ω-6 PUFAs because they share desaturases and chain-elongation enzymes (Smink et al., 2012). As a consequence, it's essential to consume LA and ALA in balance. ω-3 PUFAs like EPA and DHA are formed from the essential fatty acid ALA.

19.6 ENRICHMENT OF PUFAS

Numerous oil-producing algae have been certified as novel food sources and are currently being used in regular meals. Concurrently, related food products are already available for purchase. While some people may need to use medicines, health products, or even milk powder, others may need to augment their diet with foods and edible oils high in omega-3 PUFAs. Omega-3 PUFA raw materials with a single component and specific content are needed to make these goods. However, PUFAs in algal oil need to be purified to achieve the required product characteristics when utilized in pharmaceuticals, health products, or auxiliary medicines (Wei and Wang, 2020; Li X et al., 2019). As an illustration, even more purity is needed for some omega-3 PUFA prescription medications that the US Food and Drug Administration has approved for the prevention and treatment of excessive triglycerides (Blair and Dhillon, 2014; Hoy and Keating, 2009).

19.6.1 Hydrolysis of Lipids

Ester-forming fatty acids can be hydrolyzed and separated via fatty acid hydrolysis, and then they can be further separated and used according to their various qualities.

19.6.2 Saponification-Esterification Method

Using the variations in fatty acid metal salt solubility in specific organic solvents, PUFAs can be isolated and purified. Lipids can be saponified under heating circumstances using a certain ratio of ethanol and NaOH to produce refined oil with a greater PUFA content. After filtering the saponified mixture to produce saponified liquid and solid particles, the saponified compositions can be esterified using diluted H_2SO_4.

19.6.3 Non-Catalytic Hydrolysis Method

Triglycerides can be transesterified directly at high temperatures and pressure, via the Thermal-Fenton mechanism, or in supercritical methanol conditions. However, this method is mostly used to produce biodiesel, which has contaminants and can readily harm PUFAs.

19.6.4 Chemical Catalytic Hydrolysis Method

Lipid hydrolysis is frequently accomplished in the oleochemical sector using acid catalysis, high-pressure hydrolysis, and saponification–acidolysis. The saponification-acidification and acid-catalyzed processes offer a hydrolysis rate of more than 90% and are reasonably priced. But they also generate a lot of waste alkali or acid (Zhuang et al., 2012). Although it needs more equipment and is more expensive to operate, high-pressure steam hydrolysis is a good substitute. These techniques—especially the high-pressure hydrolysis method—are typically applied at higher temperatures (Zarli, 2020).

Neag et al. (2023) reported that alkaline and acid catalysts are two further subcategories of chemical catalysts, which are further split into homogeneous and heterogeneous catalysts. The raw materials CaO/MgO, KF/CaO, magnetic nano-scale $KOH/Fe_2O_3-Al_2O_3$ (Kazemifard et al., 2019), $CaMgO/Al_2O_3$, zirconia (ZrO_2), titanium oxide (TiO_2), zeolites, ion exchange resins, and algal carbon-based solid acid catalysts are among the raw materials that are suitable for heterogeneous catalytic processes (Cao et al., 2021).

19.6.5 Lipase-Catalyzed Method

The selective specificity of lipase, an enzyme that catalyzes a variety of processes, including alcoholysis, hydrolysis, acidolysis, transesterification, and reverse synthesis of triacylglyceride and other water-insoluble esters, is used in enzymatic enrichment. Esterification, transesterification, and hydrolysis are the three primary processes in enzymatic enrichment.

19.6.6 Enzymatic Hydrolysis

Without the requirement for pressurization or steam boiling, the enzymatic hydrolysis method—which uses lipase to break down oil into fatty acids and glycerol—can be performed under mild conditions. Lipases can produce large amounts of particular fatty acids by hydrolysis, transesterification, and esterification processes.

19.6.7 Enzymatic Esterification Method

To enhance UFAs, lipase preferentially catalyzes the reaction between alcohols and saturated fatty acids to form esters, or between alcohols and UFAs to produce esters (Jithu Paul, 2019).

19.6.8 Enzymatic Transesterification Method

The catalysis of lipase entails the catalytic exchange of acyl groups between glycerides and alcohols, free fatty acids, or other esters. These processes are known as transesterification, alcoholysis, and acidolysis, in that order. This technique works very well for making PUFA methyl esters (ethyl esters) or high-concentration PUFAs (Castejón and Señoráns, 2020; Farmani et al., 2007). The lipase hydrolysis approach doesn't require a lot of organic chemicals or further pretreatment; it just needs a single-step reaction in mild circumstances. This technique can enrich PUFAs on glycerides, and Yang et al. (1989) also reported that these fatty acid glycerides hydrolyze more quickly than methyl and ethyl ester enhanced using physiochemical techniques. For this reason, the hydrolysis approach is better suited for human absorption and digestion.

Even after ten reaction cycles, the immobilized lipase can retain 60% of its initial activity (Liu et al., 2018). Extensive screening is necessary for lipase with high specificity, which can raise expenses. Computer-aided design can be used to improve lipase's functional specificity.

19.7 INDUSTRIAL DEVELOPMENT

In addition to their efficient photosynthesis, oleaginous microalgae have a lower growth area requirement than terrestrial plants, and they are capable of channeling most of their energy into cell division, increasing biomass productivity (Mwakasege et al., 2021). As a result, they are a promising source of renewable biofuels. Among the four types of cultivation of microalgae, autotrophic, mixotrophic, heterotrophic, and photoheterotrophic, both inorganic and organic carbon sources can be used (Gunstone, 2011; Copolovici et al., 2017). Microalgae synthesize TAGs mainly within subcellular compartments like chloroplasts and endoplasmic reticulum through multiple enzyme reactions (Faraji Sarabmirza et al., 2023). The accumulation of lipids in microalgae is caused by three major steps: fatty acid synthesis in the chloroplast, glycerolipid synthesis in the endoplasmic reticulum, and TAG accumulation in the oil bodies (Anzhany et al., 2021). Various stress factors, such as environmental, chemical, or physical stresses, enable the synthesis of large amounts of lipids individually or in combination (Najar-Villarreal et al., 2019). Microalgae can shift their metabolism to produce and accumulate neutral lipids, or TAGs, under various stress conditions. TAGs are a type of carbon and energy storage (Patel et al., 2017; Arora et al., 2018). The de novo pathway, which microalgae use to synthesize lipids, begins in the chloroplast with carbon dioxide fixation into glucose (sugar). These sugars are then further metabolized to form acetyl-CoA, a precursor to the synthesis of fatty acids (Yu et al., 2011). In addition to providing a carbon source, photosynthetic reactions in autotrophic cultivation help produce the reducing power (NADH and NADPH) that is ultimately needed for lipid synthesis (Fan et al., 2011). However, the main disadvantages of using autotrophic culture on an industrial scale are its poor biomass and lipid productivity as well as the need for suitable photobioreactors (Perez-Garcia et al., 2011). Because heterotrophic cultivation of algae has numerous advantages over photoautotrophic farming, including cost-effectiveness and relative ease of cultivation with low daily maintenance, researchers have been concentrating more on it in recent decades (Guldhe et al., 2017; Perez-Garcia et al., 2011). For the heterotrophic mode of cultivation, glucose is a frequently used carbon source; however, to minimize the high production cost of feedstocks, glucose must be derived from renewable sources (Chen et al., 2017).

19.7.1 Oleaginous Microalgae Used for Nutraceuticals

Thraustochytrids and other oleaginous microorganisms with high PUFA content were investigated for the synthesis of nutraceuticals.

19.7.1.1 Oleaginous *Thraustochytrids*

While *throaustochytrids* are a heterotrophic fungus-like group of *Stramenopiles*, the term "algae" is frequently used inappropriately to describe them. When referring to photosynthetic organisms, the term "algae" is frequently used to exclude prokaryotes and embryophyta, such as cyanobacteria (Leyland et al., 2017).

A good source for the commercial production of DHA is *thaustochytrids*. Some researchers are creating new and improved technologies to produce DHA using *thraustochytrids* in response to the growing demand for DHA. Higher temperatures (25°C–30°C) often promote the best growth, while lower temperatures encourage the creation of DHA at the expense of slower growth. This ultimately leads to a low DHA output overall (Raghukumar, 2008). Several cultivation strategies can be used to get around this. For example, *thraustochytrids* can be grown at 25°C at first to promote their growth, and then they can be shifted to 15°C to increase the production of DHA (Singh et al., 1996). The species has been shown to have a broad pH tolerance, with growth and DHA production occurring in the range of 5–8. Moreover, there are strain-specific variations in the salinity optima and tolerance levels of *thraustochytrids*; certain strains can even thrive at salinities as low as 2 ppt (Burja et al., 2006). *Schizochytrium spp.* is a well-researched *thraustochytrid* that may produce between 35% and 40% DHA by weight of its total fatty acid content. Nowadays, *Schizochytrium* species are employed in the commercial synthesis of PUFA (Dewapriya and Kim, 2014). Studies on the potential for a heterotrophic culture of the marine species *Aurantiochytrium* sp. T66 (ATCC PRA-276) utilizing hydrolysates of forest biomass (30 g/L glucose) in flasks has been carried out recently. Results of the investigation showed that the total lipids and cell dry weight were 4.98 and 10.39 g/L, respectively, with DHA making up 25.98% of the total lipids (Patel et al., 2019). *Aurantiochytrium* sp. T66 was also grown in a fermenter in the same work, and the results showed increased values for docosahexaenoic acid content (35.76% of total lipids), total lipid (5.90 g/L), and cell dry weight (11.24 g/L) (Patel et al., 2019).

19.7.1.2 Oleaginous Microalgae and Diatoms

While marine microorganisms have adapted to a variety of environments, they have developed to be metabolically efficient. As a result, they have evolved to use the restricted amount of dissolved organic matter in a ratio where more metabolites are created than energy is consumed, enabling them to produce distinctive microbial metabolites (Dewapriya and Kim, 2014). Fatty acid biosynthesis is species-specific and reliant on various cultivation methods (Yodsuwan et al., 2017). By adding various specialized abiotic variables, it is possible to increase the lipid content and the generation of certain fatty acids in culture, but high lipid content circumstances may not be the most conducive to growth (Aussant et al., 2018).

The ability of certain marine microbes to synthesize fatty acids with nutraceutical value has been utilized (Dewapriya and Kim, 2014; Gong et al., 2014). Since marine species must synthesize more unsaturated fatty acids to survive in the marine environment, marine microalgae have a higher content of PUFA than water bodies' species (Dewapriya and Kim, 2014). Therefore, growing marine microalgae might be more profitable. One marine oleaginous diatom that has been shown to produce large levels of EPA is *Fistulifera solaris*, which is grown in a photoautotrophic environment. The goal of the study was to get *F. solaris* to produce EPA photoautotrophically on par with heterotrophic fungal cultures. It has been stated that *F. solaris* can produce 135.7 mg/(L • day) of EPA under ideal production circumstances (Tanaka et al., 2017).

While *Phaeodactylum tricornutum* is a photoautotrophic diatom, it can also be grown in a mixotrophic environment, which allows for both photoautotrophic and heterotrophic cultivation conditions to occur simultaneously. In this scenario, however, light is necessary for the glucose to be utilized (Zheng et al., 2013). *P. tricornutum* could grow heterotrophically in the dark on several carbon sources when genetic engineering introduced glucose transporters (De Swaaf et al., 2013). When *P. tricornutum* was cultivated mixotrophically, its EPA and DHA productivities were higher than those obtained via photoautotrophic growth. Additionally, it has been demonstrated that *P. tricornutum* produced 3.11- and 3.2-times higher eicosapentaenoic acid production when using birch and spruce hydrolysates as a glucose source in comparison to photoautotrophic cultivation. While DHA makes up nearly all of the PUFA in *C. cohnii's* lipid profile—up to 65% of all fatty acids—it is unlike any other PUFA (Ratledge et al., 2010). This feature makes the procedure of purifying DHA from *C. cohnii* highly appealing.

Martek Biosciences generated DHA-rich oil from *C. cohnii* in the late 1980s for the infant formula market, but the oil had to be free of EPA and have a high DHA content (Miller et al., 2019). Being heterotrophic organisms that may be cultivated to produce large biomass in fermenters is a benefit shared by *C. cohnii* and the previously stated *thraustochytrids*. Self-shading is a key drawback of photoautotrophic culture. For instance, pond-grown algae typically reach a culture density of around 0.5 g/L, at which point they start to shade themselves and produce less biomass (Gladue et al., 1994). Due to this, their limited production keeps them from being commercialized (Knothe, 2012). To prevent costly heating and artificial lighting, species that may be grown at the site must be considered, depending on the latitudes of interest (Schulze et al., 2019). *Leptocylindrus danicus*, a diatom strain, is capable of growing at temperatures as low as 8°C and has demonstrated a consistent biomass content over a broad temperature range, perhaps making it suitable for year-round cultivation (Yodsuwan et al., 2017). When considering the high lipid productivity that may be attained in comparison to the photoautotrophic yield, mixotrophic and heterotrophic cultivation circumstances are thought to be more favorable, even if phototrophic cultivations fix carbon dioxide and release oxygen.

REFERENCES

Amorim M.L., Soares J., Coimbra J.S.D.R., Leite M.D.O., Albino L.F.T., & Martins M.A. (2021). Microalgae proteins: Production, separation, isolation, quantification, and application in food and feed. *Critical Reviews in Food Science and Nutrition*, *61*(12), 1976–2002. https://doi.org/10.1080/10408398.2020.1768046

Anzhany D., DespalToharmat T., Nuraina N., Hamidah A.N., & Rofiah N. (2021). Effect of different altitudes on milk fatty acid and conjugated linoleic acid (CLA) profiles. *IOP Conference Series Earth and Environmental Science*, *667*, 012102.

Arora N., Patel A., Pruthi P.A., Poluri K.M., & Pruthi V. (2018). Utilization of stagnant non-potable pond water for cultivating oleaginous microalga *Chlorella minutissima* for biodiesel production. *Renewable Energy*, *126*, 30–37.

Aşkın B., & Kaya Y. (2020). Effect of deep frying process on the quality of the refined oleic/linoleic sunflower seed oil and olive oil. *Journal of Food Science and Technology*, *57*, 4716–4725. https://doi.org/10.1007/s13197-020-04655-4

Aussant J., Guihéneuf F., & Stengel D.B. (2018). Impact of temperature on fatty acid composition and nutritional value in eight species of microalgae. *Applied Microbiology Biotechnology*, *102*, 5279–5297. https://doi.org/10.10.1007/s00253-018-9001-x.

Ayerza R. (1995). Oil content and fatty acid composition of chia (*Salvia hispanica* L.) from five northwestern locations in Argentina. *Journal of American Oil Chemists' Society*, *72*, 1079–1081. https://doi.org/10.1007/BF02660727

Becker W. (2003a). Microalgae for aquaculture: The nutritional value of microalgae for aquaculture. In: Handbook Microalgal Culture: Applied Phycology (pp. 380–391). https://doi.org/10.1002/9780470995280.ch21

Becker W. (2003b). Microalgae in human and animal nutrition. In: Handbook Microalgal Culture: Applied Phycology and Biotechnology (pp. 312–351). https://doi.org/10.1002/9780470995280.ch18

Blair H.A., & Dhillon S. (2014). Omega-3 carboxylic acids (Epanova): A review of its use in patients with severe hypertriglyceridemia. *American Journal of Cardiovascular Drugs*, *14*, 393–400. https://doi.org/10.1007/s40256-014-0090-3

Brett M., & Müller-Navarra D. (1997). The role of highly unsaturated fatty acids in aquatic foodweb processes. *Freshwater Biology*, *38*, 483–499.

Burja A.M., Radianingtyas H., Windust A., & Barrow C.J. (2006). Isolation and characterization of polyunsaturated fatty acid producing Thraustochytrium species: Screening of strains and optimization of omega-3 production. *Applied Microbiology and Biotechnology*, *72*, 1161–1169. https://doi.org/10.10.1007/s00253-006-0419-1.

Cao M., Peng L., Xie Q., Xing K., & Lu M., Ji J. (2021). Sulfonated *Sargassum horneri* carbon as solid acid catalyst to produce biodiesel via esterification. *Bioresource Technology*, *324*, 124614. https://doi.org/10.1016/j.biortech.2020.124614

Capek L., Uhliarikova I., Kostalova Z., Hindakova A., & Capek P. (2023). Structural properties of the extracellular biopolymer (beta-D-xylo-alpha-D-mannan) produced by the green microalga *Gloeocystis vesiculosa* Nageli. *Carbohydrates Resources*, *525*, 108766.

Castejón N., & Señoráns F.J. (2020). Enzymatic modification to produce health-promoting lipids from fish oil, algae and other new omega-3 sources: A review. *New Biotechnology*, *57*, 45–54. https://doi.org/10.1016/j.nbt.2020.02.006

Chauhan A.S., Patel A.K., Chen C.-W., Chang J.-S., Michaud P., & Dong C.-D. et al. (2023). Enhanced production of high-value polyunsaturated fatty acids (PUFAs) from potential thraustochytrid *Aurantiochytrium* sp. *Bioresource Technology*, *370*, 128536. https://doi.org/10.1016/j.biortech.2022.128536

Cheah Y.T., Ng B.W., Tan T.L., Chia Z.S., & Chan D.J.C. (2022). Biomass and eicosapentaenoic acid production from *Amphora* sp. under different environmental and nutritional conditions. *Biotechnology and Applied Biochemistry*, *2379*. https://doi.org/10.1002/bab.2379

Chen H., Li K., Xue C., & Wang Q. (2021). A novel method for non-invasive estimation of primary productivity in aquatic ecosystems using a chlorophyll fluorescence-induced dynamic curve. *Frontiers Microbiology*, *12*, 682250. https://doi.org/10.3389/fmicb.2021.682250

Chen H., & Wang Q. (2021). Regulatory mechanisms of lipid biosynthesis in microalgae. *Biology Reviews*, *96*, 2373–2391. https://doi.org/10.1111/brv.12759

Chen H.H., & Jiang J.G. (2017). Lipid accumulation mechanisms in auto- and heterotrophic microalgae. *Journal of Agricultural and Food Chemistry*, *65*, 8099–8110.

Chen W., Li T., Ren Y., Wang J., Chen H., & Wang Q. (2023). Biological hydrogen with industrial potential: Improvement and prospection in biohydrogen production. *Journal of Cleaner Production*, *387*, 135777. https://doi.org/10.1016/j.jclepro.2022.135777

Copolovici D., Bungau S., Boscencu R., Tit D.M., & Copolovici L. (2017). The fatty acids composition and antioxidant activity of walnut cold press oil. *Revista de Chimie*, *68*, 507–509. https://doi.org/10.37358/rc.17.3.5489

Cui Y., Thomas-Hall S.R., Chua E.T., & Schenk P.M. (2021). Development of high-level omega-3 eicosapentaenoic acid (EPA) production from *Phaeodactylum tricornutum. Journal of Phycology*, *57*, 258–268. https://doi.org/10.1111/jpy.13082

Da Silva T.L., Mendes A., Mendes R.L., Calado V., Alves S.S., Vasconcelos J.M.T., & Reis A. (2006). Effect of n-dodecane on *Crypthecodinium cohnii* fermentations and DHA production. *Journal of Industrial Microbiology Biotechnology*, *33*, 408–416.

De Swaaf M.E., Sijtsma L., & Pronk J.T. (2003). High-cell-density fed-batch cultivation of the docosahexaenoic acid producing marine alga *Crypthecodinium cohnii. Biotechnology Bioengineering*, *81*, 666–672. https://doi.org/10.10.1002/bit.10513.

Dewapriya P., & Kim S. (2014). Kwon Marine microorganisms: An emerging avenue in modern nutraceuticals and functional foods. *Food Research International*, *56*, 115–125.

Djuricic I., & Calder P.C. (2021). Beneficial outcomes of omega-6 and omega-3 polyunsaturated fatty acids on human health: An update for 2021. *Nutrients*, *13*, 2421. https://doi.org/10.3390/nu13072421

Ebm N., Guo F., Brett M.T., Bunn S.E., & Kainz M.J. (2021). Polyunsaturated fatty acids in fish tissues more closely resemble algal than terrestrial diet sources. *Hydrobiologia*, *848*, 371–383. https://doi.org/10.1007/s10750-020-04445-1

Fan J., Andre C., & Xu C. (2011). A chloroplast pathway for the de novo biosynthesis of triacylglycerol in *Chlamydomonas reinhardtii. FEBS Letters*, *585*, 1985–1991. https://doi.org/10.10.1016/j.febslet.2011.05.018.

Faraji Sarabmirza R., Joolaei Ahranjani P., Rashidi L., Mousavi M., Khodaiyan F., & Rashidi Nodeh H. (2023). An investigation on conjugated linoleic acid content, fatty acid composition, and physicochemical characteristics of Iranian Kurdish butter oil. *Food Science and Nutrition*, *11*, 1051–1058. https://doi.org/10.1002/fsn3.3142

Farmani J., Hamedi M., Safari M., & Madadlou A. (2007). Trans-free Iranian vanaspati through enzymatic and chemical transesterification of triple blends of fully hydrogenated soybean, rapeseed and sunflower oils. *Food Chemistry*, *102*, 827–833.

Gladue R.M., & Maxey J.E. (1994). Microalgal feeds for aquaculture. *Journal of Applied Phycology*, *6*, 131–141. https://doi.org/10.10.1007/BF02186067.

Gong Y., Wan X., Jiang M., Hu C., Hu H., & Huang F. (2014). Metabolic engineering of microorganisms to produce omega-3 very long-chain polyunsaturated fatty acids. *Progress in Lipid Research*, *56*, 19–35. https://doi.org/10.10.1016/j.plipres.2014.07.001.

Guldhe A., Ansari F.A., Singh P., & Bux F. (2017). Heterotrophic cultivation of microalgae using aquaculture wastewater: A biorefinery concept for biomass production and nutrient remediation. *Ecological Engineering*, *99*, 47–53. https://doi.org/10.10.1016/j.ecoleng.2016.11.013.

Gunstone F. (2011). *Vegetable Oils in Food Technology: Composition, Properties and Uses.* New York: John Wiley & Sons.

Hoy S.M., & Keating G.M. (2009). Omega-3 ethylester concentrate. *Drugs*, *69*, 1077–1105. https://doi.org/10.2165/00003495–200969080-00008

Huang P.W., Wang L.R., Geng S.S., Ye C., Sun X.M., & Huang H. (2021). Strategies for enhancing terpenoids accumulation in microalgae. *Applied Microbiology and Biotechnology*, *105*, 4919–4930. https://doi.org/10.1007/s00253-021-11368-x

Jacob J.P., & Mathew S. (2017). Effect of lipases from *Candida cylinderacea* on enrichment of PUFA in marine microalgae. *Journal of Food Processing and Preservation*, *41*, e12928. https://doi.org/10.1111/jfpp.12928

Jensen G., Ginsberg D.I., & Drapeau C. (2001). Blue-green algae as an immuno-enhancer and biomodulator. *Journal of America Nutraceutical Association*, *3*, 24–30.

Jithu Paul J. (2019). Bioconcentration of marine algae using lipase enzyme. In: V. Milada (Ed.), *Microalgae*. Rijeka: IntechOpen. https://doi.org/10.5772/intechopen.87026

Kazemifard S., Nayebzadeh H., Saghatoleslami N., & Safakish E. (2019). Application of magnetic alumina-ferric oxide nanocatalyst supported by KOH for *in-situ* transesterification of microalgae cultivated in wastewater medium. *Biomass Bioenergy*, *129*, 105338. https://doi.org/10.1016/j.biombioe.2019.105338

Knothe G. (2012). Fuel properties of highly polyunsaturated fatty acid methyl esters. Prediction of fuel properties of algal biodiesel. *Energy Fuels*, *26*, 5265–5273. https://doi.org/10.10.1021/ef300700v.

Lee R E. (2018). *Phycology*, 5th edn. Cambridge: Cambridge University Press.

Lehninger A.L. (1982). *Principles of Biochemistry*. New York: Worth Publishers Inc.

Leyland B., Leu S., & Boussiba S. (2017). Are Thraustochytrids algae *Fungal Biology*, *121*, 835–840. https://doi.org/10.10.1016/j.funbio.2017.07.006.

Li X., Liu J., Chen G., Zhang J., Wang C., & Liu B. (2019). Extraction and purification of eicosapentaenoic acid and docosahexaenoic acid from microalgae: A critical review. *Algal Research*, *43*, 101619. https://doi.org/10.1016/j.algal.2019.101619

Liu J., Mao X., Zhou W., & Guarnieri M.T. (2016). Simultaneous production of triacylglycerol and high-value carotenoids by the astaxanthin-producing oleaginous green microalga *Chlorella zofingiensis*. *Bioresource Technology*, *214*, 319–327. https://doi.org/10.1016/j.biortech.2016.04.112

Liu N., Li D., Wang W., Hollmann F., Xu L., & Ma Y. et al. (2018). Production and immobilization of lipase PCL and its application in the synthesis of α-linolenic acid-rich diacylglycerol. *Journal of Food and Biochemistry*, *42*, e12574. https://doi.org/10.1111/jfbc.12574

Ljubic A., Thulesen E.T., Jacobsen C., & Jakobsen J. (2021). UVB exposure stimulates production of vitamin D-3 in selected microalgae. *Algal Research-Biomass Biofuels Bioproduction*, *59*, 102472. https://doi.org/10.1016/j.algal.2021.102472

Lu X., Liu B., He Y., Guo B., Sun H., & Chen F. (2019). Novel insights into mixotrophic cultivation of *Nitzschia laevis* for co-production of fucoxanthin and eicosapentaenoic acid. *Bioresource Technology*, *294*, 122145. https://doi.org/10.1016/j.biortech.2019.122145

Marella T.K., Bhattacharjya R., & Tiwari A. (2021). Impact of organic carbon acquisition on growth and functional biomolecule production in diatoms. *Microbial Cell Factories*, *20*, 135. https://doi.org/10.1186/s12934-021-01627-x

Miller K.K., & Alper H.S. (2019). *Yarrowia lipolytica*: More than an oleaginous workhorse. *Applied Microbiology Biotechnology*, *103*, 9251–9262.

Mwakasege E., Treydte A., Hoeglinger O., Kassim N., & Makule E. (2021). Fatty acid contents and stability of oyster nut oil (*Telfairia pedata*) compared to flaxseed and sunflower oil. *International Journal of Food Science*, 1–8. https://doi.org/10.1155/2021/9985910

Najar-Villarreal F., Boyle E., Danler R., O'quinn T., Houser T., & Gonzalez J. (2019). Fatty acid composition, proximate analysis, and consumer sensory evaluation of United States retail grass-fed ground beef. *Meat and Muscle Biology*, *3*, 18. https://doi.org/10.22175/mmb2019.06.0018

Napier J.A. (2002). Plumbing the depths of PUFA biosynthesis: A novel polyketide synthase-like pathway from marine organisms. *Trends Plant Science*, *7*, 51–54. https://doi.org/10.1016/S1360-1385(01)02191-4

Neag E., Stupar Z., Maicaneanu S.A., & Roman C. (2023). Advances in biodiesel production from microalgae. *Energies*, *16*, 1129. https://doi.org/10.3390/en16031129

Nichols D.S. (2003). Prokaryotes and the input of polyunsaturated fatty acids to the marine food web. *FEMS Microbiology Letters*, *219*, 1–7. https://doi.org/10.1016/s0378-1097(02)01200-4

Pan X., Xu L., Zhang Y., Xiao X., Wang X., & Liu Y. et al. (2012). Efficient display of active *Geotrichum sp.* lipase on Pichia pastoris cell wall and its application as a whole-cell biocatalyst to enrich EPA and DHA in fish oil. *Journal of Agricultural and Food Chemistry*, *60*, 9673–9679. https://doi.org/10.1021/jf301827y

Patel A., Matsakas L., Rova U., & Christakopoulos P. (2018). Heterotrophic cultivation of *Auxenochlorella protothecoides* using forest biomass as a feedstock for sustainable biodiesel production. *Biotechnology Biofuels*, *11*, 169. https://doi.org/10.10.1186/s13068-018-1173-1.

Patel A., Pruthi V., & Pruthi P.A. (2017). Synchronized nutrient stress conditions trigger the diversion of CDP-DG pathway of phospholipids synthesis towards de novo TAG synthesis in oleaginous yeast escalating biodiesel production. *Energy*, *1390*, 962–974. https://doi.org/10.10.1016/j.energy.2017.08.052.

Patel A., Rova U., Christakopoulos P., & Matsakas L. (2019). Simultaneous production of DHA and squalene from *Aurantiochytrium* sp. grown on forest biomass hydrolysates. *Biotechnology Biofuels*, *12*, 1–12. https://doi.org/10.10.1186/s13068-019-1593–6.

Pawar P.R., Velani S., Kumari S., Lali A.M., & Prakash G. (2021). Isolation and optimization of a novel thraustochytrid strain for DHA rich and astaxanthin comprising biomass as aquafeed supplement. *Biotechnology*, *11*, 71. https://doi.org/10.1007/s13205-020-02616-4

Perez-Garcia O., Escalante F.M.E., de-Bashan L.E., & Bashan Y. (2011). Heterotrophic cultures of microalgae: Metabolism and potential products. *Water Resource*, *45*, 11–36.

Raghukumar S. (2008). Thraustochytrid marine protists: Production of PUFAs and other emerging technologies. *Marine Biotechnology*, *10*, 631–640.

Rashid N., Park W.K., & Selvaratnam T. (2018). Binary culture of microalgae as an integrated approach for enhanced biomass and metabolites productivity, wastewater treatment, and bioflocculation. *Chemosphere*, *194*, 67–75. https://doi.org/10.1016/j.chemosphere.2017.11.108

Ratledge C. (2001). Microbial lipids. In H.-J. Rehm & G. Reed (Eds.), *Biotechnology Set*, 133–197. Wiley.

Ratledge C., Streekstra H., Cohen Z., & Fichtali J. (2010). *Single Cell Oils: Microbial and Algal Oils*, 2nd ed. San Diego, CA: Elsevier.

Schulze P.S.C., Hulatt C.J., Morales-Sánchez D., Wijffels R.H., & Kiron V. (2019). Fatty acids and proteins from marine cold adapted microalgae for biotechnology. *Algal Research*, *42*, 42. https://doi.org/10.10.1016/j.algal.2019.101604.

Simopoulos A.P., Leaf A., & Salem N. Jr (2000). Workshop statement on the essentiality of and recommended dietary intakes for omega-6 and omega-3 fatty acids. *Prostaglandins, Leukotrienes, and Essential Fatty Acids*, *63*, 119–121. https://doi.org/10.1054/plef.2000.0176

Singh A., Wilson S., & Ward O.P. (1996). Docosahexaenoic acid (DHA) production by *Thraustochytrium sp.* ATCC 20892. *World Journal of Microbiology and Biotechnology*, *12*, 76–81. https://doi.org/10.10.1007/BF00327806.

Smink W., Gerrits W.J.J., Gloaguen M., Ruiter A., & van Baal J. (2012). Linoleic and α-linolenic acid as precursor and inhibitor for the synthesis of long-chain polyunsaturated fatty acids in liver and brain of growing pigs. *Animal*, *6*, 262–270. https://doi.org/10.1017/S1751731111001479

Steinrücken P., Prestegard S.K., de Vree J.H., Storesund J.E., Pree B., Mjøs S.A., & Erga S.R. (2018). Comparing EPA production and fatty acid profiles of three *Phaeodactylum tricornutum* strains under western Norwegian climate conditions. *Algal Research*, *30*, 11–22. https://doi.org/10.10.1016/j.algal.2017.12.001.

Tanaka T., Yabuuchi T., Maeda Y., Nojima D., Matsumoto M., & Yoshino T. (2017). Production of eicosapentaenoic acid by high cell density cultivation of the marine oleaginous diatom *Fistulifera solaris*. *Bioresource Technology*, *245*, 567–572.

Valverde L.M., Moreno P.A.G., Callejón M.J.J., Cerdán L.E., & Medina A.R. (2013). Concentration of eicosapentaenoic acid (EPA) by selective alcoholysis catalyzed by lipases. *European Journal of Lipid Science and Technology*, *115*, 990–1004.

Wang J., Zhou W., Chen H., Zhan J., He C., & Wang Q. (2019). Ammonium nitrogen tolerant *Chlorella* strain screening and its damaging effects on photosynthesis. *Frontier Microbiology*, *9*, 3250. https://doi.org/10.3389/fmicb.2018.03250

Wei B., & Wang S. (2020). Separation of eicosapentaenoic acid and docosahexaenoic acid by three-zone simulated moving bed chromatography. *Journal of Chromatography*, (1625), 461326. https://doi.org/10.1016/j.chroma.2020.461326

Whalen J., Xu C., Shen F., Kumar A., Eklund M., & Yan J. (2017). Sustainable biofuel production from forestry, agricultural and waste biomass feedstocks. *Applied Energy*, *198*, 281–283. https://doi.org/10.10.1016/j.apenergy.2017.05.079.

Xia Y., Zhang Y.-T., Sun J.-Y., Huang H., Zhao Q., & Ren L.-J. (2020). Strategies for enhancing eicosapentaenoic acid production: From fermentation to metabolic engineering. *Algal Research*, *51*, 102038. https://doi.org/10.1016/j.algal.2020.102038

Yang J., Wen C., Duan Y., Deng Q., Peng D., & Zhang H. et al. (2021). The composition, extraction, analysis, bioactivities, bioavailability and applications in food system of flaxseed (*Linum usitatissimum L.*) oil: A review. *Trends Food Science and Technology*, *118*, 252–260. https://doi.org/10.1016/j.tifs.2021.09.025

Yang L.-Y., Kuksis A., & Myher J.J. (1989). Lumenal hydrolysis of menhaden and rapeseed oils and their fatty acid methyl and ethyl esters in the rat. *Biochemical Cell Biology*, *67*, 192–204. https://doi.org/10.1139/o89-030

Yang X., Li Y., Li Y., Ye D., Yuan L., & Sun Y. et al. (2019). Solid matrix-supported supercritical CO2 enhances extraction of γ-linolenic acid from the *Cyanobacterium arthrospira* (*spirulina*) platensis and bioactivity evaluation of the molecule in zebrafish. *Marine Drugs*, *17*, *203*. https://doi.org/10.3390/md17040203

Yew G.Y., Tan X., Chew K.W., Chang J.-S., Tao Y., & Jiang N. et al. (2021). Thermal-Fenton mechanism with sonoprocessing for rapid non-catalytic transesterification of microalgal to biofuel production. *Chemical Engineering Journal*, *408*, 127264.

Yodsuwan N., Sawayama S., & Sirisansaneeyakul S. (2017). Effect of nitrogen concentration on growth, lipid production and fatty acid profiles of the marine diatom *Phaeodactylum tricornutum*. *Agricultural Natural Resource*, *51*, 190–197.

Yu, W.L., Ansari, W., Schoepp, N.G., Hannon, M.J., Mayfield, S.P., & Burkart, M.D. (2011). Modifications of the metabolic pathways of lipid and triacylglycerol production in microalgae. *Microbial Cell Factories*, *10*, 1–11. https://doi.org/10.10.1186/1475–2859-10-91.

Zarli A. (2020). Oleochemicals: All time players of green chemistry. In: A. Basile, G. Centi, M.D. Falco, G. Iaquaniello (Eds.), *Studies in Surface Science and Catalysis* (pp. 77–95). Amsterdam, Netherlands: Elsevier, https://doi.org/10.1016/B978-0-444–64337-7.00006-9

Zheng S., Zou S., Feng T., Sun S., Guo X., & He M. et al. (2022). Low temperature combined with high inoculum density improves alpha-linolenic acid production and biochemical characteristics of *Chlamydomonas reinhardtii*. *Bioresources Technology*, *348*, 126746. https://doi.org/10.1016/j.biortech.2022.126746

Zhuang Y., Guo J., Chen L., Li D., Liu J., & Ye N. (2012). Microwave-assisted direct liquefaction of Ulva prolifera for bio-oil production by acid catalysis. *Bioresources Technology*, *116*, 133–139. https://doi.org/10.1016/j.biortech.2012.04.036

20 Fermentation Biotechnology and Cereal Industry
Nutritional and Economic Aspects

Sukhvinder Singh Purewal, Avneet Kaur, Raj Kumar Salar, and Anil Kumar Siroha

20.1 INTRODUCTION

Cereal grains constitute a significant component of the global diet, providing essential nutrients such as carbohydrates, proteins, vitamins, and minerals (Kaur et al. 2019; Salar et al. 2012). Nonetheless, these grains often contain non-nutrient constituents and exhibit limited bioavailability of certain nutrients. Solid-state fermentation (SSF) has emerged as a promising biotechnological strategy to enhance the nutritional profile of cereal grains and other natural resources (Zhang et al. 2022; Roasa et al. 2021; Costa-Maia et al. 2020). This chapter aims to comprehensively explore the impact of SSF on the nutrients of cereal grains. SSF stands as a promising biotechnological approach with the potential to revolutionize the nutritional, functional, and sensory properties of cereal grains. Cereal grains, including staples like wheat, rice, maize, barley, and sorghum, play a pivotal role in global diets, constituting a significant source of essential nutrients (Kaur et al. 2021; Serna-Saldivar 2016). Nonetheless, these grains often exhibit inherent limitations such as nutrient availability, digestibility challenges, and the presence of non-nutrient components. SSF presents an innovative solution to address these concerns, harnessing the dynamic interactions between microorganisms and cereal grains within a controlled, non-aqueous environment. Diverging from conventional submerged fermentation where microorganisms grow in liquid media, SSF involves the cultivation of microorganisms on the solid surface of grain substrates (Pandey 2003). This unique characteristic lends itself well to the modification of cereal grains, closely mimicking natural microbial ecosystems while facilitating the synthesis of diverse enzymes and bioactive compounds (Kaur and Purewal 2023; Bhanja et al. 2009).

Utilizing microorganisms such as fungi, bacteria, and yeast, SSF involves their colonization of the cereal grain matrix (Lizardi-Jimenez et al. 2017). This colonization prompts the release of a spectrum of enzymes that catalyze the breakdown of complex macromolecules within the grains. This enzymatic activity encompasses amylolytic

DOI: 10.1201/9781032642567-24

enzymes responsible for starch hydrolysis, proteases that degrade proteins, and phytases that target antinutritional factors such as phytates. The result is an enhanced release of nutrients and bioactive compounds, contributing to increased nutrient bioavailability, heightened protein quality, and the attenuation of antinutritional factors hindering nutrient absorption. The allure of SSF lies in its flexibility, adaptability, and capacity to create functional foods with augmented nutritional profiles. This process can be fine-tuned by adjusting microbial strains, fermentation conditions, and substrates, allowing for tailored outcomes aligned with specific nutritional objectives (Lou et al. 2023; Roasa et al. 2021; Garrido-Galand et al. 2021; Postemsky and Curvetto 2015). Moreover, SSF represents an environmentally friendly and cost-effective avenue for value augmentation of cereal grains, utilizing agricultural residues and food industry by-products as substrates, thereby promoting waste reduction and sustainability. This chapter delves into the nuances of applying SSF to cereal grains, unraveling the underlying mechanisms, the influence of microbial activities on nutritional transformation, and the broader implications for human health and food security.

20.2 SOLID-STATE FERMENTATION AND ITS MECHANISMS

SSF involves the cultivation of microorganisms, typically fungi and bacteria, on solid substrates without the presence of free-flowing liquid. This process occurs under controlled conditions, facilitating the synthesis of enzymes by microorganisms (Nighojkar et al. 2019; Salar et al. 2012). These enzymes facilitate the hydrolysis of complex compounds within cereal grains, thereby augmenting their nutritional content and bioavailability (Purewal et al. 2019). Key mechanisms underlying SSF include enzymatic degradation, synthesis of bioactive compounds, reduction of non-nutritional factors, and enhancement of microbial biomass.

20.2.1 Effects on Carbohydrates

SSF involves the cultivation of microorganisms, such as fungi and bacteria, on the surface of cereal grains. These microorganisms produce enzymes that can effectively hydrolyze and initiate breakdown of complex carbohydrates, such as starch, into simpler sugars including glucose, fructose, and maltose (Purewal et al. 2020; Ravindran and Jaiswal 2016; Salar et al. 2012; de-Souza et al. 2010). This amylolytic enzymatic activity enhances the glycemic index and digestibility of fermented grains, potentially benefiting individuals with diabetes or glucose intolerance. The glycemic index (GI) gauges the rate at which carbohydrate-containing foods elevate blood sugar levels (Demirkesen-Bicak et al. 2021; Shumoy and Raes 2017; Novotni et al. 2011). Through fermentation-induced starch hydrolysis, the GI of cereal grains can be influenced, potentially yielding cereal products with a lowered GI. Such products could prove beneficial for individuals managing diabetes or aiming to control blood sugar levels. Certain microorganisms utilized in SSF are capable of generating enzymes that transform intricate carbohydrates, like non-digestible fibers, into abbreviated carbohydrates with prebiotic properties. Prebiotics foster the growth of beneficial gut bacteria, contributing to enhanced gut health and potentially influencing the general metabolic response to carbohydrate ingestion. Another effect of fermentation is dextrinization, involving the disintegration of complex carbohydrates into smaller dextrins. These altered carbohydrate structures might possess distinct physicochemical attributes compared to the original starch, potentially affecting properties like viscosity and texture in the fermented grains. In specific scenarios, SSF can encourage the development of resistant starch, a starch variety that evades digestion in the small intestine and reaches the colon undigested. Resistant starch has been associated with various health advantages, including better gut health, heightened satiety, and the potential reduction of post-meal blood sugar spikes. However, it is pertinent to acknowledge that the consequences of SSF on carbohydrate content can vary contingent upon factors such as the microorganisms employed, the fermentation conditions, and the initial composition of the cereal grains. Moreover, while SSF holds the capability to improve the nutrients of cereal grains by modifying carbohydrate content, it is imperative to consider the broader repercussions on other nutrients, non-nutritional factors, and the overall quality of the final product.

20.2.2 Protein Enhancement and Amino Acid Profile

Microorganisms cultivated during SSF secrete proteases that facilitate the breakdown of protein structures within cereal grains. This enzymatic action contributes to the increase in free amino acids and peptides, thereby enhancing the overall protein quality. Additionally, SSF can influence the amino acid profile, leading to elevated levels of essential amino acids like lysine and methionine. These amino acids are often limiting in cereal grains and contribute to improved protein digestibility, rendering the protein content more accessible for human digestion. Darwish et al. (2012) reported that the effect of fermentation on maize stalk using starter culture *Saccharomyces cerevisiae* and *Pleurotus ostreatus* which significantly improved protein content (from 3.60% to 11.80%), surpassing the impact of *Pleurotus ostreatus*, which increased protein content only 3.60%–6.30% (7 days). Pampangouda et al. (2015) promulgated the effect of fermentation on protein and fat contents of little millet flour, which includes a shift in protein content from 9.8% to 10.9% and a decrease in fat from 3.2% to 2.6%. Purwar et al. (2016) investigated the nutritional quality of rice during SSF using *Monascus purpureus*. Their findings indicated a significant increase in moisture (13.32%) and protein content (29.62%), with a decrease in fat (27%), fiber (0.28%), and carbohydrate content (11.32%). Nelofer et al. (2018) utilized *Rhizopus oligosporus* as starter culture to check its effect on bioactive profile of barley. Fermentation resulted in a significant elevation in protein content (from 10.2% to 16.8%) and a reduction in fat content (from 2.1% to 1.6%). Bartkiene et al. (2019) harnessed a *Pediococcus acidilactici* strain to check its specific effect on barley-based industrial by-products. The process led to changes in protein content (16.7%–13.1%), crude fat (4.2%–2.5%), and crude fiber (12.3%–16.85%). Terefe et al. (2021) conducted a study to investigate the impact of SSF lasting 48 h on the nutritional composition, digestibility of protein, and antinutrients of maize. They observed a notable enhancement in protein content after fermentation with *Saccharomyces cerevisiae*, *Lactobacillus plantarum*, and their cocultures. Wu et al. (2022) reported that fermentation using *Rhizopus oryzae* and *Lactobacillus plantarum* synergistically improved the antioxidant properties and nutrients of oats. During fermentation, oats proximate composition was modulated resulting in enhancement of protein content from 89.1 to 118.5 mg/g. Influence of fermentation on the in vitro digestibility, antioxidant characteristics, and mineral composition of finger millet flour was emphasized (Omotoke-Azeez et al. 2022). Changes included an increase in protein content (from 9% to 10.7 g/100g) and a reduction in fat content (from 1.8% to 1.1 g/100g), among other alterations. In a similar vein, Zhai et al. (2023) explored the effects of SSF on water-, β-glucan, protein, and soluble dietary fibers content in black barley and yellow barley. They discovered that after SSF, the water-soluble dietary fibers, protein, and β-glucan contents were approximately 4.6 times, 1.3 times, and 1.2 times higher than those in raw materials. Li et al. (2022) investigated the impact of SSF on canola meals, utilizing two strains (*Aspergillus niger* NRRL 334 and *A. oryzae* NRRL 5590) and varying fermentation durations (24, 48, and 72 hours). The results revealed several significant changes in the nutritional characteristics of both cold-pressed (CP) and hexane-extracted (HE) canola meals following SSF. SSF led to an increase in the protein content of both CP and HE canola meals, elevating it from approximately 36% to around 40%. In CP meals, SSF caused a reduction in oil content from about 12% to 9%. Phytic acid content in both types of canola meals was significantly decreased by approximately 80% during fermentation with either of the fungal strains. The IVPD (in vitro protein digestibility) of the meals changed with fermentation duration. A shorter fermentation time (24 hours) increased IVPD compared with the controls, ranging from approximately 73% to 77%–81%, except for the CP meal fermented with *A. niger*, which experienced decreased IVPD at all fermentation times. Notably, a longer fermentation period (72 hours) decreased IVPD.

20.2.3 Augmentation of Vitamins and Minerals

SSF triggers the synthesis of specific vitamins, such as B vitamins (e.g., folate, riboflavin, niacin), by microorganisms during the fermentation process. Furthermore, minerals such as iron, zinc, and calcium can experience heightened bioavailability due to microbial enzymatic activity targeting non-nutritional factors that typically bind these minerals in less absorbable forms.

20.2.4 Reduction of Non-nutritional Factors

Cereal grains frequently contain non-nutritional compounds, such as phytates, tannins, and lectins, which impede nutrient absorption and bioavailability. Through SSF, these non-nutritional factors can undergo degradation by microbial enzymes, leading to enhanced nutrient utilization. Fermented food products, particularly those derived from grains and legumes, play a significant role in enhancing diet diversity on a global scale. These foods are crucial for ensuring food security and make substantial contributions to nutrition. Fermentation is a process that brings about favorable changes in food components, improving their taste, sensory qualities, bioavailability, and nutritional composition. There was a consistent decrease in antinutritional factors during fermentation, accompanied by an increase in nutrient bioavailability and accessibility. Key mechanisms contributing to these modifications include transamination or the creation of new compounds during fermentation, the utilization of nutrients as energy sources, and the metabolic activity of microorganisms, which could lead to the breakdown or enhancement of certain constituents (Adebo et al. 2022).

20.2.5 Bioactive Compounds and Functional Attributes

SSF-driven microbial activity gives rise to the synthesis of bioactive compounds including antioxidants, antimicrobial peptides, and phytochemicals. These compounds contribute to the functional properties of fermented grains and offer potential health benefits such as improved immune function and reduced risk of chronic ailments. SSF has emerged as a promising technique for the enhancement of phenolic compounds in cereal grains, offering numerous advantages over traditional liquid fermentation methods. This chapter explores the impact of SSF on the release of phenolic compounds from cereal grains, highlighting key findings from various studies. Researchers have employed diverse fungal strains, such as *Rhizopus oligosporus*, *M. purpureus*, *Grifola gargal*, *A. oryzae*, and *Trichoderma harzianum*, on a variety of cereal substrates including rice, wheat, brown rice, maize, and oats. Each fungal–cereal combination exhibits unique interactions, leading to distinct changes in the phenolic composition. One of the significant mechanisms by which SSF influences phenolic compounds is through fungal-mediated hydrolysis. Enzymes produced by the fungi during fermentation efficiently break down glycosidic bonds of bioactive compounds, including phenolic glucosides. This enzymatic action contributes to the release of free polyphenols, altering the nutritional profile of the cereal grains. The augmented phenolic content in SSF-treated cereal grains correlates with enhanced antioxidant properties. Fermented rice bran, wheat, and oats demonstrated superior antioxidant activity compared to their nonfermented counterparts. This suggests that SSF has the potential to produce cereal-based products with improved health-promoting properties. While SSF significantly improves the bioactive profile, the choice of appropriate extraction conditions is crucial. Researchers emphasize the importance of parameters like solvent concentration and extraction temperature for maximizing the recovery of bioactive compounds, including phenolic compounds, from SSF-treated cereal grains. Fungal-assisted SSF shows promise for industrial applications, particularly in the production of antioxidant-rich animal feeds. The scalability and efficiency of SSF processes make them attractive for large-scale production, potentially impacting the nutritional quality of food products derived from cereal grains. SSF has demonstrated a significant influence on the release of phenolic compounds from cereal grains. The enzymatic hydrolysis mediated by fungal strains during SSF enhances phenolic content and antioxidant properties, making it a valuable approach for improving the nutritional quality of cereals and their derived products. Further research is warranted to explore specific fungal–cereal combinations and optimize SSF conditions for diverse applications in the food and feed industries.

Various scientists and researchers utilized filamentous fungi, such as *R. oligosporus* and *M. purpureus*, in conjunction with rice. Details of starter strain used on different substrates are presented in Table 20.1. Additionally, *Grifola gargal* and *G. sordulenta* were applied to sunflower seed hulls and wheat grains (Abd-Razak et al. 2015; Postemsky and Curvetto 2014, 2015). *Aspergillus oryzae*, *R. oligosporus*, and *A. awamori* were used on wheat, brown rice, maize, and oat grains (Dey and Kuhad 2014). *Thamnidium elegans* was investigated on maize (Salar et al. 2012), *Cordyceps militaris* on wheat, *T. harzianum* on soybean seed matrix (Singh et al. 2010), and *A. awamori* and *A. oryzae* on wheat (Bhanja et al. 2009) to

TABLE 20.1
List of Natural Substrates Fermented with Different Microbial Strains

Substrates	*Starter Culture*	Purpose/Outcomes	References
Rice and Barley	*Aspergillus oryzae* *Rhizopus microspores*	Solid-state fungal fermentation enhances the protein quality of subpar plant-based foods	Zwinkels et al. (2023)
Corn kernels	*Aspergillus flavus*	Enhancing the degradation of aflatoxin B1 in corn using Ganoderma sinense via solid-state fermentation optimization	Lou et al. (2023)
Soybean	*Rhizopus oryzae*	The utilization of *Rhizopus oryzae* HC-1 in solid-state fermentation enhances soybeans' metabolite profiling, boosts antioxidant activity, and augments its effect on gut microbiota modulation	Chen et al. (2023)
Quinoa	*Rhizopus oligosporus*	Production and examination of metabolites generated through the solid-state fermentation of *Chenopodium formosanum* (Djulis) sprouts within a bioreactor	Hseih et al. (2023)
Apple pomace	*Zygomycetes*	Utilizing Zygomycetes fungi in solid-state fermentation as a method for enhancing the γ-linolenic acid content, enriching carotenoid pigments, and augmenting phenolic antioxidants in apple pomace biofortification	Dulf et al. (2023)
Sorghum	*Xiaoqu culture*	The impact of grain properties and fermenting culture on the solid-state fermentation of grain sorghum for Chinese liquor production	Liu et al. (2023)
Yibin Yacai	*Bacillus subtilis*	The influence of solid-state fermentation involving *Bacillus subtilis* Y4 on the *Yibin Yacai's* quality	Zou et al. (2023)
Soy protein	*R. mucilaginosa*, *L. plantarum* *M. purpureus*	Creating a plant-based meat substitute from soy proteins using a multi-strain solid-state mixing fermentation process	Ou et al. (2023)
Red kidney beans	*R. oligosporus*	Solid-state fermentation changes the behavior of red kidney bean protein during oral and gastrointestinal digestion, and this is associated with the integrity of the cotyledon cell wall	Sun et al. (2023)
Quinoa seeds	*Rhizopus chinensis*	Assessment of the saponin and phenolic composition of quinoa seeds following fungal fermentation	Starzynska-Janiszewska et al. (2023)
Barley beverage	*Lacticaseibacillus casei*	Impact of fermentation with *Lacticaseibacillus casei* on the bioactive compounds, volatile and non-volatile components, and physiological characteristics of barley-based beverage	Guo et al. (2023)

(*Continued*)

TABLE 20.1 (*Continued*)
List of Natural Substrates Fermented with Different Microbial Strains

Substrates	*Starter Culture*	Purpose/Outcomes	References
Quinoa and Lentil	*Pleurotus ostreatus*	Analysis of the volatile compounds in quinoa and lentil flour during fungal fermentation and subsequent drying	Sanchez-Garcia et al. (2024)
Brewer's spent grain	*Lactic acid bacteria and Propionibacterium*	Fermenting the liquids from brewer's spent grain to extend shelf life and create an ingredient enriched with organic acids	Shetty et al. (2023)
White and black Qingke bran	*Aspergillus niger*	Improvement in the liberation of phenolic compounds from both white and black Qingke bran through autoclaving and fermentation processes	He et al. (2023)
Quinoa flour	*P. ostreatus*	Fermented quinoa flour: Examination of the impact of fungal solid-state bioprocessing and drying on nutritional and antioxidant attributes	Sanchez-Garcia et al. (2023)
Barley grains	*Aspergillus oryzae* *Bacillus subtilis* *Lactobacillus salivarius* *Saccharomyces cerevisiae*	Enhancing the content of biologically active compounds in barley grain through microbial fermentation and characterizing them using infrared spectroscopy	Tosun and Yasar (2023)
Agricultural waste	*Burkholderis* sp.	Employing *Burkholderia* sp. ISTR5 to improve the saccharification and fermentation of agricultural waste, leading to the production and upgrading of biogas with the assistance of calcite-based bio-composite materials	Ghai et al. (2023)
Corn bran	*Lactobacillus plantarum* *Lactobacillus reuteri*	Fermented corn bran: A byproduct featuring enhanced total phenolic content and increased antioxidant activity	Akbari et al. (2023)
Oat	*S. thermophilus*	Growth patterns of starter cultures and sensory attributes of oat-based fermented beverages	Kutt et al. (2023)
Sorghum	*Lactic acid bacteria*	The possibility of elevating folate levels in traditional African fermented sorghum gruel (Motoho) through the utilization of presumptive probiotic lactic acid bacteria	Fayemi et al. (2023)
Wheat bran	*Saccharomyces cerevisiae*	Evaluation of Nile tilapia, *Oreochromis niloticus*, focusing on growth, feed utilization, blood parameters, and the histology of the intestine and liver, through the fermentation of wheat bran using Saccharomyces cerevisiae	Mohammady et al. (2023)

(*Continued*)

TABLE 20.1 (*Continued*)
List of Natural Substrates Fermented with Different Microbial Strains

Substrates	*Starter Culture*	Purpose/Outcomes	References
Wheat bran	*Lactobacillus apis*	Assessment of the influence of Lactobacillus apis on the antioxidant activity, degradation of phytic acid, nutraceutical value, and flavor characteristics of fermented wheat bran, in comparison to *Saccharomyces cerevisiae* and *Lactobacillus plantarum*	Ghamry et al. (2023)
Wheat bran	*Lactobacillus*	Fermentation of wheat bran by *Lactobacillus* led to the regulation of the bacteria-fungi composition and a reduction in fecal heavy metal concentrations in growing pigs	Zhang et al. (2023)
Guangan honey pear cider	*Saccharomyces cerevisiae*	Influence of different *Saccharomyces cerevisiae* strains on the metabolomic profiles of Guangan honey pear cider	Zhu et al. (2023)
Pearl millet	*Aspergillus sojae*	Fermented pearl millet (*Pennisetum glaucum*) exhibiting in vitro DNA damage protection activity, enriched with bioactive compounds, and showcasing antioxidant potential	Salar et al. (2017)
Pearl millet	*Aspergillus oryzae*	Enhancement of DNA damage protection and antioxidant properties in biotransformed pearl millet (*Pennisetum glaucum*) cultivar PUSA-415 through the utilization of Aspergillus oryzae MTCC 3107	Salar and Purewal (2016)
Pearl millet	*Aspergillus awamori*	Refining extraction parameters and boosting phenolic content and antioxidant capacity in pearl millet fermented with *Aspergillus awamori* MTCC-548	Salar et al. (2016)
Barley bran	*Aspergillus oryzae*	Fermented barley bran: Enhancement of phenolic compounds and antioxidant characteristics	Bangar et al. (2022)
Rice bran	*Aspergillus oryzae*	Rice bran fermented by *Aspergillus oryzae*: A byproduct enriched with increased bioactive compounds and heightened antioxidant potential	Punia et al. (2020)
Lentil	*Aspergillus awamori*	Solid-state fermentation of lentil (*Lens culinaris* L.) utilizing *Aspergillus awamori*: Impact on phenolic compounds, mineral content, and their subsequent bioavailability	Dhull et al. (2020)

(*Continued*)

TABLE 20.1 (*Continued*)
List of Natural Substrates Fermented with Different Microbial Strains

Substrates	*Starter Culture*	Purpose/Outcomes	References
Pearl millet	*Aspergillus oryzae and Rhizopus azygosporus*	Solid-state fermentation of pearl millet by *Aspergillus oryzae* and *Rhizopus azygosporus*: Influence on the bioactive profile and the activity of protecting against DNA damage	Purewal et al. (2020)
Pearl millet	*Rhizopus azygosporus*	Fermented pearl millet: A product enriched with heightened bioactive compounds and exhibiting DNA damage protection activity	Purewal et al. (2019)

enhance phenolic content and antioxidant properties through SSF. Postemsky et al. (2011) conducted a comparative analysis of the genotoxic protection provided by biotransformed wheat grains using basidiomes and mycelia from *G. gargal*. Punia et al. (2021) investigated the impact of *A. oryzae*-assisted fungal fermentation on rice bran, observing significant changes during SSF including a remarkable increase in phenolics from 1 to 8.8 mg/g (717% increment). Fermented rice bran exhibited superior antioxidant properties compared to unfermented rice bran. The fungal-assisted fermentation process holds promise for producing antioxidant-rich animal feeds on an industrial scale. Shin et al. (2019) employed a similar fungus on black rice bran, noting its effects on the bioactive profile, with significant increases in protocatechuic acid, vanillic acid, caffeic acid, and *p*-coumaric acid. Furthermore, Chen et al. (2020) observed substantial improvements in the bioactive profile of oats through SSF with *Bacillus subtilis* and *M. anka*, resulting in fermented oat fractions with higher phenolic content compared to nonfermented oats. Georgetti et al. (2009) discovered that enzymes produced by fungal strains during the fermentation process efficiently hydrolyze glycosidic bonds of bioactive compounds and phenolic glucosides, offering a promising approach to increase the levels of bound to free polyphenols and modulate the nutritional profile of various food products. Although SSF significantly enhances the bioactive profile in fermented products, the selection of experimental conditions for extraction of bioactive components is very crucial. Process parameters such as solvent concentration and extraction temperature play a pivotal role in achieving maximal recovery of bioactive compounds (Salar et al. 2016). Sandhu et al. (2016) noted the highest extractable total phenolic content in wheat fermented with *A. awamori* on the fourth day of fermentation (2.1 mg GAE/dwb). Additionally, Abd-Razak et al. (2015) reported an up to eightfold increment in total phenolic content (TPC) and a twofold increase during SSF of rice bran with *R. oligosporus* and *M. purpureus*, respectively.

20.3 OTHER USES OF FERMENTATION

20.3.1 Production of Fermented Foods

SSF is commonly employed in the production of fermented foods such as bread, cheese, sausages, and certain types of alcoholic beverages like sake and some traditional Chinese and Japanese fermented foods. In these processes, microorganisms like bacteria, yeast, and molds are utilized to transform raw materials into the final food product.

20.3.2 Bread

SSF is widely used in breadmaking. The process involves the fermentation of dough by yeast (e.g., *Saccharomyces cerevisiae*). During fermentation, yeast consumes sugars in the dough, producing carbon dioxide gas and alcohol. The gas causes the dough to rise, resulting in the characteristic texture of the bread. The alcohol evaporates during baking.

20.3.3 Cheese

Various types of cheese, such as blue cheese, Roquefort, and Camembert, are produced using SSF. In cheese production, specific molds or bacteria are introduced to milk, where they break down proteins and fats, leading to the formation of unique flavors and textures.

20.3.4 Alcoholic Beverages

SSF is employed in the production of alcoholic beverages like sake (Japanese rice wine) and certain traditional Chinese alcoholic drinks. In sake production, rice is fermented by a special yeast called Saccharomyces cerevisiae var. sake, leading to the conversion of starches into alcohol.

20.3.5 Fermented Vegetables

Foods like sauerkraut (fermented cabbage) and kimchi (fermented Korean vegetables) are made through SSF. Lactic acid bacteria naturally present in the vegetables ferment the sugars, preserving the vegetables and giving them a tangy flavor.

20.3.6 Enhancing Flavor and Aroma

SSF can enhance the flavor and aroma of foods. Microorganisms involved in fermentation can produce various compounds that contribute to the unique taste and smell of fermented products. For example, yeast in breadmaking produces alcohol and carbon dioxide, which contribute to the bread's flavor and texture. Microorganisms involved in SSF can produce a wide range of volatile compounds and flavor-enhancing molecules. These compounds can include alcohols, esters, and organic acids, which contribute to the unique taste and aroma of fermented foods. For example, the esters produced by yeast in beer and wine are crucial to their flavor profiles.

20.3.7 Improving Nutritional Value

Fermentation can also improve the nutritional value of foods. It can break down complex compounds into simpler, more digestible forms and increase the bioavailability of certain nutrients. For instance, fermentation of soybeans in the production of soy sauce and miso can increase the availability of essential amino acids. SSF can enhance the nutritional value of foods. Microorganisms can break down complex compounds in raw materials, making them more bioavailable to the human body. In some cases, SSF can reduce antinutritional factors, such as phytic acid and tannins, in grains and legumes, making them more nutritious and easier to digest.

20.3.8 Preservation

Fermentation can act as a natural preservative by creating an acidic or alcoholic environment that inhibits the growth of spoilage microorganisms and pathogens. This preservation quality has been historically crucial for extending the shelf life of various food products. Some fermented foods are associated with health benefits. For example, yogurt and other fermented dairy products contain probiotic bacteria that can promote gut health. Fermented foods can also reduce the levels of antinutritional factors in certain grains and legumes, making them more nutritious and easier to digest. Fermentation can act as a natural preservation method. The acidic or alcoholic environment created during fermentation inhibits the growth of spoilage microorganisms and pathogens. This preservation quality has been historically crucial for extending the shelf life of various food products, especially in the absence of modern refrigeration.

20.3.9 Diverse Product Development

SSF allows for the development of a wide range of food products with unique characteristics and flavors. Different microorganisms and substrates can be used to create an array of fermented products, catering to various tastes and preferences. Certain fermented foods, such as yogurt and kefir, contain probiotic bacteria that can promote gut health by maintaining a healthy balance of microorganisms in the digestive tract. These probiotics may offer various health benefits, including improved digestion and immune function.

20.3.10 Sustainable Food Production

Fermentation can be considered a sustainable food production method as it often requires fewer resources (such as water and energy) than some other food processing methods. It can also reduce food waste using raw materials that might be otherwise discarded. SSF aligns with sustainability goals in food production. It often requires fewer resources compared with other processing methods. For example, fermentation can reduce water usage and energy consumption. Additionally, it can make use of raw materials that might otherwise go to waste, such as overripe fruit or vegetable trimmings. SSF allows for a diverse range of food product development. Different microorganisms, substrates, and fermentation conditions can be employed to create an array of fermented products, each with its own unique characteristics and flavors. This diversity caters to various tastes and preferences in the food market.

20.4 CONCLUSION

SSF stands as a promising avenue to enhance the nutritional composition of cereal grains. Through enzymatic degradation, reduction of non-nutritional factors, enrichment of protein quality, and synthesis of bioactive compounds, SSF holds the potential to significantly elevate the nutritional value and functional characteristics of cereal grains. Further research should concentrate on optimizing SSF conditions, exploring diverse microbial strains, and assessing the long-term health implications of consuming fermented cereal products. The integration of SSF into food processing could potentially address nutritional deficiencies and positively impact global food security and human well-being.

REFERENCES

Abd-Razak, D.L., Abd Rashid, N.Y., Jamaluddin, A., Sharifudin, S.A., & Long, K. (2015). Enhancement of phenolic acid content and antioxidant activity of rice bran fermented with *Rhizopus oligosporus* and *Monascus purpureus*. *Biocatalysis and Agricultural Biotechnology*, 4, 33–38.

Adebo, J.A., Njobeh, P.B., Gbashi, S., Oyedeji, A.B., Ogundele, O.M., Oyeyinka, S.A., & Adebo, O.A. (2022). Fermentation of cereals and legumes: Impact on nutritional constituents and nutrient bioavailability. *Fermentation*, 8, 63.

Akbari, M., Razavi, S.H., Khodaiyan, F., Blesa, J., & Esteve, M.J. (2023). Fermented corn bran: A by-product with improved total phenolic content and antioxidant activity. *LWT*, 184, 115090. https://doi.org/10.1016/j.lwt.2023.115090

Bangar, S.P., Sandhu, K.S., Purewal, S.S., Kaur, M., Kaur, P., Siroha, A.K., Kumari, K., Singh, M., & Kumar, M. (2022). Fermented barley bran: An improvement in phenolic compounds and antioxidant properties. *Journal of Food Processing and Preservation*, *46*(10), e15543.

Bartkiene, E., Mozuriene, E., Lele, V., Zokaityte, E., Gruzauskas, R., Jakobsone, I., Juodeikiene, G., Ruibys, R., & Bartkevics, V. (2019). Changes of bioactive compounds in barley industry by-products during submerged and solid state fermentation with antimicrobial *Pediococcus acidilactici* strain LUHS29. *Food Science & Nutrition*, *8*, 340–350.

Bhanja, T., Kumari, A., & Banerjee, R. (2009). Enrichment of phenolics and free radical scavenging property of wheat koji prepared with two filamentous fungi. *Bioresource Technology*, 100, 2861–2866.

Chen, G., Liu, Y., Zeng, J., Tian, X., Bei, Q., & Wu, Z. (2020). Enhancing three phenolic fractions of oats (*Avena sativa* L.) and their antioxidant activities by solid-state fermentation with *Monascus anka* and *Bacillus subtilis*. *Journal of Cereal Science*, 93, 102940.

Chen, J., Chen, Y., Hu, J., He, C., Peng, X., Li, Z., Wang, Y., Zhu, M., & Xiao, Y. (2023) Solid-state fermentation with *Rhizopus oryzae* HC-1 improves the metabolites profiling, antioxidant activity and gut microbiota modulation effect of soybeans. *LWT*, 187, 115253.

Costa-Maia, I., Santos-D'Almeida, C.T., Guimaraes-Freire, D.M., Costa-Cavalcanti, E.A., Cameron, L.C., Furtado-Dias, J., & Larraz-Ferreira, M.S. (2020). Effect of solid-state fermentation over the release of phenolic compounds from brewer's spent grain revealed by UPLC-MS^E. *LWT*, 133, 110136.

Darwish, Galila A.M.A., Bakr, A.A., & Abdallah, M.M.F. (2012). Nutritional value upgrading of maize stalk by using *Pleurotus ostreatus* and *Saccharomyces cerevisiae* in solid state fermentation. *Annals of Agricultural Sciences*, *57*, 47–51.

Demirkesen-Bicak, H., Arici, M., Yaman, M., Karasu, S., & Sagdic, O. (2021). Effect of different fermentation condition on estimated glycemic index, in vitro starch digestibility, and textural and sensory properties of sourdough bread. *Foods*, 10, 514.

de-Souza, P.M., & Magalhaes, P.O. (2010). Application of microbial α-amylase in industry: A review. *Brazilian Journal of Microbiology*, 41, 850–861.

Dey, T.B., & Kuhad, R.C. (2014). Upgrading the antioxidant potential of cereals by their fungal fermentation under solid-state cultivation conditions. *Letters in Applied Microbiology*, 59, 493–499.

Dhull, S.B., Punia, S., Kidwai, M.K., Kaur, M., Chawla, P., Purewal, S.S., Sangwan, M., & Palthania, S. (2020). Solid-state fermentation of lentil (*Lens culinaris* L.) with *Aspergillus awamori*: Effect on phenolic compounds, mineral content, and their bioavailability. *Legume Science*, 2, e37.

Dulf, F.V., Vodnar, D.C., & Dulf, E.H. (2023). Solid-state fermentation with *Zygomycetes* fungi as a tool for biofortification of apple pomace with γ-linolenic acid, carotenoid pigments and phenolic antioxidants. *Food Research International*, 173, 113448.

Fayemi, O.E., Akanni, G.B., Sobowale, S.S., Oelofse, A., & Buys, E.M. (2023). Potential for increasing folate contents of traditional African fermented sorghum gruel (Motoho) using presumptive probiotic lactic acid bacteria. *Journal of Food Composition and Analysis*, 115, 104850.

Garrido-Galand, S., Asensio-Grau, A., Calvo-Lerma, J., Heredia, A., & Andres, A. (2021). The potential of fermentation on nutritional and technological improvement of cereal and legume flours: A review. *Food Research International*, 145, 110398.

Georgetti, S.R., Vicentini, F.T.M.C., Yokoyama, C.Y., Borin, M.F., Spadaro, A.C.C., & Fonseca, M.J.V. (2009). Enhanced in vitro and in vivo antioxidant activity and mobilization of free phenolic compounds of soybean flour fermented with different β-glucosidase-producing fungi. *Journal of Applied Microbiology*, 106(2), 459–466.

Ghai, M.K., Khatri, A., & Thakur, I.S. (2023). Utilization of Burkholderia sp. ISTR5 for enhanced saccharification and fermentation of agricultural waste for production and upgradation of biogas by calcite-based bio-composite materials. *Bioresource Technology Reports*, 24, 101608. https://doi.org/10.1016/j.biteb.2023.101608

Ghamry, M., Zhao, & W., Li, L. (2023). Impact of *Lactobacillus apis* on the antioxidant activity, phytic acid degradation, nutraceutical value and flavor properties of fermented wheat bran, compared to *Saccharomyces cerevisiae* and *Lactobacillus plantarum*. *Food Research International*, 163, 112142. https://doi.org/10.1016/j.foodres.2022.112142

Guo, W., Chen, M., Cui, S., Tang, X., Zhang, Q., Zhao, J., Mao, B., & Zhang, H. (2023). Effects of *Lacticaseibacillus casei* fermentation on the bioactive compounds, volatile and non-volatile compounds, and physiological properties of barley beverage. *Food Bioscience*, 53, 102695. https://doi.org/10.1016/j.fbio.2023.102695

He, C., Liu, X., Zhang, H., Mu, T., Zhang, Y., Ren, X., Han, L., & Wang, M. (2023). Enhancement of the release of phenolic compounds from white and black Qingke bran by autoclaving and fermentation treatments. *Food Bioscience*, 53, 102696.

Hseih, C.C., Yu, S.H., Cheng, K.W., Liou, Y.W., Hsu, C.C., Hseih, C.W., Kuo, C.H., & Cheng, K.C. (2023). Production and analysis of metabolites from solid-state fermentation of *Chenopodium formosanum* (Djulis) sprouts in a bioreactor. *Food Research International*, 168, 112707. https://doi.org/10.1016/j.foodres.2023.112707

Kaur, A., & Purewal, S.S. (2023). Modulation of Cereal biochemistry via solid-state fermentation: A fruitful way for nutritional improvement. *Fermentation*, 9, 817. https://doi.org/10.3390/fermentation9090817

Kaur, P., Purewal, S.S., Sandhu, K.S., Kaur, M., & Salar, R.K. (2019). Millets: A cereal grain with potent antioxidants and health benefits. *Journal of Food Measurement and Characterization*, 13, 793–806. https://doi.org/10.1007/s11694-018-9992-0

Kaur, P., Sandhu, K.S., Purewal, S.S., Kaur, M., & Singh, S.K. (2021). Rye: A wonder crop with industrially important macromolecules and health benefits. *Food Research International*, 150, 110769. https://doi.org/10.1016/j.foodres.2021.110769

Kutt, M.L., Orgusaar, K., Stulova, I., Priidik, R., Pismennoi, D., Vaikma, H., Kallastu, A., Zhogoleva, A., Morell, I., & Krisciunaite, T. (2023) Starter culture growth dynamics and sensory properties of fermented oat drink. *Heliyon*, 9, e15627.

Li, K., Li, X.Q., Li, G.X., Cui, L.J., Qin, X.M., Li, Z.Y., Du, Y.G., Liu, Y.T., Li, A.P., Zhao, X.Y., & Fan, X.H. (2022). Relationship between the structure and immune activity of components from the active polysaccharides APS-II of astragali radix by enzymolysis of endo α-1,4- glucanase. *Frontiers in Pharmacology*, 13, 839635. https://doi.org/10.3389/fphar.2022.839635

Liu, M., Li, S., Weiss, T., Li, Y., Wang, D., & Zheng, Y. (2023). Solid-state fermentation of grain sorghum to produce Chinese liquor: Effect of grain properties and fermenting culture. *Journal of Cereal Science*, 114, 103776. https://doi.org/10.1016/j.jcs.2023.103776

Lizardi-Jiménez, M.A., & Hernández-Martínez, R. (2017). Solid state fermentation (SSF): Diversity of applications to valorize waste and biomass. *3 Biotech*, 7(1), 44. https://doi.org/10.1007/s13205-017-0692-y

Lou, H., Yang, C., Li, Y., Li, Y., Li, Y., & Zhao, R. (2023). Optimization of aflatoxin B1 degradation in corn by *Ganoderma sinense* through solid-state fermentation. *LWT*, 183, 114959. https://doi.org/10.1016/j.lwt.2023.114959

Mohammady, E.Y., Aboseif, A.M., Soaudy, M.R., Ramadan, E.A., & Hassaan, M.S. (2023). Appraisal of fermented wheat bran by *Saccharomyces cerevisiae* on growth, feed utilization, blood indices, intestinal and liver histology of Nile tilapia, *Oreochromis niloticus*. *Aquaculture*, 575, 739755. https://doi.org/10.1016/j.aquaculture.2023.739755

Nelofer, R., Nadeem, M., Irfan, M., & Syed, Q. (2018). Nutritional enhancement of Barley in solid state fermentation by *Rhizopus oligosporus* ML-10. *Nutrition & Food Science International Journal*, 6, 1–7.

Nighojkar, A., Patidar, M.K., & Nighojkar, S. (2019). Pectinases: Production and applications for fruit juice beverages. *Processing and Sustainability of Beverages: The Science of Beverages*, 2, 235–273. https://doi.org/10.1016/B978-0-12-815259-1.00008-2

Novotni D., Ćurić D., Bituh M., Colić Barić I., Škevin D., & Čukelj N. (2011). Glycemic index and phenolics of partially-baked frozen bread with sourdough. *International Journal of Food Science and Nutrition*, 62, 26–33.

Omotoke-Azeez, S., Chinma, C.E., Bassey, S.O., Eze, U.R., Makinde, A.F., Sakariyah, A.A., Okubanjo, S.S., Danbaba, N., & Adebo, O.A. (2022). Impact of germination alone or in combination with solid-state fermentation on the physicochemical, antioxidant, in vitro digestibility, functional and thermal properties of brown finger millet flours. *LWT*, 154, 112734.

Ou, M., Lou, J., Lao, L., Guo, Y., Pan, D., Yang, H., & Wu, Z. (2023). Plant-based meat analogue of soy proteins by the multi-strain solid-state mixing fermentation. *Food Chemistry*, 414, 135671. https://doi.org/10.1016/j.foodchem.2023.135671

Pampangouda, P., Munishamanna, K.B., & Gurumurthy, H. (2015). Effect of *Saccharomyces boulardii* and *Lactobacillus acidophilus* fermentation on little millet (*Panicum sumatrense*). *Journal of Applied and Natural Science*, 7, 260–264.

Pandey, A. (2003). Solid-state fermentation. *Biochemical Engineering Journal*, 13, 81–84. https://doi.org/10.1016/S1369-703X(02)00121-3

Postemsky, P.D., & Curvetto, N.R. (2014). Enhancement of wheat grain antioxidant activity by solid state fermentation with *Grifola spp. Journal of Medicinal Food*, 17 (5), 543–549.

Postemsky, P.D., & Curvetto, N.R. (2015). Solid-state fermentation of cereal grains and sunflower seed hulls by *Grifola gargal* and *Grifola sordulenta. International Biodeterioration and Biodegradation*, 100, 52–61. https://doi.org/10.1016/j.ibiod.2015.02.016

Postemsky, P.D., Palerma, A.M., & Curvetto, N.R. (2011). Protective effects of new medicinal mushroom *Grifola gargal* Singer (higher Basidiomycetes) on induced DNA damage in somatic cells of *Drosophila melanogaster. International Journal of Medicinal Mushrooms*, 13, 6.

Punia, S., Sandhu, K.S., Grasso, S., Purewal, S.S., Kaur, M., Siroha, A.K., Kumar, K., Kumar, V., & Kumar, M. (2021). *Aspergillus oryzae* fermented rice bran: A byproduct with enhanced bioactive compounds and antioxidant potential. *Foods*, 10(1), 70. https://doi.org/10.3390/foods 10010070

Purewal, S.S., Salar, R.K., Bhatti, M.S., Sandhu, K.S., Singh, S.K., & Kaur, P. (2020). Solid-state fermentation of pearl millet with *Aspergillus oryzae* and *Rhizopus azygosporus*: Effects on bioactive profile and DNA damage protection activity. *Journal of Food Measurement and Characterization*, 14, 150–162. https://doi.org/10.1007/s11694-019-00277-3

Purewal, S.S., Sandhu, K.S., Salar, R.K., & Kaur, P. (2019). Fermented pearl millet: A product with enhanced bioactive compounds and DNA damage protection activity. *Journal of Food Measurement and Characterization*, 13, 1479–1488. https://doi.org/10.1007/s11694-019-00063-1

Purwar, S., Gupta, E., & Zaki, S. (2016). Effect of solid state fermentation on nutritive values of rice by *Monascus spp. Bioved*, 27, 185–189.

Ravindran, R., & Jaiswal, A.K. (2016). Microbial enzyme production using lignocellulosic food industry wastes as feedstock: A review. *Bioengineering*, 3, 30. https://doi.org/10.3390/bioengineering3040030

Roasa, J., Villa, R.D., Mine, Y., & Tsao, R. (2021). Phenolics of cereal, pulse and oilseed processing by-products and potential effects of solid-state fermentation on their bioaccessibility, bioavailability and health benefits: A review. *Trends in Food Science & Technology*, 116, 954–974. https://doi.org/10.1016/j.tifs.2021.08.027

Salar, R.K., Certik, M., & Brezova, V. (2012). Modulation of phenolic content and antioxidant activity of maize by solid state fermentation with *Thamnidium elegans* CCF 1456. *Biotechnology and Bioprocess Engineering*, 17, 109–116. https://doi.org/10.1007/s12257-011-0455-2

Salar, R.K., & Purewal, S.S. (2016). Improvement of DNA damage protection and antioxidant activity of biotransformed pearl millet (*Pennisetum glaucum*) cultivar PUSA-415 using *Aspergillus oryzae* MTCC 3107. *Biocatalysis and Agricultural Biotechnology*, 8, 221–227. https://doi.org/10.1016/j.bcab.2016.10.005

Salar, R.K., Purewal, S.S., & Bhatti, M.S. (2016). Optimization of extraction conditions and enhancement of phenolic content and antioxidant activity of pearl millet fermented with *Aspergillus awamori* MTCC-548. *Resource Efficient Technologies*, 2, 148–157. https://doi.org/10.1016/j.reffit.2016.08.002

Salar, R.K., Purewal, S.S., & Sandhu, K.S. (2017). Fermented pearl millet (*Pennisetum glaucum*) with in vitro DNA damage protection activity, bioactive compounds and antioxidant potential. *Food Research International*, 100, 204–210.

Sanchez-Garcia, J., Munoz-Pina, S., Garcia-Harnandez, J., Heredia, A., & Andres, A. (2023). Fermented quinoa flour: Implications of fungal solid-state bioprocessing and drying on nutritional and antioxidant properties. *LWT*, 182, 114885.

Sanchez-Garcia, J., Munoz-Pina, S., Garcia-Hernandez, J., Heredia, A., & Andres, A. (2024). Volatile profile of quinoa and lentil flour under fungal fermentation and drying. *Food Chemistry*, 430, 137082.

Sandhu, K.S., Punia, S., & Kaur, M. (2016). Effect of duration of solid state fermentation by *Aspergillus awamori nakazawa* on antioxidant properties of wheat cultivars. *LWT Food Science and Technology*, 71, 323–328. https://doi.org/10.1016/j.lwt.2016.04.008

Serna-Saldivar, S.O. (2016). Cereals: Dietary importance. reference module in food science. In *B. Caballero, P. M. Finglas & F. Toldrá* (Eds.), *Encyclopedia of Food and Health*, Academic Press, 703–711.

Shetty, R., Petersen, F.R., Podduturi, R., Molina, G.E.S., Watjen, A.P., Madsan, S.K., Zioga, E., Ozmerih, S., Hobley, T.J., & Bang-Berthelsen, C.H. (2023). Fermentation of brewer's spent grain liquids to increase shelf life and give an organic acid enhanced ingredient. *LWT*, 182, 114911. https://doi.org/10.1016/j.lwt.2023.114911

Shin, H.Y., Kim, S.M., Lee, J.H., & Lim, S.T. (2019). Solid-state fermentation of black rice bran with *Aspergillus awamori* and *Aspergillus oryzae*: Effects on phenolic acid composition and antioxidant activity of bran extracts. *Food Chemistry*, 272, 235–241.

Shumoy H., & Raes K. (2017). In vitro starch hydrolysis and estimated glycemic index of tef porridge and injera. *Food Chemistry*, 229, 381–387.

Singh, H.B., Singh, B.N., Singh, S.P., & Nautiyal, C.S. (2010). Solid-state cultivation of *Trichoderma harzianum* NBRI-1055 for modulating natural antioxidants in soybean seed matrix. *Bioresource Technology*, 101, 6444–6453.

Starzynska-Janiszewska, A., Stodolak, B., Dulinski, R., Fernandez-Fernandez, C., Martin-Garcia, B., & Gomez-Caravaca, M. (2023). Evaluation of saponin and phenolic profile of quinoa seeds after fungal fermentation. *Journal of Cereal Science*, 111, 103656.

Sun, W., He, J., Wang, H., Zhang, Q., Li, W., & Rui, X. (2023). Solid-state fermentation alters the fate of red kidney bean protein during buccal and gastrointestinal digestion: Relationship with cotyledon cell wall integrity. *Food Chemistry*, 410, 135370.

Terefe, Z.K., Omwamba, M.N., & Nduko, J.M. (2021). Effect of solid state fermentation on proximate composition, antinutritional factors and in vitro protein digestibility of maize flour. *Food Science and Nutrition*, 9, 6343–6352.

Tosun, R., & Yasar, S. (2023). Enrichment of barley grain in some biologically active molecules by microbial fermentation and their characterisation by infrared spectroscopy. *Animal Feed Science and Technology*, 303, 115713.

Wu, H., Liu, H.N., Ma, A., Zhou, J.Z., & Xia, X.D. (2022). Synergetic effects of *Lactobacillus plantarum* and *Rhizopus oryzae* on physicochemical, nutritional and antioxidant properties of whole-grain oats (*Avena sativa* L.) during solid-state fermentation. *LWT*, 154, 112687.

Zhai, X.T., Wang, H.Q., Han, L., Wang, F.Y., Zhang, Y.H., Li, J., He, C.A., Tan, B., & Wang, M. (2023). Impact of solid-state fermentation by *Aspergillus niger* 4Q and Cyberlindnerafabianii J2 on antioxidant and flavor properties of whole grain barley. *Cereal Chemistry*, 100, 914–926.

Zhang, D., Liu, H., Wang, S., Liu, Y., & Ji, H. (2023). Wheat bran fermented by *Lactobacillus* regulated the bacteria–fungi composition and reduced fecal heavy metals concentrations in growing pigs. *Science of the Total Environment*, 858, 159828. https://doi.org/10.1016/j.scitotenv.2022.159828

Zhang, D., Ye, Y., & Tan, B. (2022). Comparative study of solid-state fermentation with different microbial strains on the bioactive compounds and microstructure of brown rice. *Food Chemistry*, 397, 133735. https://doi.org/10.1016/j.foodchem.2022.133735

Zhu, C., Yang, Z., Lu, X., Yi, Y., Tian, Q., Deng, J., Jiang, D., Tang, J., & Laghi, L. (2023). Effects of *Saccharomyces cerevisiae* strains on the metabolomic profiles of Guangan honey pear cider. *LWT*, 182, 114816. https://doi.org/10.1016/j.lwt.2023.114816

Zou, Y., Zhong, Y., Zhou, Q., Jia, Z., Chen, Q., Xu, W., Wu, Y., Wei, S., Zhong, K., & Gao, H. (2023). Effects of solid-state fermentation with *Bacillus subtilis* Y4 on the quality of Yibin Yacai. *Food Bioscience*, 51, 102258. https://doi.org/10.1016/j.fbio.2022.102258

Zwinkels, J., Wolkers-Rooijackers, J., & Smid, E.J. (2023). Solid-state fungal fermentation transforms low-quality plant-based foods into products with improved protein quality. *LWT*, 184, 114979. https://doi.org/10.1016/j.lwt.2023.114979

Index

Note: **Bold** page numbers refer to tables, *italic* page numbers refer to figures.

Kolibri 360 GmbH
Ferdinand-Jühlke-Straße 7
99095 Erfurt, Deutschland
produktsicherheit@kolibri360.de